Cellular and Molecular Mechanisms in Cardiopathy and Therapeutic Strategies

Cellular and Molecular Mechanisms in Cardiopathy and Therapeutic Strategies

Guest Editors

Ying-Hsi Lin
Shengjie Lu

MDPI

Basel • Beijing • Wuhan • Barcelona • Belgrade • Novi Sad • Cluj • Manchester

Guest Editors
Ying-Hsi Lin
Cardiovascular and Metabolic Disorders Program
Duke-National University of Singapore Medical School
Singapore

Shengjie Lu
National Heart Research Institute
National Heart Centre
Singapore
Singapore

Editorial Office
MDPI AG
Grosspeteranlage 5
4052 Basel, Switzerland

This is a reprint of the Special Issue, published open access by the journal *Biomolecules* (ISSN 2218-273X), freely accessible at: https://www.mdpi.com/journal/biomolecules/special_issues/X4VV795EZV.

For citation purposes, cite each article independently as indicated on the article page online and as indicated below:

Lastname, A.A.; Lastname, B.B. Article Title. *Journal Name* **Year**, *Volume Number*, Page Range.

ISBN 978-3-7258-8321-9 (Hbk)
ISBN 978-3-7258-8322-6 (PDF)
https://doi.org/10.3390/books978-3-7258-8322-6

Contents

About the Editors

Ying-Hsi Lin

Ying-Hsi Lin is a cardiac muscle biologist investigating the molecular and biophysical basis of myocardial dysfunction in heart failure, with a particular focus on heart failure with preserved ejection fraction (HFpEF). His research centers on sarcomere-level mechanisms, examining how molecular perturbations propagate through cross-bridge kinetics to influence myofibril mechanics in both physiological and disease contexts. Using integrated experimental platforms spanning murine models, human induced pluripotent stem cell-derived cardiomyocytes, and ex vivo myocardial systems, his work aims to define mechanisms underlying impaired myocardial performance and to identify potential therapeutic strategies. Based at Duke-NUS Medical School, his research seeks to translate mechanistic insights into targeted, disease-relevant interventions for heart failure.

Shengjie Lu

Shengjie Lu is an Assistant Professor at Duke-NUS Medical School and a Junior Principal Investigator at the National Heart Centre Singapore (NHCS). He also serves as the theme lead of Translational Biomedical Imaging, Biomechanics, and Biomaterials at the College of Healthcare Engineering, SingHealth. His research focus is on developing innovative biomaterial-based therapeutic solutions for cardiovascular diseases such as myocardial infarction (MI) and heart failure. He has established a translational nanomedicine platform, leveraging nanotechnology and targeted strategies to deliver a variety of therapeutic agents, including small molecules, peptides, RNA, and genes, specifically for heart disease treatment. He is also passionate about advancing medical technologies. His interests include intravascular imaging and medical device innovations, particularly in catheter delivery systems such as drug-eluting stents (DESs) and drug-coated balloons (DCBs). Over the past few years, he has led multiple translational research projects with clinical healthcare impacts and expanded the biomaterials platform for diverse bench-to-bedside innovations.

Preface

Heart failure represents the final common manifestation of diverse cardiopathies characterized by impaired contractile function, maladaptive remodeling, electrophysiological instability, and disrupted energetic homeostasis. While current therapies primarily target systemic and neurohormonal pathways, the intrinsic molecular determinants of contractile dysfunction and maladaptive remodeling remain incompletely defined, limiting the development of precise and disease-aligned therapies.

The following Reprint, Cellular and Molecular Mechanisms in Cardiopathy and Therapeutic Strategies, compiles studies addressing key mechanistic drivers of cardiopathy, including oxidative stress, inflammation, mitochondrial dysfunction, electrophysiological instability, and sarcomere dysregulation. These processes likely interact across scales—from molecular modification to tissue-level remodeling—to impair cardiac performance and contribute to disease progression. Advances in structural biology, biophysical analysis, and translational platforms now enable more precise interrogation of these mechanisms and their potential therapeutic modulation.

We anticipate that this collection will provide a mechanistically grounded perspective on cardiopathy and help inform the development of targeted, disease-aligned strategies for heart failure.

Ying-Hsi Lin and Shengjie Lu
Guest Editors

biomolecules

MDPI

Article

Trandolapril Attenuates Pro-Arrhythmic Downregulation of Cx43 and Cx40 in Atria of Volume Overloaded Hypertensive and Normotensive Rats

Matúš Sýkora *, Katarína Ondreják Andelová , Alexandra Mrvová, Barbara Szeiffová Bačová and Narcis Tribulová *

Centre of Experimental Medicine, Institute for Heart Research, Slovak Academy of Sciences, 841 04 Bratislava, Slovakia; katarina.andelova@savba.sk (K.O.A.); mrvova.alexandra@gmail.com (A.M.); usrdbaca@savba.sk (B.S.B.)
* Correspondence: matus.sykora@savba.sk (M.S.); narcisa.tribulova@savba.sk (N.T.)

Abstract

Pressure overload in non-treated or resistant hypertension (HTN) increases the risk of heart failure (HF) as well as the occurrence of fatal ventricular arrhythmias and stroke-provoking atrial fibrillation (AF), while perturbed connexin-43 (Cx43) and Cx40 might be involved. In addition, kidney dysfunction may facilitate hemodynamic volume overload and congestive HF. We investigated the impact of volume overload on Cx43 and Cx40 in right and left heart atria of hypertensive pressure overloaded Ren-2 transgenic (TGR) strain and normotensive Hannover Sprague Dawley (HSD) rats, as well as the efficacy of renin–angiotensin blockade with trandolapril and losartan. Key novel findings revealed lower levels of Cx43 and Cx40 proteins in left as well as right heart atria in pressure overloaded hypertensive rats compared to normotensive rats. There was a significant decrease in Cx43 and Cx40 proteins due to volume overload in both atria of normotensive as well as hypertensive rats. Treatment with trandolapril increased Cx43 and Cx40 levels in right and left heart atria of normotensive as well as hypertensive volume overloaded rats. While losartan increased Cx43 and did not affect Cx40 in left and right heart atria of volume overloaded rats. Findings of this study point out that right heart atria of normotensive as well as hypertensive rats are more susceptible to volume overload comparing to the left heart atria. Trandolapril attenuated pro-arrhythmic downregulation of Cx43 and Cx40 in atria of volume overloaded normotensive as well as hypertensive rats. This fact as well as examining AF inducibility requires further investigation.

Keywords: hypertensive TGR; normotensive HSD; ACF volume overload; left and right atria; connexin-40; connexin-43; trandolapril; losartan

1. Introduction

Arterial hypertension (HTN)-induced hemodynamic pressure overload is a prevalent cardiovascular risk factor associated with the development of heart failure (HF), malignant cardiac arrhythmias, and adverse kidney events in both clinical and experimental settings [1–4]. Mechanical dysfunction and electrical instability aggravate over time and contribute significantly to morbidity and mortality [5], thereby highlighting the need for further cardioprotective investigation. Spontaneously hypertensive rats (SHR, a polygenic model imitating human primary HTN) and the hypertensive Ren-2 transgenic rats (TGR, a monogenic model of HTN) are experimental models imitating HTN associated with pressure overload [3,6]. Structural remodeling—mainly myocardial hypertrophy and interstitial collagen deposition—accompanied by disorders of electrical cardiomyocyte coupling at the connexin channels is considered a key factor in the deterioration of cardiac function and the increased vulnerability to arrhythmias [7,8]. On the other hand, renin–angiotensin–aldosterone system (RAAS) inhibition, resulting in the suppression of fibrosis, has been correlated with a reduction in arrhythmias [9]. Despite recent advances in cardiovascular pharmacotherapy, the prevention and treatment of resistant HTN and pulmonary arterial hypertension remain limited [10]. This highlights the need for novel, more effective, and safer pharmacological strategies.

Compared to pressure overload, much less is known about the impact of volume overload, which is typical in decompensated patients or those with renal disease/dysfunction [2,11]. In a rat model of acute decompensation due to volume overload induced by aorto-caval fistula (ACF), the condition—when combined with a hypertensive stimulus—progressively deteriorates over time into chronic congestive HF [12]. Myocardial stretch occurring during hemodynamic overload regulates caveolae-linked β_2-adrenergic receptor (β_2-AR) signaling [13], which contributes to the pathogenesis of HF. However, despite the high vulnerability to atrial fibrillation (AF), affecting approximately 60–80% of hypertensive patients [14], there remains a gap in comprehensive knowledge regarding atrial alterations. Chronic pressure or volume overload promotes AF [15–17] particularly through enhanced extracellular matrix accumulation and fibrosis, which contribute to the arrhythmogenic substrate. Hemodynamic overload of the atria is an important pathogenic factor in structural remodeling, including fibrosis [14,18]. The latter results from the stretch of overloaded cardiac myocytes, which directly activates fibroblasts [19]. Noteworthy, structural alterations accompanied by disturbances in the electrical coupling protein connexin-43 (Cx43) are associated with abnormal myocardial conduction, which is highly pro-arrhythmic [20], and contribute to the development of cardiac arrhythmias, including atrial fibrillation (AF). AF is further characterized by elevated levels of inflammatory biomarkers [21]. In contrast, improving intercellular electrical communication through Cx43 and Cx40 gap junction channels expressed in heart atria [22] alleviates AF [23].

However, it remains uncertain whether the use of specific antihypertensive drugs, such as those inhibiting the RAAS, can reduce pro-arrhythmic factors and thereby lower the risk of AF. Emerging antihypertensive drug classes offer new opportunities for managing HTN [2,24]. However, their long-term effects on cardiovascular, kidney, and metabolic outcomes as well as on arrhythmias, including AF, remain to be elucidated. This is particularly important given the rising prevalence of AF, which is becoming one of the major epidemics and public health challenges.

Based on the above clinical and preclinical findings, including our recently published data obtained from heart ventricles of volume overloaded rats [7], we aimed to investigate myocardial structure, enzyme histochemistry, and expression of the electrical coupling proteins Cx43 and Cx40 in the left and right atria of hypertensive as well as normotensive rats subjected to volume overload. In addition, we evaluated the efficacy of RAAS with trandolapril (an ACE inhibitor) and losartan (an angiotensin receptor blocker).

2. Materials and Methods

2.1. Experimental Design

This study was performed using male heterozygous hypertensive Ren-2 transgenic rats [(mREN2) 27, TGR; n = 40] and normotensive Hannover Sprague Dawley rats (HSD; n = 40), housed under standard laboratory conditions (22 ± 1 °C, 12:12 h light/dark cycle), with ad libitum access to chow and water. Congestive heart failure (HF) due to volume overload was induced in 8-week-old animals by creating an aortocaval fistula (ACF) using an 18-gauge needle (1.2 mm in diameter), followed by sealing with cyanoacrylate adhesive [7,11]. ACF induction was confirmed by the presence of pulsatile blood flow in the inferior vena cava. Sham-operated rats served as controls. Five weeks after ACF-induced high-volume HF in normotensive and hypertensive rats (as documented in previous study [11]) examining rat heart ventricles and kidney a pharmacological intervention was initiated for a duration of 15-week. Rats received daily either the angiotenzin II type 1 (AT_1) receptor blocker, losartan (200 mg/L), or the angiotenzin-converting enzyme inhibitor (ACEi) trandolapril (6 mg/L) in drinking water. Age-matched untreated rats were used for comparison. At 28 weeks of age, animals were euthanized by decapitation. Blood serum, as well as the left and right ventricles and atria, were collected. Biometric parameters were recorded, and heart tissue samples were snap-frozen in liquid nitrogen and stored at −80 °C.

All procedures were approved by the Animal Care and Use Committee of the Institute for Clinical and Experimental Medicine (project number 50/2017) and adhered to Directive 2010/63/EU and ARRIVE 2.0 guidelines as previously reported [7,11].

2.2. Serum Biomarker Assessment

Atrial natriuretic peptide (ANP), a biomarker of heart failure, was quantified in serum using a commercial ELISA kit (RAB00385, Sigma-Aldrich, St. Louis, MO, USA) according to the manufacturer's instructions. Serum samples and standards were incubated in antibody-coated wells, followed by incubation with HRP-conjugated streptavidin and TMB substrate. Absorbance was measured at 450 nm using a microplate reader (Synergy H1, BioTek, Winooski, VT, USA), and ANP concentrations (pg/mL) were calculated based on the standard curve.

Lipid peroxidation was assessed by measuring thiobarbituric acid reactive substances (TBARS) using a modified spectrophotometric method [25]. Serum samples or standards were treated with trichloroacetic acid and TBARS reagent, incubated at 100 °C for 70 min, and subsequently extracted with a mixture of n-butanol and pyridine (14:1). After centrifugation (5000× *g*, 10 min), the absorbance of the organic phase was measured at 535 nm using a microplate reader (Synergy H1, BioTek, Winooski, VT, USA). Malondialdehyde (MDA) concentrations were determined from a standard curve prepared with tetrabutylammonium MDA salt.

Zymographic analysis of MMP-2 gelatinolytic activity was performed as previously described by [7]. In brief, protein samples were subjected to SDS-PAGE in gelatin-containing gels under non-reducing conditions. After electrophoresis, gels were incubated in substrate buffer and subsequently stained with Coomassie Brilliant Blue. Gelatinolytic activity was visualized as clear bands against a dark background.

TIMP-2 levels were measured using a Rat TIMP-2 ELISA kit (RAB1156, Sigma-Aldrich, St. Louis, MO, USA) according to the manufacturer's instructions. Samples were sequentially incubated with a biotinylated detection antibody, HRP-conjugated streptavidin, and TMB substrate. Absorbance was measured at 450 nm using a microplate reader (Synergy H1, BioTek, Winooski, VT, USA), and TIMP-2 concentrations (pg/mL) were calculated from a standard curve.

2.3. Microscopic Examination of Left and Right Heart Atrial Tissue

Conventional hematoxylin-eosin and Masson's trichrome staining were performed on 10 μm-thick cryosections of the right and left atria to evaluate structural alterations.

Catalytic enzyme histochemistry was performed on 10 μm-thick cryosections of the right and left atria to assess the arterial and venous capillary network in situ by determining the activity of endothelial alkaline phosphatase (AP, EC 3.1.3.1), a marker of arterial capillaries, and dipeptidyl peptidase-4 (DPP4, EC 3.4.15.4), a marker of endothelial venous capillaries [26,27]. For the detection of AP activity, tissue sections were incubated in a buffered medium containing sodium β-glycerophosphate as the substrate, resulting in blue staining of arterial capillaries. DPP4 activity was visualized using glycyl-L-proline-4-methoxy-β-naphthylamide as the specific substrate, resulting in red staining of venous capillaries.

The activity of glycogen phosphorylase (GP, EC 2.4.1.1), the enzyme involved in myocardial glycogen breakdown, was determined according to [26]. Cryosections (10 μm thick) of the right and left atria were incubated in a buffer solution containing glycogen as a primer and glucose-1-phosphate as the glycosyl group donor. The product of the reaction was a purple-colored final precipitation.

Histochemical activities of the mitochondrial enzymes succinate dehydrogenase (SDH, EC 1.3.5.1) and β-hydroxybutyrate dehydrogenase (β-HBDH, EC 1.1.1.30), involved in myocardial energy metabolism, were examined according to the method of [26]. Cryosections (10 μm thick) of the right and left atria were incubated in a phosphate-buffered medium containing succinate and β-hydroxybutyrate as substrates, NAD as a coenzyme, and nitroblue tetrazolium chloride as a hydrogen acceptor, resulting in the formation of blue formazan granules indicating enzyme activity. All images were acquired and examined using a Zeiss Apotome 2 light microscope (Carl Zeiss Microscopy GmbH, Jena, Germany).

To assess the expression and localization of connexin-43 (Cx43) in the atrial myocardium, immunofluorescence staining was performed on 10 μm-thick cryosections of the right and left atria, following the protocol described by [28]. Tissue sections were incubated with a primary mouse anti-Cx43 antibody (1:700; MAB3068, CHEMICON International, Inc., Temecula, CA, USA), followed by a FITC-conjugated secondary antibody specific to mouse IgG (1:1000; Jackson ImmunoResearch Laboratories, West Grove, PA, USA). Immunofluorescence was examined using a Zeiss Apotome 2 flourescent microscope (Carl Zeiss Microscopy GmbH, Jena, Germany).

Quantitative image analysis of microscopic images was not assessed because the number of hypertensive rats suffering from volume overload HF was reduced due to mortality.

2.4. Western Blot Analysis of Left and Right Atrial Tissue

About 100 mg of frozen left and right heart atrium was homogenized in ice-cold lysis buffer containing 180 mmol/L KCl, 4 mmol/L EDTA, pH 7.4. A protease inhibitor cocktail (Sigma-Aldrich, St. Louis, MO, USA; #P8340) was also added to prevent protein degradation. Protein extracts were mixed with Laemmli sample buffer and heated prior to electrophoresis. Equal amounts of protein (6–60 μg per lane) were separated on 10% SDS–polyacrylamide gels (Mini-Protean TetraCell, Bio-Rad, Hercules, CA, USA) at a constant voltage of 90 V and transferred onto 0.2 μm nitrocellulose membranes (Advantec, Tokyo, Japan). Membranes were blocked in 5% low-fat milk (*w*/*v* in TBS-T) for 4 h at room temperature and incubated overnight at 4 °C with the appropriate primary antibodies: anti-total Cx43—1:5000, C6219, Sigma-Aldrich, MI, USA; anti-Cx40—1:2000, sc-365107, Santa Cruz Biotechnology, Dallas, TX, USA; anti-PKCε—1:1000, sc-214, Santa Cruz Biotechnology, TX, USA; anti-PKCδ—1:1000, sc-213, Santa Cruz Biotechnology, Dallas, TX, USA; anti-MMP-2—1:1000, sc-10736, Santa Cruz Biotechnology, Dallas, TX, USA; anti-Galectin-3—1:1000, #89572, Cell Signaling Technology, Denver, CO, USA; anti-ADAMTS5—1:500, ab41037, Abcam, Cambrige, UK; anti-GAPDH—1:1000, sc-25778, Santa Cruz Biotechnology, TX, USA. After washing in TBS-T, membranes were incubated with horseradish peroxidase-conjugated secondary antibodies: anti-rabbit antibody: 1:2000, 7074S, Cell Signaling Technology, Denver, CO, USA or anti-mouse antibody: 1:2000, 7076C, Cell Signaling Technology, Denver, CO, USA, for 1–1.5 h at room temperature. Protein bands were visualized using enhanced chemiluminescence (ECL) and quantified by densitometric analysis using Carestream Molecular Imaging Software (version 5.0; Carestream Health, New Haven, CT, USA). GAPDH was used as a loading control for protein normalization. All steps were previously described [2,29].

2.5. Statistical Evaluation

The Kolmogorov–Smirnov test was used to verify the normality of data distribution. Group differences were analyzed using one-way analysis of variance (ANOVA) followed by Bonferroni's multiple comparison test. Results are presented as means ± standard deviations (SD), and a p-value of less than 0.05 was considered statistically significant. All statistical analyses were performed using GraphPad Prism 8.0.1.

3. Results

3.1. Biometric and Biochemical Parameters

As shown in Table 1, body weight (BW) and heart weight (HW) were significantly higher in hypertensive TGR strain compared to normotensive HSD rats. Volume overload induced by ACF significantly increased the HW, as well as the weight of left (LVW) and right ventricles (RVW), in both HSD and TGR, while BW was reduced in TGR. For technical reasons, it was not possible to register the right and left atrial weights in this experiment. Eight weeks of ACEi treatment attenuated the effects of volume overload and reduced LVW and RVW in both HSD and TGR strain, with a concomitant increase in BW in TGR. These biometric parameters were not reduced in volume-overloaded HSD rats treated with ARB, whereas ARB treatment did attenuate them in volume-overloaded TGR. Due to technical reasons atrial weights were not possible to register in this experiment. Serum ANP and TBARS levels were significantly increased due to volume overload in both HSD and TGR rats; however, ANP did not change in HSD rats with ACEi treatment, and TBARS levels were not altered after ARB treatment. No significant alterations in serum MMP-2 were observed among the rat groups, except for a significant reduction in TGR treated with ACEi. Serum TIMP-2 levels were significantly increased in atria of volume-overloaded HSD rats and in TGR rats treated with ACEi. These data were previously published in a papers focused on heart ventricles [7] and renal functions [11], whereby HF development was monitored and confirmed by Doppler sonography.

Table 1. Biometric and biochemic parameters registered in experimental rats.

Parameters	HSD	HSD ACF	HSD ACF ACEi	HSD ACF ARB	TGR	TGR ACF	TGR ACF ACEi	TGR ACF ARB
BW (g)	543 ± 37	548 ± 45	537 ± 31	575 ± 56	652 ± 46 ^{a}	483 ± 46 ^{c}	606 ± 53 ^{d}	643 ± 38 ^{d}
HW (mg/mm)	40.5 ± 3.3	71.2 ± 5.2 ^{a}	59.2 ± 4.7 ^{ab}	63.8 ± 4.4 ^{a}	52.0 ± 4.1 ^{a}	76.2 ± 3.0 ^{c}	51.5 ± 8.7 ^{d}	65.2 ± 6.7 ^{cd}
LVW (mg/mm)	18.8 ± 1.9	27.0 ± 2.8 ^{a}	23.6 ± 3.2 ^{a}	24. 7 ± 1.3 ^{a}	22.2 ± 2.7	33.5 ± 3.0 ^{c}	17.1 ± 3.5 ^{d}	22.4 ± 1.8 ^{d}
RVW (mg/mm)	5.2 ± 0.4	12.9 ± 1.9 ^{a}	12. ± 1.2 ^{a}	13.4 ± 1.34 ^{a}	8.04 ± 1.3 ^{a}	12.4 ± 0.4 ^{c}	10.4 ± 1.6	13.5 ± 2.68 ^{c}
Serum ANP (pg/mL)	2.7 ± 0.3	3.4 ± 0.3 ^{a}	2.9 ± 0.5	2.3 ± 0.1 ^{b}	2.4 ± 0.2	4.0 ± 0.7 ^{c}	3.3 ± 0.2 ^{d}	3.3 ± 0.9 ^{d}
Serum TBARS (µmol/µg)	18.7 ± 0.8	25.0 ± 0.8 ^{a}	19.9 ± 1.9 ^{b}	20.7 ± 1.2 ^{b}	21.3 ± 0.8	24.5 ± 1.1 ^{c}	20.1 ± 0.9 ^{d}	21.9 ± 2.1
Serum MMP-2 (%)	100 ± 8	95 ± 3	83 ± 12	96 ± 4	80 ± 7	65 ± 5	60 ± 12 ^{c}	78 ± 4
Serum TIMP-2 (pg/mL)	245 ± 39	212 ± 2	341 ± 17 ^{ab}	211 ± 20	171 ± 21	144 ± 20	226 ± 43 ^{d}	134 ± 13

BW—body weight (g), HW—heart weight (g), LVW—left ventricular weight normalized to tibia length (mg/mm), RVW—right ventricular weight normalized to tibia length (mg/mm), ANP—Atrial natriuretic peptide (pg/mL), TBARS—thiobarbituric acid reactive substances (µmol/µg), MMP-2—Matrix metalloproteinase-2, TIMP-2—Tissue inhibitor of metalloproteinases-2 (pg/mL), HSD—Hannover Sprague Dawley rats; TGR—Ren-2 transgenic rats; ACF—aortocaval fistula, surgical model of volume overload; ACEi—treatment with the angiotensin-converting enzyme inhibitor, trandolapril; ARB—treatment with an angiotensin II type 1 (AT_1) receptor blocker, losartan. n = 10 per group. Data are presented as means ± SD; ^{a} $p < 0.05$ vs. HSD, ^{b} $p < 0.05$ vs. HSD ACF, ^{c} $p < 0.05$ vs. TGR, ^{d} $p < 0.05$ vs. TGR ACF. These data were previously published in [7,11].

3.2. Microscopic Findings of Left and Right Atrial Tissue of Volume-Overloaded Normotensive and Hypertensive Rats

Microscopic alterations demonstrated on images point out the differences between right and left heart atria including responsiveness to volume overload induced congestive heart failure in normotensive HSD rats and hypertensive TGR, as well as impact of treatment with trandolapril and losartan. See Figures 1–4.

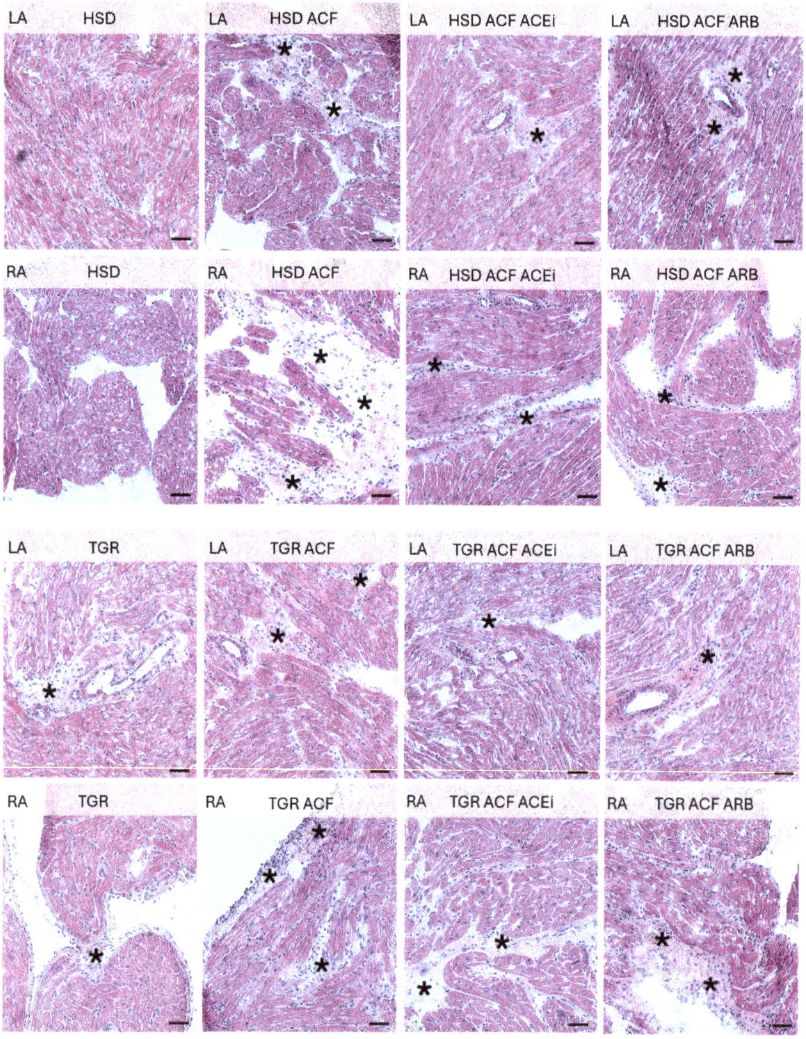

Figure 1. Haematoxylin-eosin staining of the left (LA) and right atria (RA) of experimental rats. Abundant polymorphonuclear cells (asterisk) are present in the RA of normotensive HSD rats due to ACF-induced volume overload, which was attenuated by treatment with ACE inhibitors (ACEi) and angiotensin receptor blockers (ARB). Compared to HSD rats, the atria of hypertensive TGR exhibit a higher number of polymorphonuclear cells, with volume overload further enhancing their presence, particularly in the RA. This effect was attenuated by ACEi treatment. HSD—Hannover Sprague Dawley rats; TGR—Ren-2 transgenic rats; ACF—aortocaval fistula, surgical model of volume overload; ACEi—treatment with the angiotensin-converting enzyme inhibitor, trandolapril; ARB—treatment with an angiotensin II type 1 (AT_1) receptor blocker, losartan. Scale bar: 100 μm.

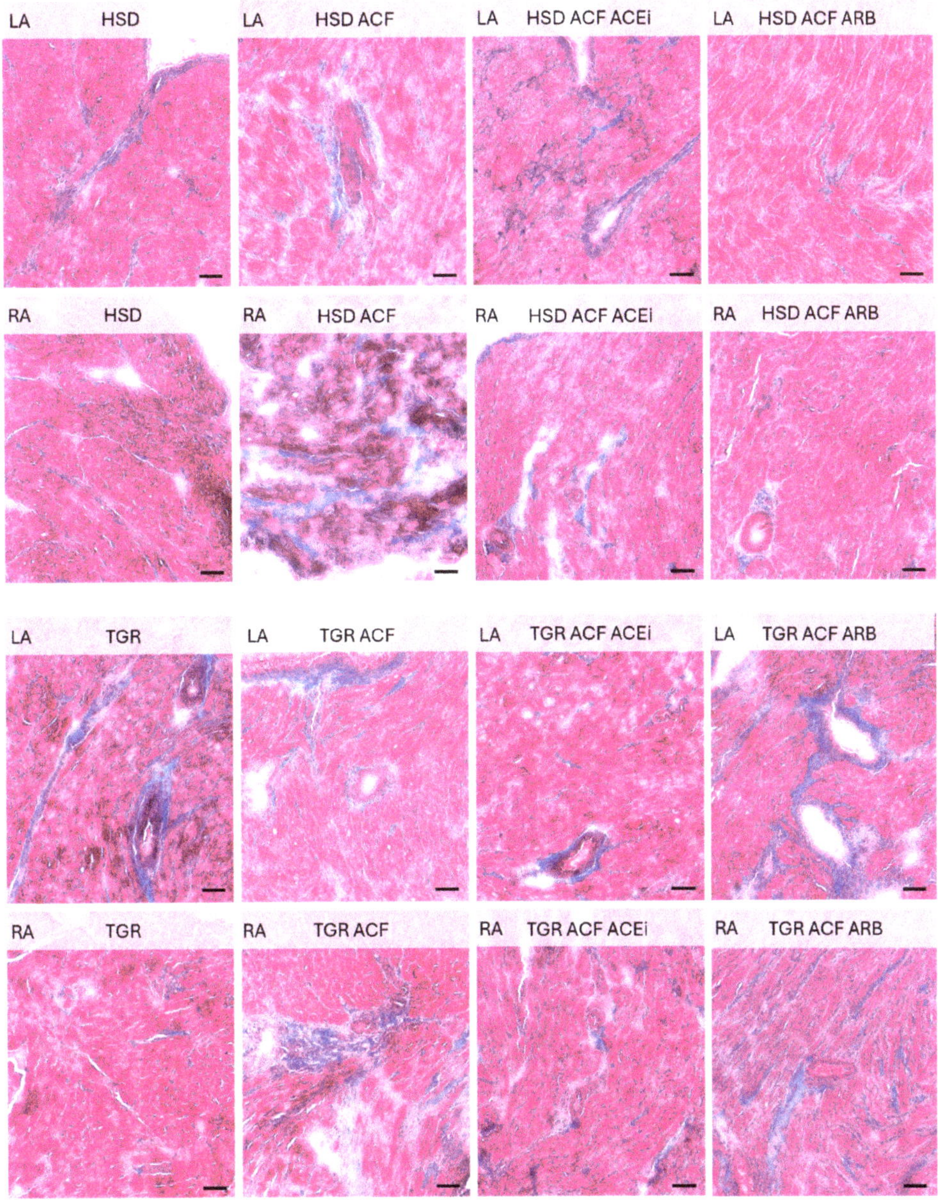

Figure 2. Masson trichrom staining of interstitial and periarterial collagen fibers (blue color) in left (LA) and right atria (RA) of experimental rats. Cardiomyocytes are stained pink. An increase in collagen deposition due to ACF-induced volume overload is evident in the left (LA) and especially in the right atria (RA) of normotensive HSD rats, which was attenuated by treatment with ACE inhibitors (ACEi) or angiotensin receptor blockers (ARB). In hypertensive TGR rats, the LA showed baseline fibrosis that was not further exacerbated by volume overload, in contrast to the RA, which exhibited massive interstitial collagen accumulation. Treatment partially suppressed extracellular matrix alterations. HSD—Hannover Sprague Dawley rats; TGR—Ren-2 transgenic rats; ACF—aortocaval fistula, surgical model of volume overload; ACEi—treatment with the angiotensin-converting enzyme inhibitor, trandolapril; ARB—treatment with an angiotensin II type 1 (AT_1) receptor blocker, losartan. Scale bar: 100 μm.

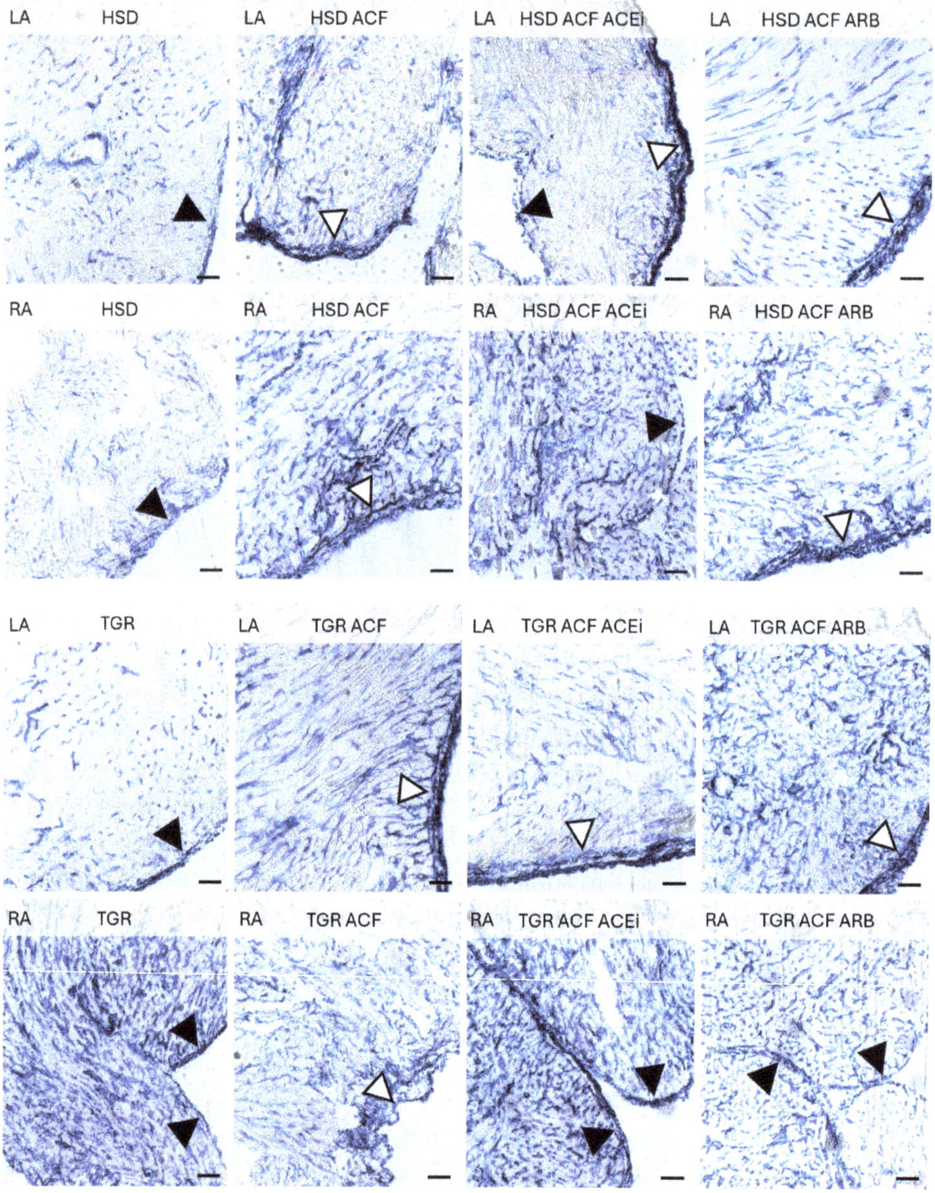

Figure 3. Alkaline phosphatase activity in the arterial capillary network (blue color) in the right (RA) and left atria (LA) of experimental rats. ACF-induced volume overload increased the density of arterial capillaries, particularly in the RA of normotensive HSD rats, along with elevated enzyme activity in the epicardium (white triangle), while the endocardium is highlighted with black triangles. These qualitative changes persisted despite treatment. In hypertensive TGR rats, the density of alkaline phosphatase-positive capillaries was higher in the RA compared to the LA and further enhanced in the LA following ACF. Treatment had no apparent effect. HSD—Hannover Sprague Dawley rats; TGR—Ren-2 transgenic rats; ACF—aortocaval fistula, surgical model of volume overload; ACEi—treatment with the angiotensin-converting enzyme inhibitor, trandolapril; ARB—treatment with an angiotensin II type 1 (AT_1) receptor blocker, losartan. Scale bar: 100 μm.

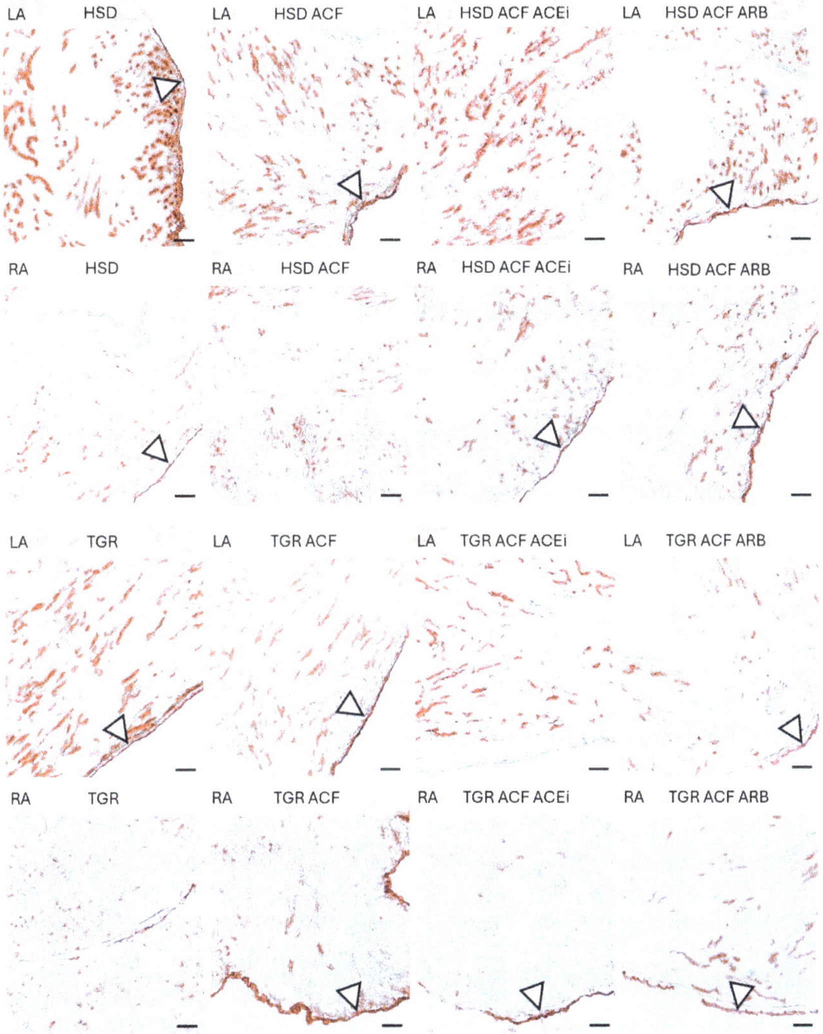

Figure 4. Dipeptidyl peptidase-4 (DPP4) activity (red color) in the venous capillary network of the left (LA) and right atria (RA) of experimental rats. Note the lower density and activity of DPP4 in the RA compared to the LA in both normotensive and hypertensive rats, regardless of intervention. White triangles indicate DPP4 activity in the epicardial region. HSD—Hannover Sprague Dawley rats; TGR—Ren-2 transgenic rats; ACF—aortocaval fistula, surgical model of volume overload; ACEi—treatment with the angiotensin-converting enzyme inhibitor, trandolapril; ARB—treatment with an angiotensin II type 1 (AT_1) receptor blocker, losartan. Scale bar: 100 µm.

Volume overload resulted in increased number of polymorphonuclears in right atria of normotensive rats and to lesser extent in TGR, while treatment with trandolapril suppressed their incidence in both normotensive and hypertensive rat strain (Figure 1).

As shown in Figure 2, an increase in interstitial collagen deposition due to volume overload was evident in the left and namely in the right atria of normotensive rats, which was attenuated by treatment with angiotensin-converting enzyme inhibitor trandorapril or angiotensin receptor blockers losartan. Peri-arterial fibrosis was observed in the let atria of hypertensive rats that was not further exacerbated by volume overload. Whilst the right atria exhibited massive interstitial and perivascular collagen accumulation. Treatment with trandorapril partially suppressed extracellular matrix formation.

Volume overload increased the density of arterial capillaries, particularly in the right atria of normotensive rats, along with elevated enzyme activity in the epicardium as demonstrated in Figure 3. These alterations persisted despite the treatment. Of interest, the density of the arterial capillary network exhibiting alkaline phosphatase activity was higher in the right versus left atria of hypertensive rats. Volume overload reduced the capillary network in the right atria of hypertensive rats while increasing it in normotensive rats. Treatment with trandolapril partially attenuated volume overload-induced alterations.

Dipeptidyl peptidase-4 activity in the venous capillary network of the left and right atria of experimental rats demonstrated in Figure 4, revealed a higher density in the left versus right atria in normotensive as well as hypertensive rats. Neither volume overload nor treatment affected this pattern.

Immunolabeling of connexin-43 revealed its dominant localization at the intercalated disks of the cardiomyocytes in the left and right atria of normotensive rats, as demonstrated in Figure 5. In contrast, lower myocardial density of Cx43 and its mis-localization to the lateral cardiomyocyte membranes were detected in atria of hyper-tensive rats. Volume overload reduced Cx43 in right and left atria of normotensive rats and to lesser extent in hypertensive rats, while abnormal lateral localization persisted. Treatment with trandolapril or losartan increased Cx43 and attenuated its mis-localization in atria of volume overloaded normotensive and partially in hypertensive rat heart.

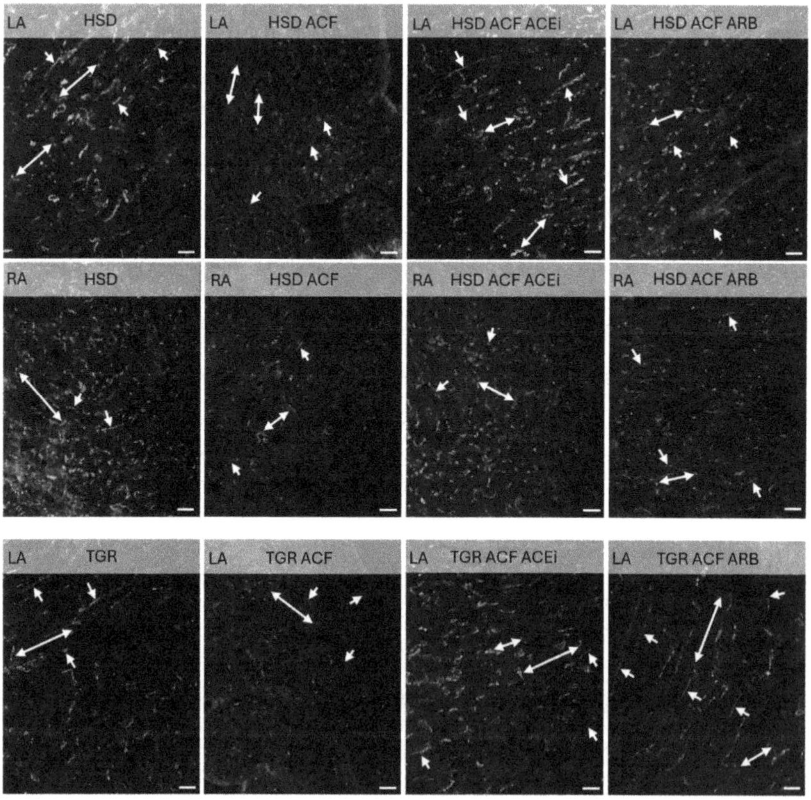

Figure 5. *Cont.*

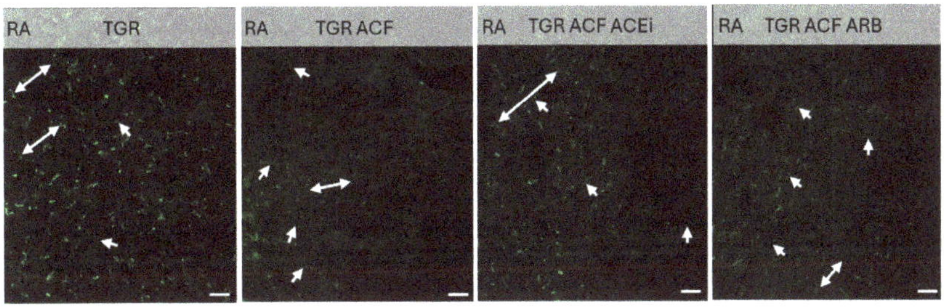

Figure 5. Immunolabeling of connexin-43 (green color) revealed higher expression predominantly at the intercalated disks (double arrows) in the left (LA) and right atria (RA) of normotensive HSD rats. In contrast, lower myocardial expression of Cx43 and its mislocalization to the lateral membranes (single arrows) were observed in hypertensive TGR. ACF-induced volume overload reduced Cx43 expression in both atria of HSD rats, while the reduction was less pronounced in TGR. Moreover, volume overload increased the localization of Cx43 to the lateral membranes of cardiomyocytes in HSD rats, whereas lateral localization persisted in hypertensive TGR. Treatment with ACEi or ARB increased Cx43 levels and attenuated its mislocalization in both atria of HSD rats, and to a lesser extent in TGR. HSD—Hannover Sprague Dawley rats; TGR—Ren-2 transgenic rats; ACF—aortocaval fistula, surgical model of volume overload; ACEi—treatment with the angiotensin-converting enzyme inhibitor, trandolapril; ARB—treatment with an angiotensin II type 1 (AT_1) receptor blocker, losartan. The scale bar indicates 20 μm.

3.3. *Findings on Examined Protein Levels in the Left and Right Atria of Volume-Overloaded Normotensive and Hypertensive Rats*

As shown in Figure 6, hypertensive TGR exhibited lower levels of Cx43 (Figure 6A) and Cx40 (Figure 6B) in both the left and right atria compared to normotensive HSD rats. A pronounced decrease in Cx43 and Cx40 protein levels was observed in both atria of normotensive HSD and hypertensive TGR strain due to volume overload. Treatment with an ACEi increased Cx43 and Cx40 levels in both atria of volume overload normotensive HSD and hypertensive TGR. In contrast, treatment with an ARB increased Cx43 but did not affect Cx40 levels in either atrium of volume-overloaded rats.

PKCε is involved in the phosphorylation of Cx43 and Cx40, which activates their function. As shown in Figure 7A, compared to normotensive HSD rats, PKCε levels were lower in both the left and right atria of hypertensive TGR. Volume overload significantly reduced PKCε protein levels only in the left atrium of normotensive HSD rats. Treatment with an ACEi increased PKCε levels in both atria of volume-overloaded normotensive HSD and hypertensive TGR strain.

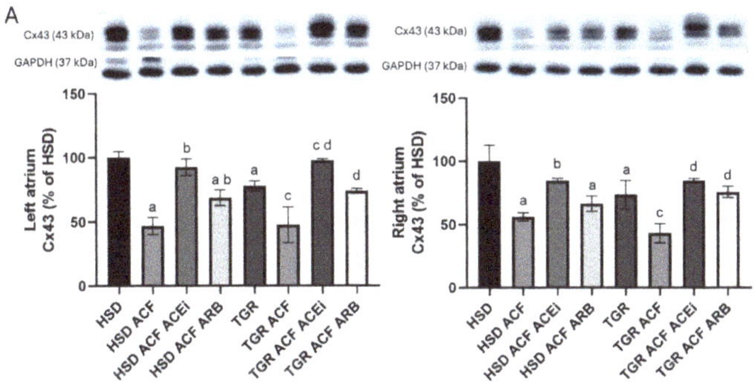

Figure 6. *Cont.*

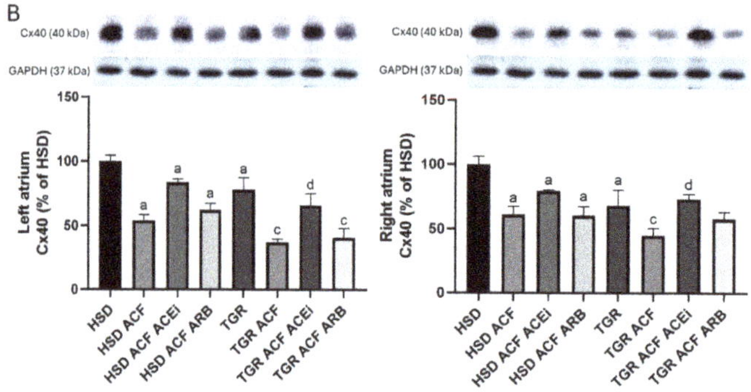

Figure 6. Protein levels of Cx43 (**A**) and Cx40 (**B**) normalized to GAPDH assessed by Western blot analysis. Hypertensive TGR rats exhibited lower levels of Cx43 and Cx40 in both the left and right atria compared to normotensive rats. A pronounced decrease in Cx43 and Cx40 protein levels was observed in both atria of normotensive HSD and hypertensive TGR rats due to volume overload. Treatment with an ACEi increased Cx43 and Cx40 levels ($p > 0.05$) in both atria of volume-overloaded normotensive and hypertensive rats. In contrast, ARB treatment increased Cx43 but did not affect Cx40 levels in either atrium of volume-overloaded rats. HSD—Hannover Sprague Dawley rats; TGR—Ren-2 transgenic rats; ACF—aortocaval fistula, surgical model of volume overload; ACEi—treatment with the angiotensin-converting enzyme inhibitor, trandolapril; ARB—treatment with an angiotensin II type 1 (AT_1) receptor blocker, losartan. n = 10 per group. Data are presented as means ± SD; ^{a} $p < 0.05$ vs. HSD, ^{b} $p < 0.05$ vs. HSD ACF, ^{c} $p < 0.05$ vs. TGR, ^{d} $p < 0.05$ vs. TGR ACF. Western blot original images can be found in Figure S1.

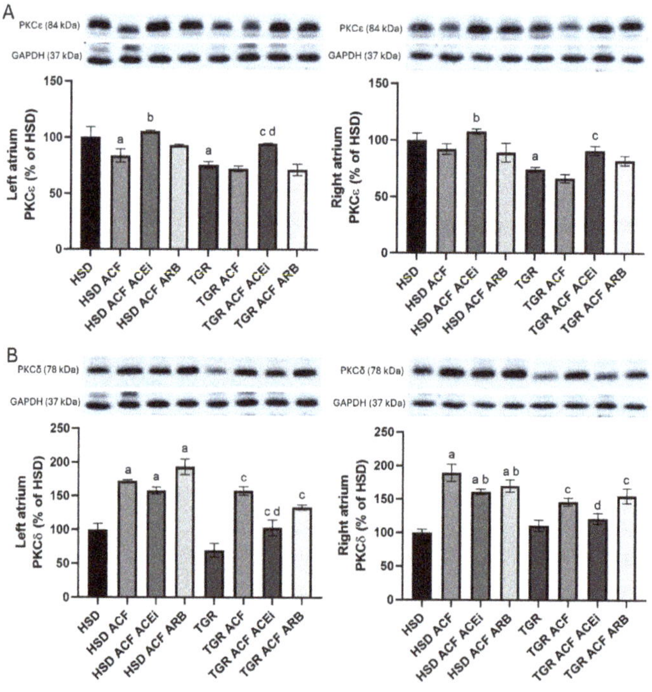

Figure 7. Protein levels of PKCε (**A**) and PKCδ (**B**) normalized to GAPDH assessed by Western blot analysis. Compared to normotensive HSD rats, PKCε levels were lower in both the left and right atria of hypertensive TGR rats. ACF-induced volume overload significantly reduced PKCε protein levels

only in the left atrium of normotensive HSD rats. Treatment with an ACEi increased PKCε levels in both atria of volume-overloaded normotensive HSD and hypertensive TGR. PKCδ protein levels were significantly increased in both the left and right atria of normotensive HSD and hypertensive TGR rats in response to volume overload. Treatment with an ACE inhibitor reduced PKCδ levels only in the left and right atria of volume-overloaded hypertensive TGR rats. HSD—Hannover Sprague Dawley rats; TGR—Ren-2 transgenic rats; ACF—aortocaval fistula, surgical model of volume overload; ACEi—treatment with the angiotensin-converting enzyme inhibitor, trandolapril; ARB—treatment with an angiotensin II type 1 (AT_1) receptor blocker, losartan. n = 10 per group. Data are presented as means ± SD; [a] $p < 0.05$ vs. HSD, [b] $p < 0.05$ vs. HSD ACF, [c] $p < 0.05$ vs. TGR, [d] $p < 0.05$ vs. TGR ACF. Western blot original images can be found in Figure S2.

PKCδ is involved in pro-hypertrophic and pro-apoptotic signaling. As shown in Figure 7B. PKCδ protein levels were significantly increased due to volume overload in both atria of normotensive HSD rats and hypertensive TGR. Treatment with an ACEi reduced PKCδ levels in the left and right atria of volume-overloaded hypertensive TGR only and did not alter the levels in the atria of normotensive rats.

Proteins involved in extracellular matrix remodeling are shown in Figure 8.

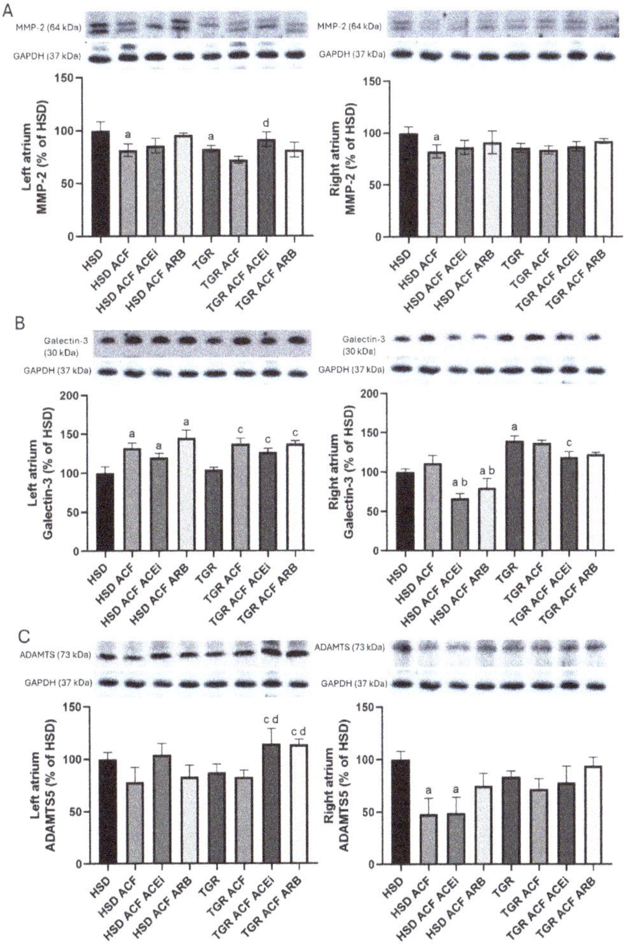

Figure 8. Protein levels of MMP-2 (**A**), Galectin-3 (**B**), and ADAMTS (**C**) normalized to GAPDH assessed by Western blot analysis. MMP-2 protein levels were significantly reduced in both the left and

right atria due to volume overload, but only in normotensive HSD rats. Treatment with an ACEi increased MMP-2 levels in the left atrium of volume-overloaded TGR rats. Galectin-3 protein levels were increased in the left atrium in response to volume overload, regardless of treatment. Interestingly, Galectin-3 levels were significantly reduced in the right atrium of volume-overloaded HSD rats treated with either an ACEi or an ARB. In contrast to the left atrium, Galectin-3 levels were significantly increased in the right atrium of volume-overloaded TGR rats. ADAMTS protein levels were not affected by volume overload in the left atrium, but were decreased in the right atrium of normotensive HSD rats. Notably, treatment with either an ACE inhibitor or an ARB increased ADAMTS protein levels in the left atrium of volume-overloaded TGR rats. Data are presented as means ± SD; ^{a} $p < 0.05$ vs. HSD, ^{b} $p < 0.05$ vs. HSD ACF, ^{c} $p < 0.05$ vs. TGR, ^{d} $p < 0.05$ vs. TGR ACF. Western blot original images can be found in Figure S3.

MMP-2 levels (Figure 8A) were significantly reduced in both the left and right atria due to volume overload, but only in normotensive HSD rats. Treatment with an ACEi increased MMP-2 protein levels in the left atrium of volume-overloaded TGR rats.

Galectin-3 levels (Figure 8B) were increased in the left and to a lesser extent in the right atria of normotensive and hypertensive rats in response to volume overload, regardless of treatment. Interestingly, Galectin-3 was significantly reduced in the right atria following treatment of volume-overloaded normotensive rats with either an ACEi or an ARB.

ADAMTS levels (Figure 8C) were not affected by volume overload in the left atrium, but were decreased in the right atrium of normotensive HSD rats. Notably, treatment with either an ACEi or an ARB increased ADAMTS proteins in the left atrium of volume-overloaded TGR rats.

4. Discussion

Compared to the heart ventricles, much less is known about atria under various pathological conditions, despite the fact that atrial fibrillation (AF), the most common arrhythmia and a major risk factor for ischemic stroke, is permanently increasing [30]. Moreover, AF and heart failure (HF) are two commonly coexisting conditions that are frequently encountered in clinical practice [14,31]. In this context, our aim was to investigate myocardial structure, enzyme histochemistry, and the expression of electrical coupling proteins Cx43 and Cx40 in the left and right atria of male hypertensive Ren-2 transgenic rats (TGR) and normotensive Hannover Sprague Dawley (HSD) rats subjected to volume overload to induce congestive HF. In addition, the efficacy of RAAS blockade using trandolapril (an ACE inhibitor) and losartan (an angiotensin receptor blocker) was evaluated.

According to the recorded biometric parameters (Table 1), which were previously published in papers focused on heart ventricles [7] and renal functions [11], hypertensive rats exhibited higher body weight (BW) and heart weight (HW) compared to normotensive rats. Volume overload increased HW, left ventricular weight (LVW), and right ventricular weight (RVW) in both rat strains, while it reduced BW in hypertensive rats. When exposed to identical volume overload, both ventricles display similar upregulation of stress and metabolic markers [32]. Relatively larger response of volume overloaded in the right ventricle compared to the left may be caused by concomitant pulmonary hypertension [33]. However, there is still minimal data on the adaptation of the right ventricle to pressure and volume overload and the transition to right ventricular failure. No evidence supports right ventricular chamber-specific regulation of protein expression in response to volume overload. Although on the cellular level, the remodeling responses of the right and left ventricles to pressure overload are largely similar, there are several key differences: the stressed right ventricle is more susceptible to oxidative stress, has a reduced angiogenic response, and is more likely to activate cell death pathways than the stressed left ventricle [33]. The need to better understand the molecular mechanisms allows for the development of right ventricle-specific heart failure therapeutics. This would be much appreciated in managing pulmonary hypertension. It is supposed that atrial dilatation due to volume overload may increase their weight as well. However, biometric parameters of atria are missing in the current study because their registration during harvesting of atrial tissue was not possible. Nonetheless, treatment of volume overload with trandolapril improved these attenuated altered ventricular biometric ventricular parameters in hypertensive rat strain [7]. While treatment with losartan did not affect biometric parameters in volume-overloaded normotensive rats, unlike in hypertensive rats. It might be explained by different mechanisms of their antihypertensive effects. Mean arterial pressure was higher in hypertensive versus normotensive rats, and volume overload significantly reduced this parameter only in hypertensive rats. Treatment with either trandolapril or losartan suppressed this parameter only in hypertensive volume-overloaded rats, as previously published [11]. In addition, cardiac output was significantly increased due to volume overload in normotensive as well as hypertensive rats, and even more after treatment with losartan. In addition, treatment reduced serum levels of atrial natriuretic peptide (ANP), thiobarbituric acid reactive substances (TBARS), and tissue inhibitor of metalloproteinases-2 (TIMP-2), all of which were elevated due to volume overload in the atria of both rat strains. Matrix metalloproteinases (MMPs), a family of proteolytic enzymes, and TIMP regulate the extracellular matrix turnover in cardiac tissue and are involved in cardiac fibrosis [34]. Inhibitors of matrix MMP 2 may minimize cardiac structural and functional alterations, thereby fighting heart failure [35]. We detected that volume overload decreased MMP 2 protein level in both atria of normotensive as well as hypertensive rat strain (Figure 8), while only treatment with trandolapril abolished this decrease in left atria of hypertensive strain. An increase in collagen deposition due to volume overload was detected in the right atria of normotensive rats, which was attenuated by treatment with either trandolapril or losartan, as shown in Figure 2. In the left atria of hypertensive rats, periarterial and interstitial fibrosis was not further aggravated by volume overload, in contrast to the right atrium, where interstitial collagen deposition was increased. Treatment partially attenuated extracellular matrix alterations in atria of volume-overloaded hypertensive rats, likewise in the ventricles, as we previously reported in the left ventricle [7]. Moreover, as shown in microscopic images in Figure 1, treatment with trandolapril or losartan reduced the abundance of polymorphonuclear cells induced by volume overload, particularly in the right atrium of normotensive rats. Compared to normotensive rats, the hypertensive strain exhibited more polymorphonuclear cells in the left atrium, with a further increase observed in the right atrium following volume overload. This increase was attenuated by trandolapril. Polymorphonuclears, in particular myeloid and lymphoid-derived immune cells such as macrophages, neutrophils, T-cells, and B-cells reflecting inflammation, are very active and may be involved in arrhythmia pathophysiology [36]. This issue requires further study regarding malignant cardiac arrhythmia prevention.

Adaptive changes in the myocardial capillary network were examined by in situ activity of endothelial alkaline phosphatase (Figure 3), a marker of arteriolar capillaries, and endothelial dipeptidyl peptidase-4 activity (Figure 4), a marker of the venular capillary network [27,37]. Volume overload induced a higher density of arteriolar capillaries, particularly in the right atrium of normotensive rats, along with increased enzyme activity in the epicardium. These qualitative changes were not affected by treatment with either trandolapril or losartan. In hypertensive rats, arteriolar capillaries were more abundant in the right compared to the left atrium, and their density was enhanced by volume overload in the left while reduced in the right atrium. There was no apparent effect of treatment. Venular capillary activity, indicated by dipeptidyl peptidase-4 red colored reaction, was higher in the left atrium compared to the right in both normotensive and hypertensive rats, regardless of intervention. Neither volume overload nor treatment with trandolapril or losartan altered this pattern. Of interest, the involvement of dipeptidyl peptidase-4 in the regulation of mitochondrial function and oxidative stress in cardiomyocytes has been recently reported [38] as well as in the development of cardiac fibrosis and atrial arrhythmias [39]. It should be emphasized that coronary microvascular dysfunction across the spectrum of heart pathologies contributes to the development and progression of HF [40].

Atrial energetic metabolism was assessed by in situ activity of cytoplasmic glycogen phosphorylase (Figure S4), mitochondrial succinic dehydrogenase (Figure S5), and β-hydroxybutyrate dehydrogenase (Figure S6). As shown in the supplementary data, no alterations indicating atrial ischemia were observed regardless of rat strain or volume overload conditions.

The most interesting novel findings of this study include alterations in the expression of the electrical coupling proteins Cx43 (Figure 6A) and Cx40 (Figure 6B). Compared to the normotensive rat strain, hypertensive rats exhibited lower levels of Cx43 and Cx40 in the left and right atria. Volume overload significantly suppressed the expression of Cx43 and Cx40 in both atria, regardless of rat strain. Changes in Cx43 expression were accompanied by altered localization, specifically a redistribution from intercalated disks to lateral cardiomyocyte membranes, as demonstrated in Figure 5, which impairs electrical conduction and is considered proarrhythmic [3,22,41]. It appears that atrial remodeling due to volume overload, characterized by structural changes and downregulation of Cx43 and Cx40, may impair myocardial electrical conduction and contribute to an increased propensity for AF [22], likewise in humans with congestive HF [42]. On the other hand, treatments, particularly with trandolapril, attenuated the downregulation of Cx43 and Cx40 in both the right and left atria, likewise previously found in the heart ventricles of volume-overloaded rats [7]. Trandolapril also reduced proarrhythmic electrical remodeling and mortality, indicating its beneficial therapeutic effects in volume overload HF [8]. This suggests pleiotropic antiarrhythmic effects, whose mechanisms require further investigation. Of interest, the multi-ion channel inhibitor acehytisine has shown benefits in preventing and terminating persistent AF in dilated atria caused by volume overload [17].

In addition, trandolapril increased PKCε protein levels in both right and left heart atria (Figure 7A), which were suppressed by volume overload in both normotensive and hypertensive rat strains. Considering that phosphorylation of Cx43 and Cx40 by PKCε is involved in the modulation of electrical conduction, it may therefore be expected to influence arrhythmogenesis [43]. However, treatment did not affect the volume overload-induced elevation of pro-hypertrophic PKCδ in either the left or right heart atria (Figure 7B), regardless of rat strain.

Pro-arrhythmic ventricular myocardial fibrosis during the progression of HF has been associated with increased MMP-2 levels [35]. In contrast, HF due to volume overload reduced MMP-2 protein levels in both left and right atria of normotensive rats (Figure 8A), while trandolapril unexpectedly increased MMP-2 in the left atria of hypertensive rats. Galectin-3 has emerged as a critical mediator involved in cardiac inflammation, vascular remodeling, and fibrotic processes [44]. Galectin-3 protein levels were increased in the left heart atria due to volume overload, regardless of treatment. Interestingly, Galectin-3 levels (Figure 8B) were significantly reduced in the right heart atria of volume-overloaded normotensive rats treated with either trandolapril or losartan. In contrast to the left atria, Galectin-3 was significantly increased in the right heart atria of the TGR strain. ADAMTS5 is a large extracellular matrix proteoglycan involved in extracellular matrix remodeling [45]. ADAMTS5 protein levels (Figure 8C) were not affected by volume overload in the left heart atria but were altered in the right atria of normotensive HSD rats. Of interest, treatment with either trandolapril or losartan increased ADAMTS protein levels in the left atria of volume-overloaded TGR strain.

Finally, it should be noted that the difference in action of Trandolapril and Losartan is primarily associated with their mechanisms of action. Trandolapril, an ACE inhibitor, inhibits the activity of the angiotensin-converting enzyme, a key component of the renin-angiotensin system that converts angiotensin I into angiotensin II. Thus, ACE inhibitors lead to a reduction in circulating levels of angiotensin II, which can itself directly activate or inactivate pathways of interest. For example, angiotensin II is involved in the down-regulation of Cx43 [46,47]. In contrast, Losartan, an angiotensin receptor blocker (ARB), does not reduce the level of circulating angiotensin II; it acts only by blocking the AT_1 receptors. Any non-receptor-mediated functions of angiotensin II remain unchanged.

5. Conclusions

Novel findings of this study indicate that the right atria of normotensive HSD as well as hypertensive TGR are more susceptible to volume overload compared to the left atria. The downregulation of Cx43 and Cx40 caused by volume overload strongly suggests their participation in the development of atrial fibrillation, as these gap junction proteins are essential for electrical coupling and action potential propagation among cardiomyocytes. Noteworthy, trandolapril attenuated the pro-arrhythmic downregulation of Cx43 and Cx40 in the atria of volume-overloaded normotensive and hypertensive rats, while losartan was less efficient. Based on these findings, further studies should investigate the impact of volume overload on the incidence of atrial fibrillation in both normotensive and hypertensive rats, as well as explore the molecular mechanisms underlying Cx43 and Cx40 upregulation by trandolapril.

Limitations of This Study

The main limitation of this study is the lack of examination of electrocardiographic parameters and testing propensity to atrial arrhythmias, as well as investigation of proposed antiarrhythmic mechanisms of pharmacotherapy that require further research and investigations. Another limitation of this study is the absence of quantitative analysis for microscopic findings. As the conclusions are based on representative images, they should be interpreted with caution. Further studies, including quantitative assessment, are required to validate these observations.

Supplementary Materials: The following supporting information can be downloaded at: https://www.mdpi.com/article/10.3390/biom15101457/s1, Figure S1: Protein levels of Cx43 and Cx40 normalized to GAPDH assessed by Western blot analysis; Figure S2: Protein levels of PKCε and PKCδ normalized to GAPDH assessed by Western blot analysis; Figure S3: Protein levels of MMP-2, Galectin-3, and ADAMTS normalized to GAPDH assessed by Western blot analysis. Figure S4: Microscopic images of glycogen phosphorylase activity in the left (LA) and right atria (RA) of normotensive (HSD) and hypertensive (TGR) rats in response to volume overload (ACF) and treatment with trandolapril (ACEi) or losartan (ARB). Figure S5: Microscopic images of mitochondrial succinate dehydrogenase activity in the left (LA) and right atria (RA) of normotensive (HSD) and hypertensive (TGR) rats in response to volume overload (ACF) and treatment with trandolapril (ACEi) or losartan (ARB). Figure S6. Microscopic images of mitochondrial beta-hydroxybutyrate dehydrogenase activity in left (LA) and right heart atria (RA) of normotensive (HSD) and hypertensive (TGR) rats were detected in response to volume overload (ACF) and to treatment with trandolapril (ACEi) and losartan (ARB).

Author Contributions: Methodology, M.S., K.O.A., and A.M.; validation, M.S. and N.T.; resources, M.S. and B.S.B.; data curation, M.S., K.O.A., N.T.; writing—original draft preparation, M.S., N.T.; writing—review and editing, M.S., K.O.A., B.S.B. and N.T.; visualization, M.S., N.T.; supervision, M.S., N.T.; project administration, M.S. and B.S.B.; funding acquisition, M.S. and B.S.B. All authors have read and agreed to the published version of the manuscript.

Funding: This research was funded by VEGA grants 2/0006/23 Slovak Research and Development Agency under Contract no. APVV-21-0410 and VV-MVP-24-0278; European Regional Development Fund: ITMS2014+: 313011AVG3. This research was also partially supported by the Ministry of Health of the Czech Republic within the project for the development of research organization [Grant Number 00023001 (IKEM)—institutional support] and also supported by the project National Institute for Research of Metabolic and Cardiovascular Diseases (Program EXCELES, Project No. LX22NPO5104)—Funded by the European Union—Next Generation EU.

Institutional Review Board Statement: All animal experiments were approved on 26 June 2017 by the Animal Care and Use Committee of the Institute for Clinical and Experimental Medicine, Prague; project number 50/2017; in accordance with guidelines and practices established by the Directive 2010/63/EU of the European Parliament on the Protection of Animals Used for Scientific Purposes.

Informed Consent Statement: Not applicable.

Data Availability Statement: Data are contained within the article and Supplementary Materials.

Conflicts of Interest: The authors declare no conflicts of interest.

Abbreviations

The following abbreviations are used in this manuscript:

HTN	Atrial hypertension
HF	Heart failure
SHR	Spontaneously hypertensive rats
TGR	Hypertensive Ren-2 transgenic rats
RAAS	Renin–angiotensin–aldosterone system
ACF	Aorto-caval fistula
AF	Atrial fibrillation
Cx43	Connexin-43
ACEi	Angiotensin-converting enzyme inhibitor
ARB	Angiotensin II type 1 receptor blocker
AP	Alkaline phosphatase
DPP-4	Dipeptidyl peptidase-4
GP	Glycogen phosphorylase
SDH	Succinate dehydrogenase
β-HBDH	β-hydroxybutyrate dehydrogenase
IOD	Integral optical density
BW	Body weight
HW	Heart weight
LVW	Left ventricular weight
RWV	Right ventricular weight
ANP	Atrial natriuretic peptide
TBARS	Thiobarbituric acid reactive substances
TIMP-2	Tissue inhibitor of metalloproteinases-2
HSD	Hannover Sprague Dawley
ACF	Aortocaval fistula, surgical model of volume overload
Cx40	Connexin-40
PKCε	Protein Kinase C epsilon
PKCδ	Protein Kinase C delta
MMP-2	Matrix Metalloproteinase-2
ADAMTS	A Disintegrin And Metalloproteinase with Thrombospondin Motifs

References

1. Buckley, L.F.; Baker, W.L.; Van Tassell, B.W.; Cohen, J.B.; Alkhezi, O.; Bress, A.P.; Dixon, D.L. Systolic Blood Pressure Time in Target Range and Major Adverse Kidney and Cardiovascular Events. *Hypertension* **2023**, *80*, 305–313. [CrossRef]
2. Gawrys, O.; Husková, Z.; Škaroupková, P.; Honetschlägerová, Z.; Vaňourková, Z.; Kikerlová, S.; Melenovský, V.; Bačová, B.S.; Sykora, M.; Táborský, M.; et al. The treatment with sGC stimulator improves survival of hypertensive rats in response to volume-overload induced by aorto-caval fistula. *Naunyn. Schmiedebergs. Arch. Pharmacol.* **2023**, *396*, 3757–3773. [CrossRef] [PubMed]
3. Sykora, M.; Andelova, K.; Szeiffova Bacova, B.; Egan Benova, T.; Martiskova, A.; Knezl, V.; Tribulova, N. Hypertension Induces Pro-arrhythmic Cardiac Connexome Disorders: Protective Effects of Treatment. *Biomolecules* **2023**, *13*, 330. [CrossRef]
4. Bueno, H.; Moura, B.; Lancellotti, P.; Bauersachs, J. The year in cardiovascular medicine 2020: Heart failure and cardiomyopathies. *Eur. Heart J.* **2021**, *42*, 657–670. [CrossRef]
5. Kitai, T.; Kohsaka, S.; Kato, T.; Kato, E.; Sato, K.; Teramoto, K.; Yaku, H.; Akiyama, E.; Ando, M.; Izumi, C.; et al. JCS/JHFS 2025 Guideline on Diagnosis and Treatment of Heart Failure. *Circ. J.* **2025**, *89*, CJ-25-0002. [CrossRef]
6. Langheinrich, M.; Lee, M.A.; Böhm, M.; Pinto, Y.M.; Ganten, D.; Paul, M. The hypertensive Ren-2 transgenic rat TGR(mREN2)27 in hypertension research: Characteristics and functional aspects. *Am. J. Hypertens.* **1996**, *9*, 506–512. [CrossRef]
7. Sykora, M.; Kratky, V.; Cervenka, L.; Kopkan, L.; Tribulova, N.; Szeiffova Bacova, B. The treatment with trandolapril and losartan attenuates pressure and volume overload alternations of cardiac connexin-43 and extracellular matrix in Ren-2 transgenic rats. *Sci. Rep.* **2023**, *13*, 20923. [CrossRef]
8. Jarkovská, D.; Miklovič, M.; Švíglerová, J.; Červenka, L.; Škaroupková, P.; Melenovský, V.; Štengl, M. Effects of Trandolapril on Structural, Contractile and Electrophysiological Remodeling in Experimental Volume Overload Heart Failure. *Front. Pharmacol.* **2021**, *12*, 729568. [CrossRef]
9. Stein, M.; Boulaksil, M.; Jansen, J.A.; Herold, E.; Noorman, M.; Joles, J.A.; Van Veen, T.A.B.; Houtman, M.J.C.; Engelen, M.A.; Hauer, R.N.W.; et al. Reduction of fibrosis-related arrhythmias by chronic renin-angiotensin- aldosterone system inhibitors in an aged mouse model. *Am. J. Physiol. Heart Circ. Physiol.* **2010**, *299*, 310–321. [CrossRef] [PubMed]
10. Tamargo, J.; Agewall, S.; Ambrosio, G.; Borghi, C.; Cerbai, E.; Dan, G.A.; Drexel, H.; Ferdinandy, P.; Grove, E.L.; Klingenberg, R.; et al. New pharmacological agents and novel cardiovascular pharmacotherapy strategies in 2024. *Eur. Heart J. Cardiovasc. Pharmacother.* **2025**, *11*, 292–317. [CrossRef] [PubMed]
11. Kratky, V.; Vanourkova, Z.; Sykora, M.; Bacova, B.S.; Hruskova, Z.; Kikerlova, S.; Huskova, Z.; Kopkan, L. AT1 receptor blocker, but not an ACE inhibitor, prevents kidneys from hypoperfusion during congestive heart failure in normotensive and hypertensive rats. *Sci. Rep.* **2021**, *11*, 4271. [CrossRef] [PubMed]
12. Thiele, A.; Bartolomaeus, H.; Anandakumar, H.; Haase, N.; Heuser, A.; Müller, D.N.; Dechend, R.; Wilck, N. A novel approach to modelling acute cardiac decompensation in rats. *Cardiovasc. Res.* **2025**, *121*, 379–381. [CrossRef] [PubMed]
13. Fu, J.; Mansfield, C.; Diakonov, I.; Judina, A.; Delahaye, M.; Bhogal, N.; Sanchez-Alonso, J.L.; Kamp, T.; Gorelik, J. Stretch regulation of β2-Adrenoceptor signalling in cardiomyocytes requires caveolae. *Cardiovasc. Res.* **2025**, *121*, 440–453. [CrossRef]
14. Verdecchia, P.; Angeli, F.; Reboldi, G. Hypertension and atrial fibrillation: Doubts and certainties from basic and clinical studies. *Circ. Res.* **2018**, *122*, 352–368. [CrossRef] [PubMed]
15. Aimoto, M.; Yagi, K.; Ezawa, A.; Tsuneoka, Y.; Kumada, K.; Hasegawa, T.; Kuze, T.; Chiba, T.; Nagasawa, Y.; Tanaka, H.; et al. Chronic Volume Overload Caused by Abdominal Aorto-Venocaval Shunt Provides Arrhythmogenic Substrates in the Rat Atrium. *Biol. Pharm. Bull.* **2022**, *45*, 635–642. [CrossRef] [PubMed]
16. Cao, X.; Aimoto, M.; Fukumoto, M.; Nagasawa, Y.; Tanaka, H.; Takahara, A. Influence of chronic volume overload-induced atrial remodeling on electrophysiological responses to cholinergic receptor stimulation in the isolated rat atria. *J. Pharmacol. Sci.* **2018**, *136*, 73–78. [CrossRef] [PubMed]
17. Cao, X.; Nagasawa, Y.; Zhang, C.; Zhang, H.; Aimoto, M.; Takahara, A. Acehytisine suppresses atrial fibrillation in rats with dilated atria caused by chronic volume overload. *J. Pharmacol. Sci.* **2020**, *142*, 34–40. [CrossRef]
18. Boixel, C.; Fontaine, V.; Rücker-Martin, C.; Milliez, P.; Louedec, L.; Michel, J.B.; Jacob, M.P.; Hatem, S.N. Fibrosis of the left atria during progression of heart failure is associated with increased matrix metalloproteinases in the rat. *J. Am. Coll. Cardiol.* **2003**, *42*, 336–344. [CrossRef]
19. Li, X.; Garcia-Elias, A.; Benito, B.; Nattel, S. The effects of cardiac stretch on atrial fibroblasts: Analysis of the evidence and potential role in atrial fibrillation. *Cardiovasc. Res.* **2022**, *118*, 440–460. [CrossRef]
20. Kléber, A.G.; Rudy, Y. Basic Mechanisms of Cardiac Impulse Propagation and Associated Arrhythmias. *Physiol. Rev.* **2004**, *84*, 431–488. [CrossRef]
21. Li, J.; Solus, J.; Chen, Q.; Rho, Y.H.; Milne, G.; Stein, C.M.; Darbar, D. Role of inflammation and oxidative stress in atrial fibrillation. *Heart Rhythm* **2010**, *7*, 438–444. [CrossRef]
22. Dhein, S.; Salameh, A. Remodeling of cardiac gap junctional cell–cell coupling. *Cells* **2021**, *10*, 2422. [CrossRef]
23. Zhang, L.; Zhou, J.; Kong, W. Extracellular matrix in vascular homeostasis and disease. *Nat. Rev. Cardiol.* **2025**, *22*, 333–353. [CrossRef]

24. Le, N.N.; Tran, T.Q.B.; Mcclure, J.; Gill, D.; Padmanabhan, S. Emerging antihypertensive therapies and cardiovascular, kidney, and metabolic outcomes: A Mendelian randomization study. *Eur. Heart J. Cardiovasc. Pharmacother.* **2025**, *11*, 264–274. [CrossRef]
25. Shlafer, M.; Shepard, B.M. A method to reduce interference by sucrose in the detection of thiobarbituric acid-reactive substances. *Anal. Biochem.* **1984**, *137*, 269–276. [CrossRef]
26. Lojda, Z.; Gossrau, R.; Schiebler, T.H. *Enzymhistochemische Methoden*; Springer: Berlin/Heidelberg, Germany, 1976.
27. Koyama, T.; Xie, Z.; Gao, M.; Suzuki, J.; Batra, S. Adaptive changes in the capillary network in the left ventricle of rat heart. *Jpn. J. Physiol.* **1998**, *48*, 229–241. [CrossRef]
28. Andelova, K.; Szeiffova Bacova, B.; Sykora, M.; Pavelka, S.; Rauchova, H.; Tribulova, N. Cardiac Cx43 Signaling Is Enhanced and TGF-β1/SMAD2/3 Suppressed in Response to Cold Acclimation and Modulated by Thyroid Status in Hairless SHRM. *Biomedicines* **2022**, *10*, 1707. [CrossRef]
29. Sykora, M.; Kratky, V.; Kopkan, L.; Tribulova, N.; Szeiffova Bacova, B. Anti-Fibrotic Potential of Angiotensin (1-7) in Hemodynamically Overloaded Rat Heart. *Int. J. Mol. Sci.* **2023**, *24*, 3490. [CrossRef]
30. Lippi, G.; Sanchis-Gomar, F.; Cervellin, G. Global epidemiology of atrial fibrillation: An increasing epidemic and public health challenge. *Int. J. Stroke* **2021**, *16*, 217–221. [CrossRef] [PubMed]
31. Ben Morrison, T.; Jared Bunch, T.; Gersh, B.J. Pathophysiology of concomitant atrial fibrillation and heart failure: Implications for management. *Nat. Clin. Pract. Cardiovasc. Med.* **2009**, *6*, 46–56. [CrossRef]
32. Havlenova, T.; Skaroupkova, P.; Miklovic, M.; Behounek, M.; Chmel, M.; Jarkovska, D.; Sviglerova, J.; Stengl, M.; Kolar, M.; Novotny, J.; et al. Right versus left ventricular remodeling in heart failure due to chronic volume overload. *Sci. Rep.* **2021**, *11*, 17136. [CrossRef]
33. Reddy, S.; Bernstein, D.; Alto, P. Molecular Mechanisms of RV failure. *Circulation* **2016**, *132*, 1734–1742. [CrossRef]
34. Díez, J.; De Boer, R.A. Management of cardiac fibrosis is the largest unmet medical need in heart failure. *Cardiovasc. Res.* **2022**, *118*, E20–E22. [CrossRef]
35. Gonçalves, P.R.; Nascimento, L.D.; Gerlach, R.F.; Rodrigues, K.E.; Prado, A.F. Matrix Metalloproteinase 2 as a Pharmacological Target in Heart Failure. *Pharmaceuticals* **2022**, *15*, 920. [CrossRef]
36. Keefe, J.A.; Wang, J.; Song, J.; Ni, L.; Wehrens, X.H.T. Immune cells and arrhythmias. *Cardiovasc. Res.* **2025**, *121*, 382–395. [CrossRef]
37. Andelova, K.; Sykora, M.; Farkasova, V.; Stankovicova, T.; Szeiffova Bacova, B.; Knezl, V.; Egan Benova, T.; Pravenec, M.; Tribulova, N. Acclimation of Hairless Spontaneously Hypertensive Rat to Ambient Temperature Attenuates Hypertension-Induced Pro-Arrhythmic Downregulation of Cx43 in the Left Heart Ventricle of Males. *Biomolecules* **2024**, *14*, 1509. [CrossRef]
38. Lee, S.Y.; Wu, S.T.; Du, C.X.; Ku, H.C. Potential Role of Dipeptidyl Peptidase−4 in Regulating Mitochondria and Oxidative Stress in Cardiomyocytes. *Cardiovasc. Toxicol.* **2024**, *24*, 1090–1104. [CrossRef]
39. Lendeckel, U.; Arndt, M.; Wrenger, S.; Nepple, K.; Huth, C.; Ansorge, S.; Klein, H.U.; Goette, A. Expression and activity of ectopeptidases in fibrillating human atria. *J. Mol. Cell. Cardiol.* **2001**, *33*, 1273–1281. [CrossRef]
40. La Vecchia, G.; Fumarulo, I.; Caffè, A.; Chiatto, M.; Montone, R.A.; Aspromonte, N. Microvascular Dysfunction across the Spectrum of Heart Failure Pathology: Pathophysiology, Clinical Features and Therapeutic Implications. *Int. J. Mol. Sci.* **2024**, *25*, 7628. [CrossRef]
41. Tribulova, N.; Egan Benova, T.; Szeiffova Bacova, B.; Viczenczova, C.; Barancik, M. New aspects of pathogenesis of atrial fibrillation: Remodeling of intercalated discs. *J. Physiol. Pharmacol.* **2015**, *66*, 625–634.
42. Sanders, P.; Morton, J.B.; Davidson, N.C.; Spence, S.J.; Vohra, J.K.; Sparks, P.B.; Kalman, J.M. Electrical remodeling of the atria in congestive heart failure: Electrophysiological and electroanatomic mapping in humans. *Circulation* **2003**, *108*, 1461–1468. [CrossRef]
43. Cruciani, V.; Mikalsen, S.O. Connexins, gap junctional intercellular communication and kinases. *Biol. Cell* **2002**, *94*, 433–443. [CrossRef]
44. Seropian, I.M.; El-Diasty, M.; El-Sherbini, A.H.; González, G.E.; Rabinovich, G.A. Central role of Galectin-3 at the cross-roads of cardiac inflammation and fibrosis: Implications for heart failure and transplantation. *Cytokine Growth Factor Rev.* **2024**, *80*, 47–58. [CrossRef]
45. Barallobre-Barreiro, J.; Radovits, T.; Fava, M.; Mayr, U.; Lin, W.Y.; Ermolaeva, E.; Martínez-López, D.; Lindberg, E.L.; Duregotti, E.; Daróczi, L.; et al. Extracellular Matrix in Heart Failure: Role of ADAMTS5 in Proteoglycan Remodeling. *Circulation* **2021**, *144*, 2021–2034. [CrossRef] [PubMed]
46. Chen, Y.; Qiao, X.; Zhang, L.; Li, X.; Liu, Q. Apelin-13 Regulates Angiotensin II-Induced Cx43 Downregulation and Autophagy via the AMPK/mTOR Signaling Pathway in HL-1 Cells. *Physiol. Res.* **2020**, *69*, 813–822. [CrossRef] [PubMed]
47. Cai, W.; Ruan, L.; Wang, Y.; Chen, J. Effects of angiotensin II on connexin 43 of VSMCs in arteriosclerosis. *J. Zhejiang Univ. Sci. B* **2006**, *7*, 648–653. [CrossRef] [PubMed]

biomolecules

MDPI

Article

Supercomplex Restructuring in Heart Mitochondria of COX7A1-Deficient Mice

Lauren Pavelich [1,2], Lucynda Pham [1], Paul Stemmer [3], Icksoo Lee [4], Lawrence I. Grossman [1], Maik Hüttemann [1,2,*] and Tasnim Arroum [1,*]

1 Center for Molecular Medicine and Genetics, Wayne State University, Detroit, MI 48201, USA; laurenpavelich@wayne.edu (L.P.); lucynda.pham@med.wayne.edu (L.P.); lgrossman@wayne.edu (L.I.G.)
2 Department of Biochemistry, Microbiology, and Immunology, Wayne State University, Detroit, MI 48201, USA
3 Department of Pharmaceutical Sciences, Wayne State University, Detroit, MI 48201, USA; pmstemmer@wayne.edu
4 College of Medicine, Dankook University, Cheonan 31116, Republic of Korea; icksoolee@dankook.ac.kr
* Correspondence: mhuttema@med.wayne.edu (M.H.); ho0066@wayne.edu (T.A.); Tel.: +1-313-577-9150 (M.H.); +1-313-577-5219 (T.A.)

Abstract

The role of electron transport chain supercomplexes and factors that regulate their composition in a tissue- and species-specific manner are not fully understood. Tissue-specific isoforms have been reported for cytochrome *c* oxidase (COX), which may contribute to such regulation. Therefore, we here investigated COX activity and structural organization in wild-type (WT) and COX7A1 knockout (KO) mice, which lack the heart/skeletal muscle isoform of COX subunit VIIa. COX7A1 KO mice showed a 30% reduction in total COX activity in the heart. Although the activity of COX in the monomers and I+III_2+IV_n supercomplexes (SCs) remained unchanged, a marked reduction in COX dimers and unknown COX-containing species IV_x and IV_y contributed to the overall reduction in COX activity. Furthermore, we observed that COX7A2 substituted for COX7A1 in COX monomers, dimers, and all COX-containing SCs in the KO mice, indicating a compensatory mechanism to preserve COX functionality. Collectively, these results suggest that COX7A1 plays an important role in maintaining structural stability; however, they also suggest that loss of COX7A1 is compensated by its replacement with COX7A2.

Keywords: cytochrome *c* oxidase; isoforms; COX7aH; COX7aL; heart

1. Introduction

Cytochrome *c* oxidase (COX), the terminal oxidase of the mitochondrial electron transport chain (ETC), is a crucial enzyme involved in aerobic respiration and ATP synthesis. The enzyme's complex assembly comprises multiple subunits that are expressed in a tissue-specific manner, each playing an integral role in the enzyme's functional activity and regulatory mechanisms. Tissue-specific COX subunit isoforms could serve as a foundation for fine-tuning COX activity to meet the specific metabolic needs of different tissues. Among these subunits is COX subunit 7a (COX7A), which is comprised of distinct isoforms, with COX7A1 (the "heart-type" isoform, also called COX7AH) demonstrating tissue-specific expression, predominantly within cardiac and skeletal muscle tissue, as high-energy-demanding tissues [1]. Initial investigations concentrated on elucidating the characteristics of the COX7A1 gene [2], delineating its genomic architecture and promoter regulation while revealing its structural homology to the ubiquitously expressed COX7A2 (the "liver-type" isoform, also known as COX7AL) and emphasizing their distinct gene regulation [3–5]. Despite their structural similarity, these isoforms are encoded by separate genes and controlled by distinct transcriptional mechanisms, reflecting their specialized roles. Comparative genomic investigations further highlighted the evolutionary divergence of these isoforms, indicating functional specialization influenced by adaptive evolutionary pressure in primates, with an accelerated nonsynonymous substitution rate in the COX7A gene family—COX7A1, COX7A2, and COX7AR (also known as COX7A2L or SCAF1), aligning with the evolution of tissue-specific functions [6].

We have previously characterized the role of *Cox7a1* in a knockout mouse model, highlighting its important role in maintaining cardiac functionality. COX7A1 loss caused a decrease in total COX activity and manifestation of dilated cardiomyopathy in 6-week-old mice, which improved over time [7]. In addition, we found that COX7A1 knockout mice incorporate the liver-type isoform (COX7A2) as a compensatory mechanism. In addition, we found that COX7A1 loss leads to reduced skeletal muscle bioenergetics and causes exercise intolerance [8]. Investigations into isoform switching between COX7A1 and COX7A2 have further illuminated the role of COX7A1 in cardiac and skeletal muscle, showing a developmental transition from liver-type to heart/muscle-specific isoforms of both COX6A and COX7A during early postnatal muscle development in both human and mouse skeletal muscle, with the heart/muscle-specific isoform becoming predominant by 3 months of age in infants and 4 weeks in mice [9]. However, the authors further show that biopsy studies in patients with reversible infantile respiratory chain deficiency showed that this isoform switch does not directly contribute to the onset or recovery of the disease [9]. It should be noted that commonly used in vitro models, such as the C_2C_{12} myoblast differentiation model, fail to fully recapitulate this switch, exhibiting only limited expression of the muscle-specific isoform, thus making them suboptimal for studying the isoform switch and adaptation of muscle cells [10,11].

In this study, we aimed to investigate how deletion of the COX7A heart-specific isoform subunit affects COX activity and mitochondrial supercomplex assembly, with a particular focus on isoform switching of COX7A1 and two subunits in cardiac mitochondria in female mice.

2. Materials and Methods

2.1. Generation of Cytochrome c Oxidase Subunit 7a Isoform 1 (Heart-Type) Knockout Mice

The generation of knockout mice by replacement of all three *Cox7a1* exons with the *neo* cassette was described in detail in [7].

2.2. Heart Tissue Homogenization and Mitochondria Isolation

Heart tissue was homogenized using gentleMACs Dissociator (Miltenyi Biotec, Bergisch Gladbach, Germany, Cat. No 130-093-235) with protocol m_lung_02.01 in 2 mL of mitochondrial isolation buffer (MIB: 200 mM sucrose, 10 mM Tris, 1 mM EGTA, pH 7.4 adjusted with 1 M HEPES solution, Gibco, cat. No. 15630080) in gentleMACs C Tubes (Miltenyi Biotec, Cat. No. 130-093-237). Heat-activated sodium vanadate was added to the homogenization buffer at a final conentration of 5mM. The entire process was performed on ice to help preserve protein integrity and prevent dephosphorylation. To further break down the tissue, 0.2 mg of Subtilisin A (Protease) (Sigma, Burlington, MA, USA, Cat. No. P5380) was added to the tissue homogenate and incubated at 4 °C for 10 min with gentle shaking. To enrich the mitochondrial fraction, we used a series of filtration steps. The filters were pre-wet with MIB. First, a 40 µm filter (pluriSelect, Cat. No. 43-50040-50) was used to remove larger cellular debris and nuclei. The flowthrough containing mitochondria and other organelles was then collected and subjected to two centrifugations at medium speed (9000× *g*, 15 min, 4 °C) to precipitate the crude mitochondrial fraction and further remove contaminations. Mitochondrial fractions were washed with MIB supplemented with 2 mM phenylmethylsulphonylfluoride, 5 mM heat-activated sodium vanadate, and Protease Inhibitor Cocktail (Sigma, Cat. No. P8849). Protein concentration was determined using the DC protein assay kit (Bio-Rad, Hercules, CA, USA, cat. no. 5000111), according to the manufacturer's protocol.

2.3. Outer Mitochondrial Protein Digestion with Proteinase K

Isolated mitochondria were digested with Proteinase K solution (QIAGEN, Hilden, Germany, cat. No. 1018332) to remove the cytosolic and outer mitochondrial membrane proteins. Eighty µg of mitochondria were resuspended in 1 mL of MIB plus Proteinase K solution at a final concentration of 1mM and incubated at room temperature for 45 min, with gentle shaking. To remove the Proteinase K, Protease Inhibitor Cocktail (1:100) was added, and the solution was centrifuged at 16,900× *g* for 15 min at 4 °C to collect the mitochondrial pellet. The pellet was then washed 3 times with fresh MIB. Western blot analysis of the proteinase K treatment is presented in Supplementary Figure S3.

2.4. Mitochondrial Membrane Solubilization for Blue Native-PAGE

The Blue Native-PAGE (BN-PAGE) protocol was adapted from [12,13]. All components were used from the NativePAGE™ Sample Prep Kit (ThermoFisher Scientific, Waltham, MA, USA, cat. no. BN2008) including NativePAGE™ 4× sample buffer, NativePAGE™ 5% G-250 sample additive, except the digitonin (ThermoFisher Scientific, cat. no. 407565000), which was diluted to 5%. Briefly, for solubilization of crude mitochondria, we used NativePAGE™ 4× sample buffer supplemented with digitonin (5%) to reach a final concentration of 1.5% (digitonin/protein ratio of 6 *g*/*g*). This mixture was incubated on ice for 20 min, mixing every 5 min to ensure complete solubilization of the mitochondria. The mixture was then centrifuged at 4 °C at 17,000× *g* for 20 min. Following centrifugation, 0.5% G-250 sample additive was added to the supernatant. Subsequently, 75 µg of these samples were loaded onto 3–12% native gradient gels (Invitrogen, Carlsbad, CA, USA cat. no. BN1001BOX). Electrophoresis was conducted at 4 °C, applying 150 V until the proteins entered 1/3 into the gel (~30 min); at this point, the light-blue cathode buffer was changed to clear cathode buffer (50 mM Tricine, 15 mM Bis-Tris, pH 7.0). The voltage was then increased to 200 V until the desired separation was achieved (~1 h). The gels to be used for Western blot analyses were stained with GEL Blue (G-Biosciences, St. Louis, MO, USA, cat. no. 786-35G) for 30 min. The gel was washed 2 × 5 min with water before being imaged. The mitochondrial proteins on the gel were then transferred from the native gel to a PVDF membrane (BioRad, cat. no. 1620177) via a wet transfer at 10 V overnight. To help transfer large proteins, 0.05% SDS was added to the transfer buffer (25 mM Tris base 192 mM glycine, 20% methanol, pH adjusted to 8.3). Thirty µg of the mitochondrial sample from BN-PAGE were run on an SDS PAGE, followed by Western blot, to determine the purity of the inner mitochondrial membrane proteins (Supplementary Figure S3).

2.5. Mitochondrial Native Protein In-Gel Activity Assay (IGA)

After the BN-PAGE, we conducted an in-gel activity assay (IGA), as described [13]. Briefly, gels were incubated in 10 mL of the respective complex assay buffer at room temperature. COX IGAs were completed first, followed by complex I (CI) activity assays performed on the same gels. Each reaction was carried out at room temperature for 1 h with gentle shaking. Gels were washed 2 × 10 min in water and then imaged. The gels were washed in water overnight at 4 °C and then imaged again. To avoid saturated signals, pictures obtained after one hour of incubation were used for quantification.

2.6. Western Blot Analysis

To remove the Coomassie from the membrane, a 15 min methanol wash step was performed. Membranes were washed 2 × 5 min with water and 2 × 5 min with TBST to remove methanol. Membranes were blocked with a blocking reagent (5% bovine serum albumin (BSA), 1% Tris-buffered saline with Tween-20: TBST, 0.1% Tween- 20) for 45 min at room temperature. Next, we added the primary antibody (Table 1) with gentle shaking for 90 min at room temperature or overnight at 4 °C. The membranes were then washed two times for 5 min with TBST. Finally, membranes were incubated with the secondary antibody for 1 h at room temperature (Table 1), overlaid with Pierce™ ECL Western Blotting Substrate (ThermoFisher, cat. no. 32106), and signals were recorded on a ChemiDoc MP imaging system.

Table 1. List of antibodies used in this study.

Antibody	Dilution	μg/mL Amount	Company	Product ID
MT-CO1 (COX1)	1:1000	1	Thermofisher	PA5-68016
UQCRC1	1:2000	0.8	Proteintech	14742-1-AP
NDUFS3	1:1000	1	Abcam	17D95
COX7A2L (SCAF1)	1:1000	0.350	Proteintech	11416-1-AP
Anti-Rabbit IgG HRP-linked	1:5000	0.077	Cell Signaling Technology	7074S
Anti-Mouse IgG HRP-linked	1:8000	0.184	Cell Signaling Technology	7076S
Anti-Rabbit IgG StarBright Blue 700	1:2500	-	BioRad	12004162
Anti-Mouse IgG StarBright Blue 520	1:5000	-	BioRad	12005867

2.7. Mass Spectrometry and Proteomic Analysis

Following BN-PAGE, gel pieces were excised and peptides were prepared using the in-tryptic digestion method, following the manufacturer's protocol (Thermo Scientific™ Waltham, MA USA, cat. no. 89871). Peptide samples were analyzed on a Thermo Scientific Orbitrap Eclipse Tribrid mass spectrometer, operated in data-dependent acquisition (DDA) mode. Raw data files were processed using the MaxQuant platform (v2.6.8.0) and searched against the UniProt database for Mus musculus (https://www.uniprot.org/proteomes/UP000000589, accessed on 25 February 2025). For stoichiometric analysis, the peptide intensity signals for each band were used, normalized to a core subunit in the studied complex.

2.8. Statistical Analyses

All data are presented as means ± SEM (standard error of the mean) values. We conducted an unpaired *t*-test. Statistical analyses were conducted with GraphPad Prism version 10.0.0, GraphPad Software, Boston, MA, USA, with significance set at $p \leq 0.05$.

3. Results

3.1. COX7A1 KO Decreases COX Activity and Amount in Select Supercomplexes

To study the effect of knockout of the heart/skeletal muscle-specific COX subunit 7A1 on supercomplex composition, we investigated female WT and *Cox7a1* KO mice to analyze COX activity and supercomplex assembly changes in the heart. To explore potential alterations in the activity and composition of COX monomers, dimers, and supercomplexes (SCs), we employed Blue-Native PAGE (BN-PAGE) followed by in-gel activity (IGA) assays for COX and complex I (CI), along with immunoblotting of various ETC complexes. BN-PAGE, in combination with mild solubilization of mitochondrial membrane proteins using digitonin, preserves protein–protein interactions and maintains native complex structure and activity. This enabled us to assess structural and functional rearrangements of COX-containing SCs. Interestingly, IGA analysis of COX revealed that COX monomer activity remained unchanged (Figure 1B,D). However, activity within COX dimers (IV_2) and unknown CIV species that run at similar electrophoretic mobility as the III_2+IV and III_2+IV_2 SCs (annotated as IV_x and IV_y) exhibited significant reductions (Figure 1E–G). When considering overall COX activity across all complexes, we observed a total decrease of 30% (Figure 1I). Interestingly, this reduction was not evenly distributed across COX-containing complexes; rather, select subpopulations, such as COX dimers and IV_x and IV_y species, contributed disproportionately to the observed decline in COX activity. We also investigated CI in-gel activity (Figure 1C). Our results showed that CI monomer activity and CI-associated SC activity remained unchanged (Figures 1J and S1B,C). The only exception was SC I+III_2+IV_n, which exhibited a ~30% increase in activity (Figure 1K). Notably, this was the only SC where COX activity was preserved and did not show a reduction (Figure 1H). This suggests that in the absence of COX7A1, respiratory chain complexes undergo selective rearrangements, with certain SCs compensating to maintain respiratory chain integrity.

3.2. Proteomic Profiling Shows Distinct Distribution of the COX7A1 and COX7A2 in WT and KO Mice

Mass spectrometry, following BN-PAGE and band excision, was used to investigate the distribution profiles of COX7A1 and COX7A2 across different mitochondrial complexes in COX7A1 WT and KO mice. In WT mice, COX7A1 is primarily found in monomers (44%) and dimers (11%), in addition to its presence in IV_x, IV_y, and I+III_2+IV SCs. In contrast, COX7A2 was notably reduced in WT mice, accounting for only 5% of the total COX7A levels found solely in COX-dimeric assemblies (Figure 2A–G). Conversely, in COX7A1 KO mice, COX7A1 was absent as anticipated, while COX7A2 appeared in monomers (19%), dimers (12%), IV_x and IV_y species (15% and 25%), and 29% in the I+III_2+IV SC (Figure 2A–G). This distribution suggests that COX7A2 compensates for the lack of COX7A1 in the KO animals, leading to a reorganization of COX. Although the redistribution of COX7A2 highlights its structural importance in stabilizing COX, these structural adaptations were insufficient to fully restore COX activity (Figure 1I).

We did not detect SCAF1 in our COX complexome analysis, despite observing a signal at the same level as III_2+$IV_{1\text{-}2}$ (Figures 1A,B, 3A,B and S2). This signal was notably more evident when using COX antibodies compared to CIII antibodies, and when the image was overexposed. The mice used have a C57BL/6 background that is known to lack SCAF1. Immunoblot with SCAF1 and UQCRC1 is presented in Supplementary Figure S2.

We next assessed the stoichiometric distribution of complex III (*bc*$_1$-complex; Figure 2B) by examining representative subunits indicative of specific assembly stages. UQCRC1, a core subunit involved in all stages of complex III biogenesis, served as an indicator for the presence of complex III in the selected bands, while UQCRFS1 was used as a marker of mature dimeric $CIII_2$. We differentiated between pre-III_2 (immature) and mature III_2 complexes in both WT and KO animals. Specifically, the absence of UQCRFS1 signified pre-III_2 structures, whereas its presence confirmed complete dimerization. In Figure 2B, we can see dimerization of CIII in III_2 and supercomplex bands, while immature pre-III_2 complexes were identified running at the same level as the IV_x and IV_y bands, where no UQCRFS1 signal was detected. However, due to the lack of SCAF1 in these animals, these signals are not indicative of an assembled III_2+IV complex, and may be due to disassembly of higher I+III_2+IV SC due to detergent treatment.

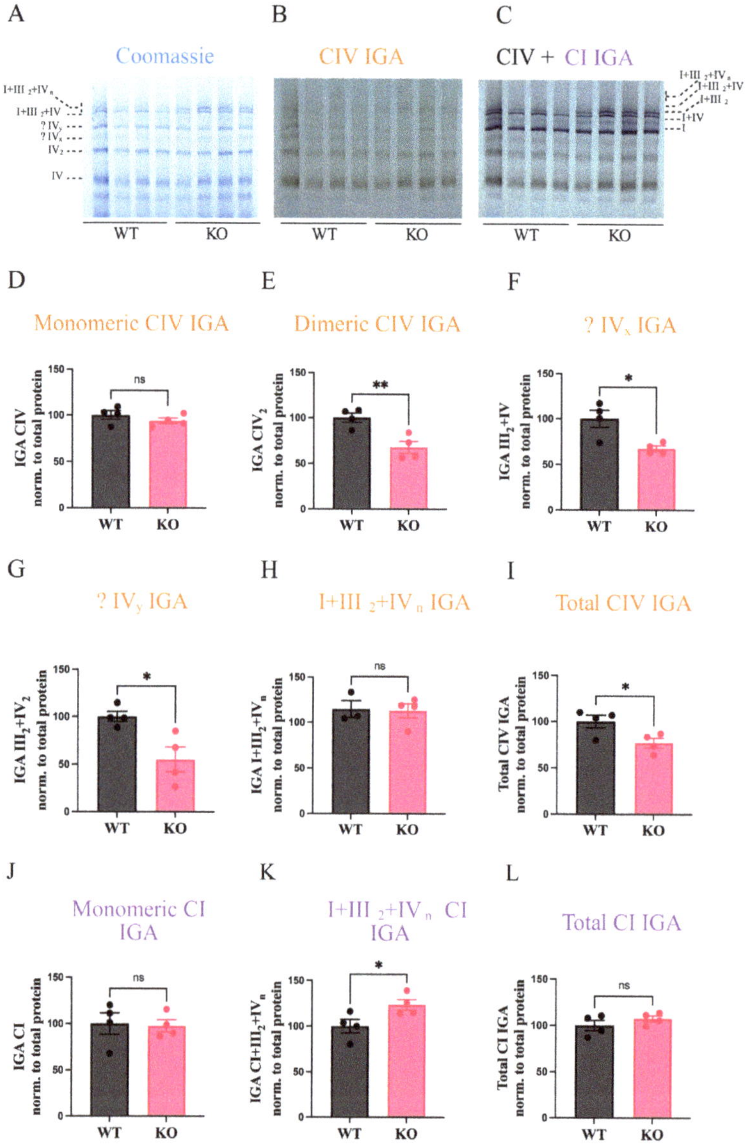

Figure 1. COX7A1 knockout decreases COX activity. BN-Page was performed and was followed by in-gel activity reactions for CIV and CI. Brown and violet bands indicate CIV and CI activity, respectively. (**A**) Coomassie gel before IGA was performed. (**B**) CIV IGA. (**C**) CIV+CI IGA. Uncropped images for (**A–C**) are found in Supplementary Figures S4 and S5. (**D**) Quantification of monomeric CIV IGA. (**E**) Quantification of CIV_2 IGA. (**F**) Quantification of IV_x CIV-IGA. (**G**) Quantification of IV_y CIV-IGA. (**H**) Quantification of I+III_2+IV_n CIV-IGA. (**I**) Quantification of total CIV IGA. (**J**) Quantification of monomeric CI IGA. (**K**) Quantification of I+III_2+IV_n CI-IGA. (**L**) Quantification of total CI IGA normalized to total protein. Data was from four female animals per group. Data are represented as means ± standard error. **, $p < 0.01$; *, $p < 0.05$, based on unpaired *t*-test. The orange color is for complex IV in-gel activity and violet color for complex I in-gel activity.

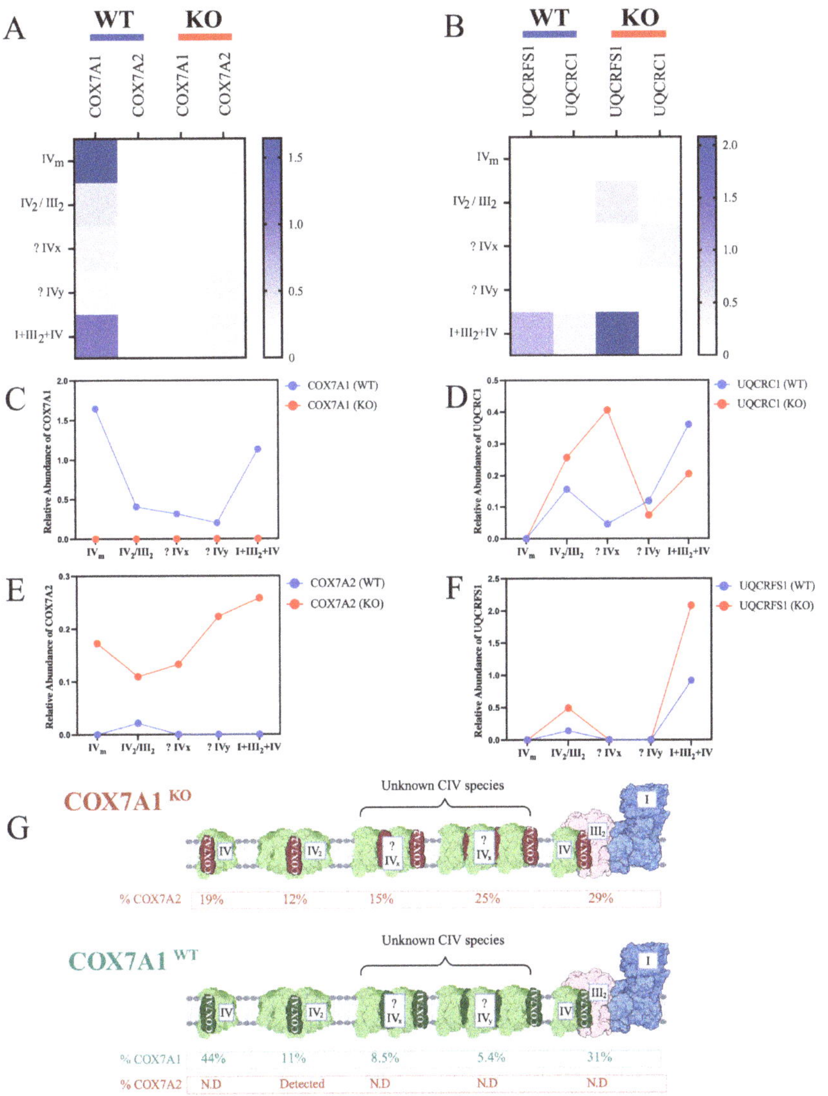

Figure 2. Structural compensation and redistribution of COX7A2 in COX7A1 knockout mice. (**A**) BN-PAGE followed by proteomic analysis, illustrating the relative distribution of COX7A1 and COX7A2. The total intensity was normalized to the COXIV subunit within each gel band. (**B**) BN-PAGE followed by proteomic analysis, illustrating the relative distribution of UQCRFS1 and UQCRC1. The total intensity was normalized to the UQCRC2 subunit within each gel band. (**C**,**E**) Relative abundance comparison of COX7A1 and COX7A2, based on data shown in (**A**). (**D**,**F**) Relative abundance comparison of UQCRC1 and UQCRFS1, based on data shown in (**B**). (G) Summary schematic of data derived from (**A**–**F**). Not detected annotated as N.D.

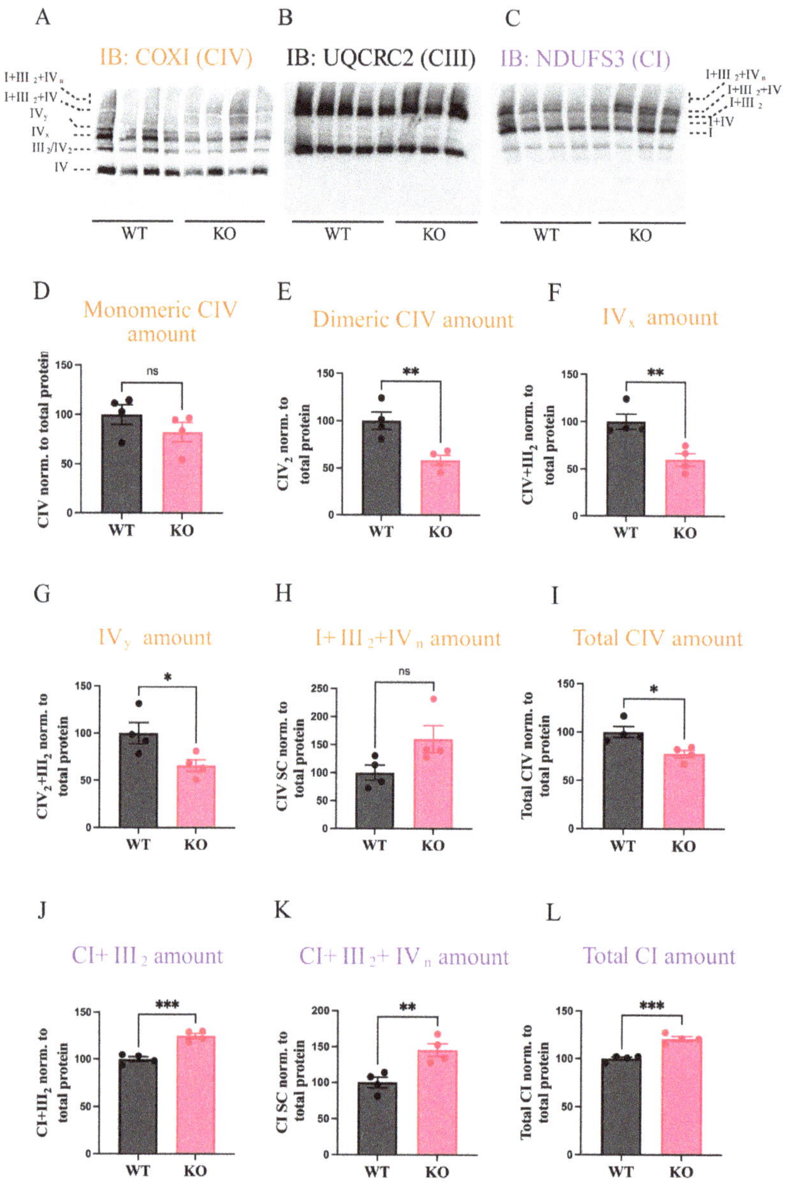

Figure 3. COX7A1 KO decreases COX amount. BN-Page was performed and was followed by immunoblotting. (**A**) Western blot following anti-COXI (CIV). (**B**) Western blot following anti-UQCRC2 (CIII). (**C**) Western blot following anti-NDUFS3 (CI). Uncropped western blots for (**A–C**) are found in Supplementary Figures S6 and S7. (**D**) Quantification of monomeric CIV amount. (**E**) Quantification of CIV_2 amount. (**F**) Quantification of CIV amount in IV_x. (**G**) Quantification of CIV amount in IV_y. (**H**) Quantification of CIV amount in I+III_2+IV_n. (**I**) Quantification of total CIV amount. (**J**) Quantification of CI amount in I+III_2. (**K**) Quantification of CI amount in I+III_2+IV_n. (**L**) Quantification of total CI. Data was from four female animals per group. Data are represented as means ± standard error. ***, $p < 0.001$; **, $p < 0.01$; *, $p < 0.05$, based on unpaired *t*-test. Orange is for complex IV immunoblot quantification and violet is for complex I immunoblot quantification.

We next examined potential changes in ETC content using immunoblotting. Following BN-PAGE separation, we performed immunoblotting for COX (COXI), CIII (UQCRC2), and CI (NDUFS3) (Figure 3A–C). Consistent with COX IGA findings, there was no significant change in monomeric COX levels or I+III_2+IV_n SCs (Figure 3D,H). However, decreases were observed in dimeric COX, IV_x, and IV_y (Figure 3E–G). Additionally, we investigated complex I assembly, noting an increase in CI within I+III_2 and I+III_2+IV_n supercomplexes (Figure 3J–L). Quantitative analysis revealed a 20% increase in total CI levels (Figure 3L) and a 30% decrease in total COX levels (Figure 3I). Further examination of complex III showed no alteration in III_2, I+III_2, I+III_2+IV_n, or total CIII content (Supplementary Figure S1G,H,K).

4. Discussion

In this study, we investigated the assembly and redistribution of mitochondrial complexes and SCs in the hearts of COX7A1 knockout mice. Using in-gel activity analysis of COX, we gained new insights into how a selective reduction of activity in COX-containing complexes took place in COX7A1-deficient mouse hearts. Recently, García-Poyatos et al. [14] showed that loss of *Cox7a1* in zebrafish leads to a severe reduction in COX homodimerization and the assembly of higher-order supercomplexes, without affecting the levels of monomeric COX. This disruption was accompanied by a marked decrease in total COX activity in *Cox7a1*-deficient zebrafish skeletal muscle. In our study, the enzymatic activity of monomeric COX remained unchanged; however, the activity associated with COX dimers and COX containing IV_x and IV_y species was significantly reduced. This suggests that the loss of COX7A1 may not impact the catalytic function of monomeric COX; rather, it could play a role in stabilizing COX homodimerization and higher-order structures involving COX. However, the decrease observed in the mouse model was not as severe as reported in the zebrafish knockout model [14]. Moreover, homodimeric COX exhibits higher enzymatic activity in the zebrafish model than its monomeric counterpart. The latter contradicts previous biochemical and structural studies conducted in mammalian tissue and in vitro enzymatic studies; we and others have shown that COX inhibition through allosteric regulation and post-translational modifications is disrupted under stress conditions, likely by Ca^{2+}-activated dephosphorylation of COX, which causes COX to transition into its monomeric form—proposed as the more active form of COX [13,15–17].

Our results show that the activity of monomeric CI and CI-containing supercomplexes remained largely unchanged between WT and *Cox7a1* knockouts. Interestingly, the only notable exception was SCI+III_2+IV_n, which exhibited a ~30% increase in CI activity. I+III_2+IV_n is the COX-containing SC in which COX activity remained preserved and did not decline. These findings suggest that, in the absence of COX7A1, a compensatory rearrangement within the ETC takes place, favoring the assembly of CI-containing SCs, potentially to rescue some of the respiratory dysfunction. This stabilization of higher-order CI-containing SCs was observed in other reports as a response to ETC defects [18,19]. When COX7A1 is absent, selective rearrangements happen within respiratory chain complexes, allowing specific supercomplexes to compensate and maintain respiratory chain integrity. Previously, we demonstrated that *Cox7a1* knockout mice showed increased incorporation of the "liver-type" isoform COX7A2 into the cardiac COX holoenzyme [8]. Rodents, like humans, undergo a developmental switch from "liver-type" COX7A2 to the "heart-type" COX7A1 during postnatal maturation [9]. In the adult stage, COX7A1 is the predominant form. However, in *Cox7a1* knockout mice, we see that COX7A2 is upregulated and is compensating for COX7A1 loss. In the present study, we expanded upon these findings using mass spectrometry following BN-PAGE analysis, revealing that COX7A2 is incorporated into all COX populations, including monomers, dimers, and higher-order supercomplexes in the mitochondria lacking COX7A1. The redistribution of COX in the KO model may help increase the enzymatic turnover rate through monomerization, leading to increased respiration that compensates for COX7A1 loss, resulting in the preference of the more active monomeric COX after the loss of COX7A1.

Previously, we found that COX7A1 KO was accompanied by an unexpected elevation in tissue ATP levels [7]. Notably, the overall reduction in COX dimerization observed in the knockout mice may help explain the elevated ATP levels, as dimeric COX has been associated with increased inhibitory phosphorylation and lower activity in vitro [20] and in pathological conditions in liver tissue [13]. COX phosphorylation was linked to increased dimer formation and decreased enzymatic activity, and potentially caused allosteric constraints that limit the turnover rate of the enzyme. Moreover, elevated cAMP levels, which promote COX phosphorylation [17], are known to favor dimer formation and suppress respiratory activity [17,21], thus limiting mitochondrial output and reactive oxygen species (ROS) generation [20]. In this context, dynamic regulation of COX dimerization may serve as a protective mechanism in high-energy-demanding tissues like the heart and skeletal muscle, helping to fine-tune respiration and minimize oxidative stress and cellular damage in tissues lacking COX7A1.

The interplay between the different isoforms needs to be further investigated and validated through additional techniques not reliant on detergent enrichment, as previous studies have explored how different detergents influence the dimerization of COX. Biochemical and structural studies have shown, for instance, that sodium cholate promotes the formation of COX dimers by binding to its dimer interface and thereby stabilizing its dimeric form [20,22]. In addition, cardiolipin was found to promote COX dimer formation [23]. This suggests that removing the cardiolipin-COX interface through detergents could destabilize dimers, increase the monomeric form of COX, and affect the dimeric turnover rate. Therefore, changes in the lipid composition and their interaction with mitochondrial ETC subunits could influence the complex and SC populations and their activity. In this study, we used digitonin, the mildest non-ionic detergent that produces a heterogeneous mix of monomeric, dimeric, and higher-order structures of COX. It frees COX from the lipid bilayer, which by itself may decrease dimerization of COX observed in the native gel analysis. However, at the moment, it is the best detergent used for mitochondrial SC enrichment.

Another subunit of the COX7A family is SCAF1 (also known as COX7AR or COX7A2L). Due to the mouse C57BL/6 background, we did not detect SCAF1 in our mass spectrometry data, confirming previous reports [24,25]. Interestingly, we observed the presence of bands with a similar electrophoretic mobility of III_2+IV_{1-2}. This signal corresponds to unknown higher-order COX species annotated as IV_x and IV_y in this report. The lack of SCAF1 leads to loss of III_2+IV SC formation. Also, the III_2+IV_2 SC was not reported in mammals, and it has been shown that a single copy of subunit 9 spans both CIII monomers, blocking the binding of a second SCAF1 molecule and thereby preventing the assembly of the III_2+IV_2 supercomplex [25]. Therefore, the observed COX species above the COX dimers are not characterized yet. These COX-containing species were also observed previously, in [24].

We also detected CIII at the same position as the IV_x and IV_y bands, but CIII was not fully mature and, notably lacked UQCRFS1, which is found in the mature III_2 complex [26]. Full-length SCAF1 is essential for promoting the formation of the III_2+IV SC by effectively tethering complexes III and IV [27], which is important for I+III_2+IV_n diversity [28], and it has been shown to regulate CIII biogenesis. However, SCAF1 did not affect mitochondrial respiration in the human cell culture system [29]. Additionally, SCAF1 and COX7A1/2 have been shown to form distinct types of respirasomes, with SCAF1-respirasome being preferred under glycolytic conditions and COX7A1/2 being more favorable for promoting OXPHOS when pyruvate dehydrogenase is activated [30].

Finally, previous studies using *Cox7a1* knockout mice have only examined male animals, reporting a reduction of approximately 50% in COX activity [7]. Although we observe a similar reduction in COX activity in female animals (this study), it is not as pronounced, with an overall reduction of 30% in *Cox7a1* knockout, suggesting that there are sex-specific modulators of ETC function that merit further investigation. Previous research investigating mitochondrial biogenesis and function in cardiac tissue has yielded varied findings regarding sex-related differences in mitochondrial function. Some studies have reported reduced mitochondrial DNA content and decreased mitochondrial function in female hearts compared to males [31]. In contrast, others have found that female hearts typically exhibit higher mitochondrial state 3 respiration, compared to male hearts [32]. Further investigations into hormonal regulation and other sex-specific mechanisms that influence cardiac mitochondrial function are warranted.

5. Conclusions

In conclusion, our study demonstrates that loss of COX7A1 in female mouse hearts specifically impairs COX activity within dimeric and higher-order supercomplexes, while monomeric COX remains functionally intact. The observed compensatory increase in COX7A2 incorporation in *Cox7a1* knockout, along with selective rearrangements favoring CI-containing supercomplexes, appears to help mitigate respiratory malfunction. Moreover, sex-specific differences in COX activity reduction underscore the need for further investigation into hormonal and genetic modulators of mitochondrial function.

Supplementary Materials: The following supporting information can be downloaded at: https://www.mdpi.com/article/10.3390/biom15091209/s1, Figure S1: Quantification of IGA and BN-Page; Figure S2: BN-Page with SCAF1; Figure S3: Proteinase K treatment quality control; Figure S4. Uncropped images for Figure 1B,C. CIV IGA, and CIV+CI IGA. Images were taken after overnight incubation and on a phone. Total of 4 biological replicates per group. Sample group and lane assignments: WT 1-4, COX7aH KO 5-8; Figure S5. Uncropped images for Figure 1A–C. Uncropped Coomassie and CIV IGA, and CIV+CI IGA. Images were taken after 1 h of incubation and on a Chemidoc imager and used for IGA quantification and normalization. Total of 4 biological replicates per group. Sample group and lane assignments: WT 1-4, COX7aH KO 5-8; Figure S6. Uncropped images for Figure 3A–C. Uncropped immunoblots of COXI, UQCRC2, and NDUFS3. Total of 4 biological replicates per group. Sample group and lane assignments: WT 1-4, COX7aH KO 5-8. Lane 9 was a rat heart used as a positive control for the Q-respirasome; Figure S7. Uncropped images for Figure 3. Uncropped Coomassie blot used for total protein normalization of UQCRC2 (left), COXI (right), and NDUFS3 (right). WT 1-4, COX7aH KO 5-8. Lane 9 was a rat heart used as a positive control for the Q-respirasome; Figure S8. Uncropped images for Supplementary Figure S2A,B uncropped UQCRC1 and SCAF1. Sample group and lane assignments: WT 1-4, COX7aH KO 5-8. Lane 9 was a rat heart used as a positive control for III2+IV SC; Figure S9. Uncropped images for Supplementary Figure S3.

Author Contributions: L.P. (Lauren Pavelich), T.A., L.P. (Lucynda Pham), P.S. and I.L. were responsible for investigation, formal analysis, validation, and visualization. T.A. and L.P. (Lauren Pavelich) were responsible for writing—original draft. T.A., M.H. and L.I.G. were responsible for conceptualization and project administration. M.H. and L.I.G. were responsible for funding acquisition and resources. T.A. and M.H. were responsible for methodology and supervision. All authors have read and agreed to the published version of the manuscript.

Funding: This work was supported by Wayne State University, Detroit, the Center for Molecular Medicine and Genetics, Wayne State University, Detroit, and a grant supplement from the National Institutes of Health (GM48517). Opinions, interpretations, conclusions, and recommendations are those of the authors, and are not necessarily endorsed by the funding agencies, including the National Institutes of Health.

Institutional Review Board Statement: Institutional Animal Care and Use Committee, Wayne State University; Approval Code: # A 02-08-14, Approval Date: 1 May 2014.

Informed Consent Statement: Not applicable.

Data Availability Statement: The original contributions presented in this study are included in the article and Supplementary Materials. Further inquiries can be directed to the corresponding authors.

Conflicts of Interest: The authors declare no conflicts of interest.

Abbreviations

COX	cytochrome *c* oxidase
COX7A1/COX7AH	cytochrome *c* oxidase subunit 7a, heart/skeletal muscle-type
COX7A2/COX7AL	cytochrome *c* oxidase subunit 7a, liver-type (ubiquitous)
SCAF1/COX7AR/COX7A2L	supercomplex associated factor 1
ETC	electron transport chain
SCs	supercomplexes
BN-PAGE	blue native polyacrylamide gel electrophoresis
IGA	in-gel activity assay
CI/CIII/CIV	complexes I, III, and IV of the ETC
KO	knockout
OCR	oxygen consumption rate
ROS	reactive oxygen species
SDS-PAGE	sodium dodecyl sulfate–polyacrylamide gel electrophoresis
TCA	tricarboxylic acid cycle
WT	wild-type

References

1. Pham, L.; Arroum, T.; Wan, J.; Pavelich, L.; Bell, J.; Morse, P.T.; Lee, I.; Grossman, L.I.; Sanderson, T.H.; Malek, M.H.; et al. Regulation of mitochondrial oxidative phosphorylation through tight control of cytochrome c oxidase in health and disease—Implications for ischemia/reperfusion injury, inflammatory diseases, diabetes, and cancer. *Redox Biol.* **2024**, *78*, 103426. [CrossRef]
2. Seelan, R.S.; Grossman, L.I. Structure and organization of the heart isoform gene for bovine cytochrome c oxidase subunit VIIa. *Biochemistry* **1992**, *31*, 4696–4704. [CrossRef]
3. Seelan, R.S.; Grossman, L.I. Cytochrome c oxidase subunit VIIa isoforms. Characterization and expression of bovine cDNAs. *J. Biol. Chem.* **1991**, *266*, 19752–19757. [CrossRef]
4. Seelan, R.S.; Grossman, L.I. Structural organization and evolution of the liver isoform gene for bovine cytochrome c oxidase subunit VIIa. *Genomics* **1993**, *18*, 527–536. [CrossRef]
5. Yu, M.; Jaradat, S.A.; Grossman, L.I. Genomic organization and promoter regulation of human cytochrome c oxidase subunit VII heart/muscle isoform (COX7AH). *Biochim. Biophys. Acta Gene Struct. Expr.* **2002**, *1574*, 345–353. [CrossRef]
6. Schmidt, T.R.; Goodman, M.; Grossman, L.I. Molecular evolution of the COX7A gene family in primates. *Mol. Biol. Evol.* **1999**, *16*, 619–626. [CrossRef]
7. Hüttemann, M.; Klewer, S.; Lee, I.; Pecinova, A.; Pecina, P.; Liu, J.; Lee, M.; Doan, J.W.; Larson, D.; Slack, E.; et al. Mice deleted for heart-type cytochrome c oxidase subunit 7a1 develop dilated cardiomyopathy. *Mitochondrion* **2012**, *12*, 294–304. [CrossRef] [PubMed]
8. Lee, I.; Hüttemann, M.; Liu, J.; Grossman, L.I.; Malek, M.H. Deletion of heart-type cytochrome c oxidase subunit 7a1 impairs skeletal muscle angiogenesis and oxidative phosphorylation. *J. Physiol.* **2012**, *590*, 5231–5243. [CrossRef] [PubMed]
9. Boczonadi, V.; Giunta, M.; Lane, M.; Tulinius, M.; Schara, U.; Horvath, R. Investigating the role of the physiological isoform switch of cytochrome c oxidase subunits in reversible mitochondrial disease. *Int. J. Biochem. Cell Biol.* **2015**, *63*, 32–40. [CrossRef] [PubMed]
10. Taanman, J.W.; Herzberg, N.H.; De Vries, H.; Bolhuis, P.A.; Van den Bogert, C. Steady-state transcript levels of cytochrome c oxidase genes during human myogenesis indicate subunit switching of subunit VIa and co-expression of subunit VIIa isoforms. *Biochim. Biophys. Acta Mol. Basis Dis.* **1992**, *1139*, 155–162. [CrossRef]

11. Ye, X.; Xie, Y.; Shi, Y.; Wang, B.; Han, X.; Zhou, X.; Pan, K.; Wang, M.; Fang, H. Switching ubiquitous and muscle-specific isoforms of mitochondrial respiratory complex IV in skeletal muscle fine-tunes complex IV activity. *FASEB J.* **2023**, *37*, e22891. [CrossRef]
12. Jha, P.; Wang, X.; Auwerx, J. Analysis of Mitochondrial Respiratory Chain Supercomplexes Using Blue Native Polyacrylamide Gel Electrophoresis (BN-PAGE). *Curr. Protoc. Mouse Biol.* **2016**, *6*, 1–14. [CrossRef] [PubMed]
13. Arroum, T.; Pham, L.; Raisanen, T.E.; Morse, P.T.; Wan, J.; Bell, J.; Lax, R.; Saada, A.; Hüttemann, M.; Weksler-Zangen, S. High Sucrose Diet-Induced Subunit I Tyrosine 304 Phosphorylation of Cytochrome c Oxidase Leads to Liver Mitochondrial Respiratory Dysfunction in the Cohen Diabetic Rat Model. *Antioxidants* **2024**, *13*, 19. [CrossRef] [PubMed]
14. Garcia-Poyatos, C.; Arora, P.; Calvo, E.; Marques, I.J.; Kirschke, N.; Galardi-Castilla, M.; Lembke, C.; Meer, M.; Fernandez-Montes, P.; Ernst, A.; et al. Cox7a1 controls skeletal muscle physiology and heart regeneration through complex IV dimerization. *Dev. Cell* **2024**, *59*, 1824–1841e10. [CrossRef] [PubMed]
15. Arnold, S.; Kadenbach, B. The intramitochondrial ATP/ADP-ratio controls cytochrome c oxidase activity allosterically. *FEBS Lett.* **1999**, *443*, 105–108. [CrossRef] [PubMed]
16. Kadenbach, B.; Hüttemann, M. The subunit composition and function of mammalian cytochrome c oxidase. *Mitochondrion* **2015**, *24*, 64–76. [CrossRef]
17. Helling, S.; Vogt, S.; Rhiel, A.; Ramzan, R.; Wen, L.; Marcus, K.; Kadenbach, B. Phosphorylation and kinetics of mammalian cytochrome c oxidase. *Mol. Cell. Proteom.* **2008**, *7*, 1714–1724. [CrossRef]
18. Tropeano, C.V.; Aleo, S.J.; Zanna, C.; Roberti, M.; Scandiffio, L.; Loguercio Polosa, P.; Fiori, J.; Porru, E.; Roda, A.; Carelli, V.; et al. Fine-tuning of the respiratory complexes stability and supercomplexes assembly in cells defective of complex III. *Biochim. Biophys. Acta Bioenerg.* **2020**, *1861*, 148133. [CrossRef]
19. Liang, C.; Padavannil, A.; Zhang, S.; Beh, S.; Robinson, D.R.L.; Meisterknecht, J.; Cabrera-Orefice, A.; Koves, T.R.; Watanabe, C.; Watanabe, M.; et al. Formation of I(2)+III(2) supercomplex rescues respiratory chain defects. *Cell Metab* **2025**, *37*, 441–459.e11. [CrossRef]
20. Ramzan, R.; Rhiel, A.; Weber, P.; Kadenbach, B.; Vogt, S. Reversible dimerization of cytochrome c oxidase regulates mitochondrial respiration. *Mitochondrion* **2019**, *49*, 149–155. [CrossRef]
21. Lee, I.; Salomon, A.R.; Ficarro, S.; Mathes, I.; Lottspeich, F.; Grossman, L.I.; Hüttemann, M. cAMP-dependent tyrosine phosphorylation of subunit I inhibits cytochrome c oxidase activity. *J. Biol. Chem.* **2005**, *280*, 6094–6100. [CrossRef]
22. Shinzawa-Itoh, K.; Muramoto, K. Biochemical and crystallographic studies of monomeric and dimeric bovine cytochrome c oxidase. *Biophys. Physicobiol.* **2021**, *18*, 186–195. [CrossRef]
23. Mende, P.; Huther, F.J.; Kadenbach, B. Specific and reversible activation and inactivation of the mitochondrial phosphate carrier by cardiolipin and nonionic detergents, respectively. *FEBS Lett.* **1983**, *158*, 331–334. [CrossRef]
24. Cogliati, S.; Calvo, E.; Loureiro, M.; Guaras, A.M.; Nieto-Arellano, R.; Garcia-Poyatos, C.; Ezkurdia, I.; Mercader, N.; Vazquez, J.; Enriquez, J.A. Mechanism of super-assembly of respiratory complexes III and IV. *Nature* **2016**, *539*, 579–582. [CrossRef]
25. Vercellino, I.; Sazanov, L.A. Structure and assembly of the mammalian mitochondrial supercomplex CIII(2)CIV. *Nature* **2021**, *598*, 364–367. [CrossRef]
26. Signes, A.; Fernandez-Vizarra, E. Assembly of mammalian oxidative phosphorylation complexes I-V and supercomplexes. *Essays Biochem.* **2018**, *62*, 255–270. [CrossRef] [PubMed]
27. Calvo, E.; Cogliati, S.; Hernansanz-Agustin, P.; Loureiro-Lopez, M.; Guaras, A.; Casuso, R.A.; Garcia-Marques, F.; Acin-Perez, R.; Marti-Mateos, Y.; Silla-Castro, J.C.; et al. Functional role of respiratory supercomplexes in mice: SCAF1 relevance and segmentation of the Q(pool). *Sci. Adv.* **2020**, *6*, eaba7509. [CrossRef] [PubMed]
28. Vercellino, I.; Sazanov, L.A. SCAF1 drives the compositional diversity of mammalian respirasomes. *Nat. Struct. Mol. Biol.* **2024**, *31*, 1061–1071. [CrossRef] [PubMed]
29. Lobo-Jarne, T.; Nyvltova, E.; Perez-Perez, R.; Timon-Gomez, A.; Molinie, T.; Choi, A.; Mourier, A.; Fontanesi, F.; Ugalde, C.; Barrientos, A. Human COX7A2L Regulates Complex III Biogenesis and Promotes Supercomplex Organization Remodeling without Affecting Mitochondrial Bioenergetics. *Cell Rep.* **2018**, *25*, 1786–1799.e4. [CrossRef]
30. Fernandez-Vizarra, E.; Lopez-Calcerrada, S.; Sierra-Magro, A.; Perez-Perez, R.; Formosa, L.E.; Hock, D.H.; Illescas, M.; Penas, A.; Brischigliaro, M.; Ding, S.; et al. Two independent respiratory chains adapt OXPHOS performance to glycolytic switch. *Cell Metab.* **2022**, *34*, 1792–1808.e6. [CrossRef]
31. Cao, Y.; Vergnes, L.; Wang, Y.C.; Pan, C.; Chella Krishnan, K.; Moore, T.M.; Rosa-Garrido, M.; Kimball, T.H.; Zhou, Z.; Charugundla, S.; et al. Sex differences in heart mitochondria regulate diastolic dysfunction. *Nat. Commun.* **2022**, *13*, 3850. [CrossRef] [PubMed]
32. Fliegner, D.; Ellieva, A.; Angelov, A.; Petrov, G.; Regitz-Zagrosek, V. Sex differences and estrogen effects in cardiac mitochondria in human aortic stenosis and in the mouse heart. *Front. Endocrinol.* **2023**, *14*, 1181044. [CrossRef] [PubMed]

biomolecules

MDPI

Review

Oxidative Stress in Diabetic Cardiomyopathy: Molecular Mechanisms and Emerging Therapeutic Targets

Umberto Capece [1,†], Davide Nilo [2,†], Cassandra Morciano [1], Roberto Nilo [3], Serenella Spiezia [4], Marta Chiara Sircana [5], Vincenzo Russo [6,7], Marco Alfonso Perrone [8], Leonilde Bonfrate [9,10], Carlo Acierno [11], Ferdinando Carlo Sasso [2] and Alfredo Caturano [9,*]

1 Centro Malattie Endocrine e Metaboliche, Dipartimento di Scienze Mediche e Chirurgiche, Fondazione Policlinico Universitario A. Gemelli IRCCS and Università Cattolica del Sacro Cuore, 00168 Roma, Italy; capeceumberto@gmail.com (U.C.); morcianocassandra@gmail.com (C.M.)
2 Department of Advanced Medical and Surgical Sciences, University of Campania "Luigi Vanvitelli", 80138 Naples, Italy; nilodavide@gmail.com (D.N.); ferdinandocarlo.sasso@unicampania.it (F.C.S.)
3 Meditrial Europe, 00198 Roma, Italy; robertonilo98@gmail.com
4 Diagnostic and Therapeutic Medicine Department, Fondazione Policlinico Universitario Campus Bio-Medico, Via Alvaro del Portillo, 200, 00128 Rome, Italy; serenella.spiezia@gmail.com
5 Department of Medical, Surgical and Pharmacology, University of Sassari, 07100 Sassari, Italy; m.sircana4@phd.uniss.it
6 Division of Cardiology, Department of Medical Translational Sciences, University of Campania "Luigi Vanvitelli", 80138 Naples, Italy; v.p.russo@libero.it
7 Department of Biology, College of Science and Technology, Sbarro Institute for Cancer Research and Molecular Medicine, Temple University, Philadelphia, PA 19122, USA
8 Division of Cardiology and CardioLab, Department of Clinical Sciences and Translational Medicine, University of Rome Tor Vergata, 00133 Rome, Italy; marco.perrone@uniroma2.it
9 Department of Human Sciences and Promotion of the Quality of Life, San Raffaele Roma University, 00166 Rome, Italy; leonilde.bonfrate@uniroma5.it
10 Center of Nutrition for the Research and the Care of Obesity and Metabolic Diseases, National Institute of Gastroenterology IRCCS "Saverio de Bellis", 70013 Castellana Grotte, Italy
11 Azienda Ospedaliera Regionale San Carlo, 85100 Potenza, Italy; carlo894@gmail.com
* Correspondence: alfredo.caturano@uniroma5.it
† These authors contributed equally to this work.

Abstract

Diabetic cardiomyopathy (DCM) is a distinct myocardial disorder that develops independently of coronary artery disease and hypertension and represents a major contributor to heart failure in patients with diabetes. Beyond hemodynamic alterations, DCM is driven by complex molecular mechanisms involving metabolic dysregulation, mitochondrial dysfunction, inflammation, and fibrotic remodeling. Increasing evidence identifies oxidative stress as a central integrative process linking these pathogenic pathways in the diabetic heart. Chronic hyperglycemia, insulin resistance, and altered substrate utilization promote excessive generation of reactive oxygen species, overwhelming endogenous antioxidant defenses and disrupting myocardial redox homeostasis. Oxidative stress induces direct damage to lipids, proteins, and DNA while simultaneously activating redox-sensitive signaling pathways that amplify inflammation, endothelial dysfunction, cardiomyocyte apoptosis, and fibrosis. In addition, epicardial and visceral adipose tissue have emerged as active contributors to myocardial oxidative stress through paracrine and systemic mechanisms, reinforcing inflammatory and fibrotic crosstalk. This review provides a comprehensive overview of the molecular sources and targets of oxidative damage in DCM, examines the impairment of antioxidant defense systems, and discusses emerging therapeutic strategies aimed at restoring redox balance.

 https://doi.org/10.3390/biom16030470

Keywords: diabetic cardiomyopathy; oxidative stress; mitochondrial dysfunction; reactive oxygen species; epicardial adipose tissue; myocardial fibrosis; redox signaling; antioxidant defense; heart failure; diabetes mellitus

1. Introduction

Diabetes mellitus represents a major global health burden and is one of the leading causes of cardiovascular morbidity and mortality worldwide [1]. In addition to accelerated atherosclerosis and ischemic heart disease, diabetes is associated with a distinct myocardial disorder known as diabetic cardiomyopathy (DCM), characterized by structural and functional alterations of the myocardium that occur independently of coronary artery disease, hypertension, or valvular abnormalities [2,3]. Clinically, DCM manifests as progressive left ventricular dysfunction, initially dominated by diastolic impairment with preserved ejection fraction and, in advanced stages, by systolic dysfunction and overt heart failure. Despite its clinical relevance and growing prevalence, DCM remains underdiagnosed and lacks disease-specific therapeutic strategies [3].

The pathogenesis of diabetic cardiomyopathy is complex and multifactorial, reflecting the chronic metabolic stress imposed by hyperglycemia, insulin resistance, and altered substrate utilization. These disturbances lead to profound changes in myocardial energy metabolism, lipid handling, and cellular homeostasis [4]. Rather than acting as independent mechanisms, these processes converge on oxidative stress, which has emerged as a central and unifying driver of myocardial injury in the diabetic heart [5]. Persistent metabolic overload promotes excessive generation of reactive oxygen species (ROS), overwhelming endogenous antioxidant defenses and disrupting redox balance across multiple cardiac cell types [6]. Oxidative stress plays a dual pathogenic role in DCM. On the one hand, excess ROS directly damage lipids, proteins, and nucleic acids, impairing cellular integrity and contractile function. On the other hand, oxidative stress acts as a key signaling mediator that amplifies maladaptive pathways involved in inflammation, fibrosis, and cell death [7]. Through activation of redox-sensitive kinases and transcription factors, oxidative stress contributes to endothelial dysfunction, microvascular impairment, cardiomyocyte apoptosis, and fibroblast activation, thereby promoting adverse myocardial remodeling and progressive ventricular dysfunction [7–10].

Beyond cardiomyocytes and endothelial cells, increasing evidence highlights the contribution of extracardiac tissues to myocardial oxidative stress in diabetes. In particular, epicardial adipose tissue (EAT) and other visceral fat depots have emerged as active pathogenic players through paracrine and systemic mechanisms [11,12]. In the diabetic milieu, adipose tissue dysfunction, macrophage infiltration, and altered adipokine secretion further intensify local inflammation and oxidative burden within the cardiac microenvironment [6]. These observations support the concept that DCM is not a purely cardiomyocyte-centric disease but rather the result of a complex multicellular and inter-organ crosstalk driven by metabolic and redox imbalance. Despite substantial progress in delineating the molecular mechanisms underlying DCM, current clinical management largely relies on glycemic control and conventional heart failure therapies, which do not specifically target the underlying myocardial pathology [4,13]. This limitation has fueled growing interest in oxidative stress and redox-regulated pathways as potential disease-modifying targets, with the aim of interrupting the progression of myocardial damage at earlier stages.

The aim of this review is to provide a comprehensive overview of the role of oxidative stress in the development and progression of diabetic cardiomyopathy. Specifically, we

https://doi.org/10.3390/biom16030470

(i) summarize the key pathophysiological features of DCM, (ii) discuss the major cellular sources and molecular targets of oxidative damage in the diabetic heart, (iii) explore the interplay between oxidative stress, inflammation, and fibrotic remodeling, and (iv) highlight emerging therapeutic strategies and future perspectives aimed at restoring redox homeostasis and improving cardiac outcomes in patients with diabetes. Unlike reviews that focus on individual molecular pathways or isolated antioxidant mechanisms, this article provides an integrated synthesis that positions oxidative stress as a central organizing process of this condition. In addition, we will focus on recent evidence particularly regarding adipose tissue, which makes an update of this nosological entity necessary. Furthermore, most studies in the current literature focus on the effects of GLP-1 receptor agonists (GLP-1RA) [14] and SGLT2 inhibitors (SGLT2-I) [15], which act on diabetic cardiomyopathy through multiple mechanisms that are still under investigation and mainly independent of oxidative stress. Our objective, instead, is to focus specifically on this pathogenetic mechanism. This review emphasizes the emerging shift from non-specific antioxidant supplementation toward precision, mechanism-based redox therapies that target specific sources, intracellular compartments, and disease stages. This integrated framework aims to support a more mechanistic and translational understanding of redox-targeted interventions in diabetic cardiomyopathy.

2. Pathophysiology of Diabetic Cardiomyopathy

DCM represents a distinct pathological entity that differs from other forms of heart failure by its unique metabolic, structural, and molecular features. Unlike ischemic or hypertensive cardiomyopathy, DCM develops in the absence of overt coronary artery disease or pressure overload and is primarily driven by chronic metabolic stress associated with diabetes, which converges on sustained oxidative stress as a central mediator of myocardial injury [4,16]. The hallmark of DCM is a progressive impairment of myocardial structure and function, initially characterized by diastolic dysfunction with preserved ejection fraction and subsequently evolving toward systolic dysfunction and overt heart failure [17].

At the structural level, diabetic hearts exhibit a constellation of histopathological changes, including cardiomyocyte hypertrophy, interstitial and perivascular fibrosis, microvascular rarefaction, and increased cardiomyocyte apoptosis. These alterations lead to increased myocardial stiffness, impaired relaxation, and reduced ventricular compliance, which are key determinants of diastolic dysfunction [16]. Fibrotic remodeling, in particular, reflects excessive deposition of extracellular matrix components, predominantly collagen types I and III, driven by persistent activation of cardiac fibroblasts and profibrotic signaling pathways [18]. Over time, this maladaptive remodeling compromises not only mechanical efficiency but also electrical stability of the myocardium [19,20].

From a metabolic standpoint, the diabetic heart undergoes a profound shift in substrate utilization. Insulin resistance and reduced glucose uptake force cardiomyocytes to rely predominantly on fatty acid oxidation for ATP generation [21]. While fatty acids constitute an efficient energy source under physiological conditions, their chronic overutilization in diabetes results in mitochondrial overload, incomplete β-oxidation, and accumulation of toxic lipid intermediates such as ceramides and diacylglycerols [22,23]. This metabolic inflexibility not only reduces cardiac energetic efficiency but also promotes lipotoxicity, mitochondrial dysfunction, and excessive generation of ROS, thereby directly linking altered substrate metabolism to myocardial redox imbalance [24,25].

Mitochondrial abnormalities are a central feature of DCM pathophysiology and represent a critical source of oxidative stress. Diabetic cardiomyocytes display altered mitochondrial morphology, impaired oxidative phosphorylation, reduced ATP production,

https://doi.org/10.3390/biom16030470

and enhanced electron leak from the respiratory chain [26]. These defects amplify ROS generation and initiate a cascade of downstream events, including mitochondrial DNA damage, opening of the mitochondrial permeability transition pore, activation of apoptotic pathways, and disruption of calcium handling [27,28]. Impaired calcium homeostasis further contributes to contractile dysfunction and increases arrhythmogenic susceptibility, reinforcing the progression toward heart failure [29]. In parallel, diabetes induces profound alterations in myocardial signaling networks. Chronic hyperglycemia, insulin resistance, and oxidative stress activate redox-sensitive pathways such as protein kinase C (PKC), mitogen-activated protein kinases (MAPKs), nuclear factor-κB (NF-κB), and transforming growth factor-β (TGF-β) [30]. These pathways orchestrate inflammatory responses, promote fibroblast activation, and sustain maladaptive myocardial remodeling. Importantly, oxidative stress functions as a key amplifier of these signaling cascades, integrating metabolic derangements with inflammation, fibrosis, and cardiomyocyte loss [31,32].

Collectively, these structural, metabolic, and molecular alterations define the complex pathophysiological landscape of diabetic cardiomyopathy. Rather than resulting from a single pathogenic insult, DCM emerges from the convergence of metabolic inflexibility, mitochondrial dysfunction, oxidative stress, inflammation, and fibrotic remodeling. Within this framework, oxidative stress should be viewed not merely as a downstream consequence of diabetes but as a central integrative mechanism linking these processes. This concept provides the rationale for a focused analysis of the cellular and molecular sources of ROS in diabetic cardiomyopathy, which are discussed in the following section.

3. Sources of Oxidative Stress in Diabetic Cardiomyopathy

3.1. Cellular Types Involved

The development of oxidative stress in DCM involves multiple cardiac and extracardiac cell types, including cardiomyocytes, endothelial cells of the coronary microvasculature, cardiac fibroblasts, and adipocytes from epicardial and other visceral fat depots [6,12,16]. Cardiomyocytes are directly exposed to hyperglycemia, insulin resistance, and altered substrate metabolism, leading to mitochondrial dysfunction, impaired calcium handling, and disruption of intracellular redox balance [32]. Endothelial cells respond to the diabetic milieu with increased ROS production, reduced nitric oxide (NO) bioavailability, and endothelial dysfunction, contributing to microvascular impairment [33]. In parallel, EAT has emerged as an active contributor to myocardial oxidative stress. Epicardial adipocytes can release free fatty acids, pro-inflammatory cytokines, and adipokines that act locally through paracrine signaling or systemically via the circulation, thereby promoting oxidative stress, inflammation, and fibrotic remodeling within the myocardium [12,34,35]. Adipose depots located outside the heart may further amplify this process by supplying circulating pro-oxidant and pro-inflammatory mediators [36]. The coordinated dysfunction of these cellular compartments results in a sustained oxidative burden that drives myocardial injury and functional deterioration in DCM.

3.2. Mitochondrial Dysfunction and ROS Overproduction

Mitochondrial dysfunction represents a primary source of oxidative stress in DCM. In diabetic cardiomyocytes, mitochondria exhibit marked morphological and functional abnormalities, including fragmentation, altered fission–fusion dynamics that may involve MFN1/2, OPA1, and DRP1, impaired electron transport chain efficiency, and increased electron leakage [37–39]. All these processes ultimately lead to excessive superoxide production [39,40]. Hyperglycemia and fatty-acid overload increase the supply of reducing equivalents (NADH and $FADH_2$) to the respiratory chain, further enhancing mitochondrial ROS generation and overwhelming intramitochondrial antioxidant systems such as

https://doi.org/10.3390/biom16030470

manganese superoxide dismutase and mitochondrial catalase [40]. This redox imbalance induces mitochondrial DNA (mtDNA) damage, opening of the mitochondrial permeability transition pore, cytochrome c release, and activation of apoptotic pathways, thereby contributing to cardiomyocyte loss, contractile dysfunction, and fibrotic remodeling [37,41]. Importantly, mitochondrial ROS generation establishes a self-perpetuating cycle, whereby oxidative damage further impairs mitochondrial function and amplifies ROS production [38,39]. In addition to electron leak from respiratory chain complexes, cytochrome c has been identified as a context-dependent contributor to ROS generation. Under physiological conditions, cytochrome c functions as a mobile electron carrier between complexes III and IV [42]. However, in the presence of elevated NADH and hydrogen peroxide concentrations, conditions frequently observed in the diabetic myocardium, cytochrome c may participate in redox cycling reactions capable of generating superoxide radicals [43,44]. Moreover, partial release of cytochrome c from mitochondria during early mitochondrial membrane destabilization can further disrupt electron transport chain efficiency, amplifying ROS production [45]. This dual role of cytochrome c, both as an essential component of oxidative phosphorylation and as a potential amplifier of oxidative stress under pathological conditions, underscores the complexity of mitochondrial redox regulation in diabetic cardiomyopathy.

3.3. NADPH Oxidases (NOX Enzymes)

In addition to mitochondria, NADPH oxidases (NOX) constitute a major enzymatic source of ROS in DCM. NOX isoforms (particularly NOX2, NOX4, and NOX5) are expressed in cardiomyocytes, endothelial cells, and cardiac fibroblasts. In the diabetic setting, NOX activity is stimulated by hyperglycemia, activation of diacylglycerol-protein kinase C pathways, angiotensin II signaling, and inflammatory mediators [46,47]. NOX-derived ROS promote superoxide formation, reduce NO bioavailability, and activate downstream inflammatory cascades, including NF-κB signaling, thereby amplifying oxidative injury and myocardial inflammation [25,48]. Experimental studies suggest that selective inhibition of NOX enzymes may attenuate oxidative stress and pathological remodeling in DCM [49].

3.4. Xanthine Oxidoreductase and Aldehyde Oxidase

Xanthine oxidoreductase (XOR) represents an additional enzymatic source of reactive oxygen species in the diabetic heart. XOR exists in two interconvertible forms, xanthine dehydrogenase and xanthine oxidase, the latter preferentially generating superoxide anion and hydrogen peroxide during purine metabolism [50]. Under physiological conditions, XOR activity is tightly regulated; however, in pathological states such as diabetes mellitus, ischemia–reperfusion injury, and chronic inflammation, increased expression and post-translational conversion to the oxidase form enhance myocardial ROS production [7,10,51]. In patients with type 2 diabetes, elevated plasma xanthine oxidoreductase activity has been significantly associated with insulin resistance, adverse metabolic parameters, and vascular complications, supporting the concept that XOR-derived oxidative stress may contribute to endothelial dysfunction and cardiovascular risk in the diabetic milieu [50,52].

Aldehyde oxidase (AO), a molybdenum-containing enzyme structurally related to XOR, has also emerged as a potential contributor to cardiac oxidative stress [53]. Although primarily studied in hepatic metabolism, AO is expressed in cardiovascular tissues and is capable of generating superoxide and hydrogen peroxide during oxidation of endogenous aldehydes and xenobiotics [54,55]. In diabetic conditions, increased availability of reactive aldehydes and enhanced inflammatory signaling [56] may favor greater AO-dependent redox activity, potentially contributing to additional intracellular ROS generation. Although direct evidence in diabetic cardiomyopathy remains limited, the biochemical properties of

https://doi.org/10.3390/biom16030470

aldehyde oxidase support a plausible contributory role under conditions of metabolic and inflammatory stress, thus representing a still insufficiently characterized source of ROS in this context.

3.5. Uncoupled Nitric Oxide Synthase (NOS)

Endothelial nitric oxide synthase (eNOS) normally produces NO, exerting vasodilatory and cardioprotective effects. However, under conditions of oxidative stress, hyperglycemia, and deficiency of essential cofactors such as tetrahydrobiopterin (BH_4), eNOS may become uncoupled, shifting from NO production to superoxide generation. This process further reduces NO bioavailability and exacerbates oxidative damage within both the endothelium and myocardium [10,41]. Experimental studies in diabetic models have demonstrated eNOS uncoupling in vascular tissues and endothelial progenitor cells, characterized by increased superoxide production and impaired endothelial function [57,58]. These alterations contribute to endothelial dysfunction and impaired NO signaling, mechanisms widely recognized as key contributors to diabetic cardiovascular complications, including diabetic cardiomyopathy [10,41]. Although eNOS uncoupling is well documented in vascular and endothelial contexts, its direct contribution to myocardial oxidative injury in DCM remains incompletely defined and should be interpreted as an extrapolated mechanism.

3.6. Advanced Glycation End-Products (AGEs) and RAGE Signaling

Chronic hyperglycemia promotes the formation and accumulation of advanced glycation end-products (AGEs), which interact with their receptor RAGE to activate pro-oxidant and pro-inflammatory signaling pathways. AGE–RAGE engagement enhances ROS generation through stimulation of NADPH oxidases, activates NF-κB signaling, and induces expression of inflammatory and fibrotic genes [59,60]. In the diabetic myocardium, this axis significantly contributes to oxidative stress accumulation, myocardial fibrosis, and ventricular dysfunction [61].

3.7. Endoplasmic Reticulum Stress and the Unfolded Protein Response (UPR)

Endoplasmic reticulum (ER) stress represents an additional contributor to oxidative stress in DCM [37]. Hyperglycemia, accumulation of misfolded or glycated proteins, and increased metabolic demand activate the unfolded protein response (UPR), mediated by PERK/eIF2α, IRE1α, and CHOP signaling pathways. Persistent UPR activation promotes ROS generation and exacerbates intracellular redox imbalance [37,62]. In diabetic cardiomyocytes, ER stress contributes to apoptosis, secondary mitochondrial dysfunction, and interstitial fibrosis, further reinforcing the oxidative and structural remodeling characteristic of DCM [63].

4. Molecular Targets of Oxidative Damage in the Heart

4.1. Lipids: Lipid Peroxidation and Membrane Damage

Excessive ROS production drives lipid peroxidation of membrane lipids, generating reactive aldehydes such as 4-hydroxy-2-nonenal (4-HNE) and malondialdehyde (MDA), which impair membrane fluidity, receptor function, ion-channel behavior and ion-pump activity [64,65]. In DCM, elevated MDA and 4-HNE levels have been correlated with the severity of myocardial dysfunction [66]. Damage to mitochondrial or plasma membranes of cardiomyocytes directly undermines cell integrity, compromises signal transduction and activates apoptotic pathways [32].

https://doi.org/10.3390/biom16030470

4.2. Proteins: Carbonylation and Enzymatic Inactivation

ROS can oxidize amino-acid residues (e.g., lysine, arginine, proline) leading to formation of carbonyl groups that alter protein tertiary structure, enzyme activity and protein stability [67]. The marker 3-nitrotyrosine (3-NT) reflects protein nitration by peroxynitrite and is elevated in cardiomyopathies [68]. Within DCM, protein carbonylation contributes to dysfunction of contractile proteins, ion-pump proteins and mitochondrial proteins, thereby disturbing energetic metabolism and increasing susceptibility to apoptosis [66].

4.3. DNA: Mitochondrial and Nuclear Damage

ROS can damage mitochondrial DNA (mtDNA) and nuclear DNA, causing single-strand or double-strand breaks, base oxidation (e.g., 8-oxo-2′-deoxyguanosine, 8-OH-dG) and activation of DNA-repair pathways [69–71]. In diabetic hearts, increased mtDNA damage has been documented and correlates with mitochondrial dysfunction and cardiac fibrosis [38,72]. DNA damage promotes p53 activation, cardiomyocyte senescence and apoptosis, thus contributing to DCM progression toward heart failure [73].

4.4. Redox-Sensitive Signaling Pathways

Oxidative stress activates several redox-sensitive signaling cascades that play a central role in translating molecular damage into maladaptive cellular responses. Among these, MAPKs, including ERK1/2, p38, and JNK, are critically involved in the regulation of cardiomyocyte hypertrophy, apoptosis, and fibrotic responses. Sustained activation of these pathways in the diabetic myocardium promotes adverse structural remodeling and functional impairment [74,75].

The transcription NF-κB represents a key downstream effector of oxidative stress. ROS-mediated NF-κB activation induces the expression of pro-inflammatory cytokines, adhesion molecules, and extracellular matrix-related genes, thereby establishing a molecular link between oxidative injury and inflammatory priming of the myocardium [76]. Importantly, NF-κB signaling acts as an upstream trigger that initiates inflammatory and fibrotic programs, which are further amplified through cellular crosstalk and tissue-level responses [77] discussed in subsequent sections. In contrast to these pro-oxidant pathways, nuclear factor erythroid 2-related factor 2 (Nrf2) serves as the master regulator of the endogenous antioxidant response. Upon activation, Nrf2 translocates to the nucleus and induces the transcription of antioxidant and cytoprotective genes involved in ROS detoxification and redox homeostasis. However, in diabetic conditions, Nrf2 signaling is often impaired or insufficiently activated, resulting in reduced antioxidant capacity and increased vulnerability of cardiomyocytes to oxidative damage [78,79]. The imbalance between persistent activation of pro-oxidant signaling pathways (MAPKs and NF-κB) and inadequate activation of compensatory antioxidant mechanisms mediated by Nrf2 facilitates the progression from oxidative stress to sustained myocardial injury. This molecular dysregulation primes inflammatory and fibrotic responses, thereby contributing to myocardial remodeling and functional decline in diabetic cardiomyopathy [80].

Collectively, oxidative damage to lipids, proteins, and nucleic acids, together with the dysregulation of redox-sensitive signaling pathways, translates molecular redox imbalance into maladaptive cellular responses in the diabetic myocardium. These intracellular mechanisms are tightly interconnected and converge to impair cardiomyocyte survival, contractile function, and stress resilience. A schematic overview of the principal molecular targets and intracellular consequences of oxidative stress in diabetic cardiomyopathy is provided in Figure 1.

https://doi.org/10.3390/biom16030470

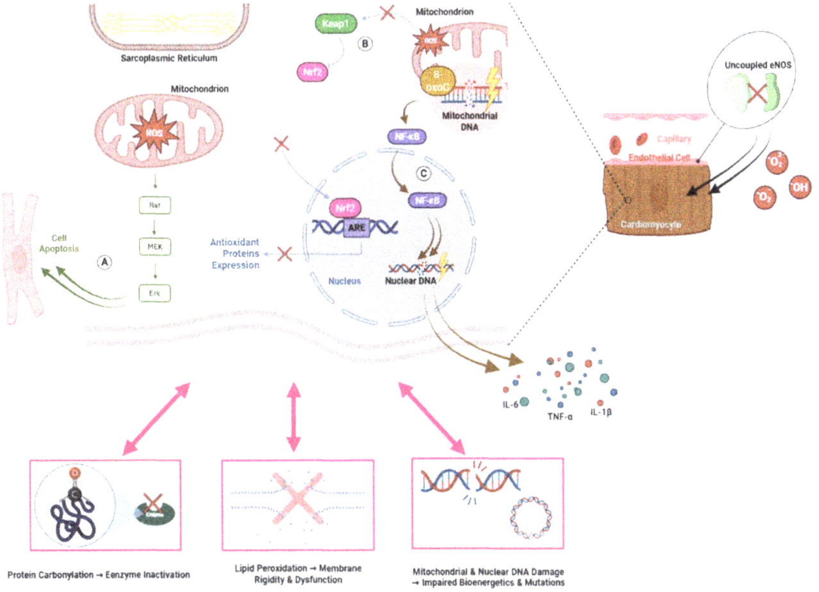

Figure 1. Molecular targets and intracellular consequences of oxidative damage in diabetic cardiomyopathy. Excessive mitochondrial production of reactive oxygen species (ROS) in cardiomyocytes induces oxidative damage to multiple intracellular targets. ROS activate pro-apoptotic signaling pathways, including the Raf–MEK–ERK cascade (**A**), and promote oxidative lesions in mitochondrial and nuclear DNA, such as 8-oxo-7,8-dihydroguanine (8-oxoG), leading to impaired bioenergetics and genomic instability. Oxidative stress disrupts the Keap1–Nrf2 axis, reducing antioxidant response element (ARE)-dependent transcription of antioxidant enzymes and weakening endogenous defense mechanisms (**B**). In parallel, ROS activate NF-κB signaling and induce transcriptional programs that exacerbate inflammatory responses and nuclear DNA damage (**C**). Uncoupling of endothelial nitric oxide synthase (eNOS) further contributes to ROS accumulation and endothelial dysfunction, amplifying oxidative injury in adjacent cardiomyocytes. At the molecular level, oxidative stress results in protein carbonylation with enzyme inactivation, lipid peroxidation leading to membrane rigidity and dysfunction, and cumulative mitochondrial and nuclear DNA damage that ultimately compromise cardiomyocyte survival and function [32,38,64–80].

5. Inflammatory and Fibrotic Crosstalk

The pathophysiology of diabetic cardiomyopathy involves, beyond the myocardium itself, a complex interaction with other organs and tissues through the release of hormones, cytokines, and circulating metabolic mediators. In recent years, increasing attention has been directed toward the role of EAT as a key contributor to myocardial inflammation and oxidative stress. EAT is anatomically contiguous with the myocardium and is distributed along the atrioventricular and interventricular grooves surrounding the coronary arteries (peri-coronary EAT) and overlying the myocardial surface (myocardial EAT) [81,82]. Due to its anatomical position, EAT shares the same microcirculation as the myocardium, in the absence of interposed muscle fascia or connective tissue layers [83]. Consequently, alterations in EAT physiology can directly influence myocardial structure and function.

Under physiological conditions, EAT exerts several cardioprotective effects. It acts as a local energy reservoir, supplying free fatty acids to the myocardium, and secretes adipokines with anti-inflammatory and anti-atherogenic properties [84]. Moreover, EAT exhibits features of both white and brown adipose tissue, indicating a metabolically active phenotype [85]. However, under pathological conditions such as obesity and diabetes mellitus, EAT undergoes marked structural and functional remodeling [86]. EAT thickness increases and a phenotypic shift toward a pro-inflammatory and pro-fibrotic state occurs, characterized by enhanced secretion of cytokines such as IL-6, TNF-α, and MCP-1, together with reduced release of protective mediators including adiponectin [87–89]. Importantly, this transition reflects not merely an expansion of adipose mass but a profound metabolic and structural reprogramming of EAT [90]. Mitochondrial dysfunction in EAT enhances local oxidative stress and promotes the accumulation of lipotoxic intermediates, which may influence the adjacent myocardium through paracrine mechanisms facilitated by the absence of a fascial barrier and the shared microcirculation between these tissues, although direct evidence in DCM remains limited [81,90]. These paracrine signals may promote oxidative stress, endothelial activation, and recruitment of pro-inflammatory macrophages within the myocardial and perivascular compartments [91]. The resulting inflammatory microenvironment induces cardiomyocyte stress and interstitial remodeling, ultimately favoring myocardial fibrosis [92]. Persistent cytokine exposure and altered lipid handling from dysfunctional EAT further promote lipotoxicity and oxidative stress, contributing to increased myocardial stiffness, diastolic dysfunction, and heart failure with preserved ejection fraction [93]. Owing to its direct contact with atrial and ventricular myocardium, EAT dysfunction may also alter local electrophysiological properties, creating a substrate for atrial and ventricular arrhythmias and increasing the incidence of atrial fibrillation [94–96].

The mechanisms underlying the development of a dysfunctional, pro-inflammatory EAT phenotype are not fully elucidated. Under physiological conditions, EAT exhibits beige-like features and retains a certain thermogenic potential. However, in the diabetic milieu, EAT may undergo a phenotypic shift toward a white adipocyte-dominant profile ("EAT whitening"), characterized by reduced thermogenic capacity, mitochondrial dysfunction, and impaired oxidative efficiency [85,94]. Specifically, this process consists of the transformation of mitochondria-rich, multilocular brown adipocytes into large, unilocular white adipocytes [97]. During sustained caloric excess, whitened adipocytes undergo to increased triglyceride storage [98] and progressive hypertrophy until a critical size threshold is reached [99], beyond which cellular stress and adipocyte death occur [100]. This critical size is lower in visceral fat depots, such as EAT, compared with subcutaneous adipose tissue [101]. Adipocyte death promotes macrophage accumulation and the formation of crown-like structures, further amplifying local inflammation [102]. Moreover, macrophage density is generally higher in visceral adipose tissue, reinforcing its pro-inflammatory profile [103]. Collectively, chronic low-grade inflammation, macrophage polarization, and lipotoxic infiltration contribute to fibrotic remodeling, microvascular impairment, and electrical instability, thereby linking EAT dysfunction to adverse myocardial structural and functional changes in diabetic cardiomyopathy, although direct mechanistic evidence remains limited and this association is largely supported by indirect and translational data. These processes are highly interconnected and likely overlap in both diabetes and obesity [104]. Supporting this concept, EAT from patients with type 2 diabetes exhibits distinct transcriptional profiles characterized by increased expression of inflammatory genes and cytokines, reinforcing the view of EAT as an active pathogenic mediator rather than a passive fat depot [105].

Pharmacological interventions such as GLP-1 receptor agonists and SGLT2 inhibitors have demonstrated promising effects in reducing EAT thickness and improving cardiovascular health. These effects are likely mediated by indirect mechanisms related to weight loss and systemic metabolic improvement for SGLT2 inhibitors, and by both direct and indirect actions for GLP-1 receptor agonists, given the expression of GLP-1 receptors on EAT and their ability to modulate adipose tissue metabolism [106–108]. The cardioprotective properties of these agents are well recognized, and attenuation of EAT-derived oxidative stress may contribute to their beneficial cardiovascular effects [109].

https://doi.org/10.3390/biom16030470

Beyond EAT, other visceral adipose depots, including perirenal, supraclavicular, and paravertebral brown adipose tissue, may influence myocardial structure and function through analogous mechanisms involving adipose whitening, macrophage infiltration, and cytokine release [110,111]. These depots have been linked to cardiovascular health, suggesting that inflammatory and fibrotic crosstalk in diabetic cardiomyopathy likely involves multiple visceral fat compartments, among which EAT exerts the most pronounced effects due to its anatomical proximity to the heart [112]. Through direct paracrine signaling [113] and systemic dissemination of pro-inflammatory mediators from distant fat depots [114], adipose tissue dysfunction generates a "distal inflammatory storm" that reinforces myocardial oxidative stress. This systemic-local axis establishes a self-perpetuating cycle of oxidative injury, accelerating the transition from metabolic compensation to overt heart failure [115].

Within this oxidative and inflammatory milieu, a central role is played by the NLRP3 inflammasome, which functions as a redox-sensitive sensor in cardiomyocytes, fibroblasts, and endothelial cells [116]. Chronic hyperglycemia-induced metabolic stress promotes NLRP3 overactivation [117], leading to caspase-1 activation, pyroptosis, and excessive release of IL-1β and IL-18 [118,119]. These cytokines propagate inflammation and stimulate fibroblast activation and collagen deposition, thereby linking inflammatory signaling to fibrotic remodeling of the diabetic heart [120]. IL-1β also interacts with transforming growth factor-β (TGF-β) pathways, promoting extracellular matrix synthesis and altering fibroblast contractile activity [121]. TGF-β signaling mobilizes Smad2 and Smad3 transcription factors, promoting the expression of fibrosis-related genes and contributing to extracellular matrix accumulation [122]. Reactive oxygen species further amplify this process by enhancing TGF-β activation [123]. In parallel, activation of the NLRP3 inflammasome contributes to endothelial dysfunction, as IL-1β promotes endothelial-to-mesenchymal transition and loss of barrier integrity, thereby exacerbating inflammation and fibrotic remodeling [124]. These mechanisms position NLRP3 as a critical molecular link between metabolic stress, inflammation, and fibrosis in the diabetic heart. Moreover, sustained ROS generation driven by mitochondrial dysfunction and NADPH oxidases activate TGF-β signaling, enhance matrix-remodeling enzyme expression, and promote fibroblast proliferation and differentiation [123,125]. ROS further exacerbate AGE formation and RAGE activation, contributing to extracellular matrix accumulation [126]. Counter-regulatory mechanisms such as sirtuin-1 activation may attenuate oxidative stress, improve mitochondrial function, and promote adipose tissue browning, representing potential modulators of these pathogenic processes [123,127–130].

Overall, inflammatory and fibrotic crosstalk in diabetic cardiomyopathy arises from the integration of metabolic stress, oxidative imbalance, adipose tissue dysfunction, and redox-sensitive inflammatory signaling (Figure 2). These interconnected mechanisms contribute to myocardial remodeling, microvascular dysfunction, and progressive ventricular impairment, highlighting oxidative stress as a central driver linking metabolic disease to cardiac structural and functional deterioration.

https://doi.org/10.3390/biom16030470

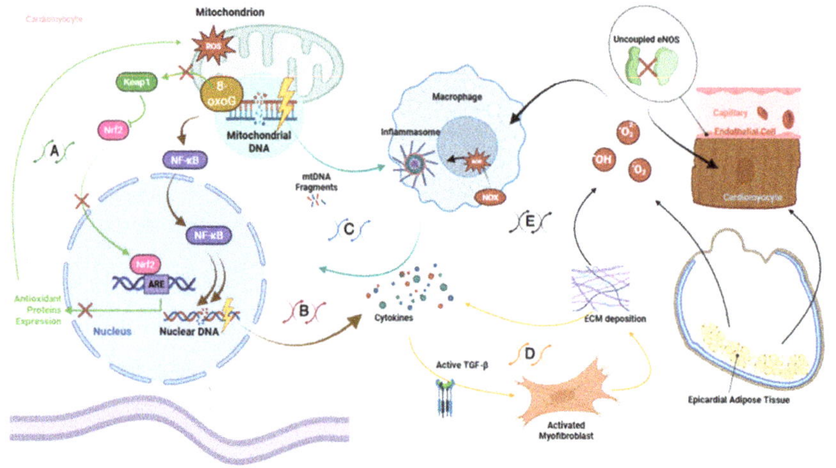

Figure 2. Vicious cycles linking oxidative stress, inflammation, and fibrosis in diabetic cardiomyopathy. (**A**) Oxidative DNA damage–antioxidant imbalance loop. Mitochondrial reactive oxygen species (ROS) induce accumulation of 8-oxo-7,8-dihydroguanine (8-oxoG) in mitochondrial and nuclear DNA, impairing mitochondrial function and further increasing ROS production. Excess ROS also disrupts the Keap1–Nrf2 axis, reducing antioxidant response element (ARE)-dependent expression of antioxidant enzymes (e.g., SOD, GPx, catalase, HO-1), thereby reinforcing oxidative stress. (**B**) NF-κB–cytokine–DNA damage loop. ROS and metabolic stress activate NF-κB signaling, leading to increased expression of pro-inflammatory cytokines such as IL-6 and TNF-α. Cytokine signaling enhances nuclear oxidative damage and further NF-κB activation, establishing a self-amplifying inflammatory circuit. (**C**) mtDNA–NLRP3 inflammasome loop. Oxidatively damaged mitochondrial DNA (mtDNA) is released as damage-associated molecular patterns (DAMPs), activating NADPH oxidase-derived ROS production and the NLRP3–caspase-1 inflammasome in macrophages. This promotes IL-1β and IL-18 secretion, which feeds back to cardiomyocytes, exacerbating mitochondrial dysfunction and ROS generation. (**D**) Cytokine–TGF-β–fibrosis loop. IL-1β, IL-18, TNF-α, and ROS activate latent transforming growth factor-β (TGF-β), driving fibroblast-to-myofibroblast differentiation and excessive extracellular matrix (ECM) deposition. Progressive fibrotic remodeling increases mechanical stress and local inflammation, sustaining cytokine release and tissue stiffening. (**E**) Integrated extrinsic drivers. Uncoupled endothelial nitric oxide synthase (eNOS), epicardial adipose tissue-derived cytokines and free fatty acids, macrophage NADPH oxidase activity, and paracrine ROS converge on cardiomyocytes and stromal cells, amplifying oxidative stress, inflammatory signaling, and fibrotic progression [81–130].

6. Antioxidant Defense Mechanisms and Their Impairment

In physiological conditions, antioxidant defenses comprise a tightly regulated network of enzymatic and non-enzymatic systems that neutralize ROS and preserve cellular redox homeostasis. The main enzymatic antioxidants include superoxide dismutase (SOD), glutathione peroxidase (GPx), and catalase. SOD catalyzes the conversion of superoxide into hydrogen peroxide, which is subsequently detoxified by catalase and GPx, thereby preventing propagation of harmful oxidative cascades. In diabetes, persistent hyperglycemia and mitochondrial overproduction of ROS chronically activate pro-oxidant pathways and progressively impair the activity of these enzymes. As a result, SOD, catalase, and GPx often become insufficient or dysfunctional in diabetic tissues, contributing to sustained oxidative stress [131–133].

This enzymatic deficit is not merely a consequence of increased antioxidant consumption but reflects a deeper disruption of the cellular transcriptional antioxidant network governed by specific proteins, particularly Nrf2 and Nrf1 [134]. Nrf2 is the master regulator

https://doi.org/10.3390/biom16030470

of the antioxidant response element pathway. Upon activation by oxidative stress, Nrf2 dissociates from Keap1, translocates to the nucleus, and induces the expression of multiple antioxidant and cytoprotective genes involved in ROS scavenging and redox balance [135]. In diabetes, impaired Nrf2 signaling is associated with reduced induction of these defense genes and increased susceptibility of cardiomyocytes and vascular cells to ROS-mediated injury [136].

Although less extensively studied, Nrf1 also plays a relevant role in maintaining redox homeostasis by regulating subsets of antioxidant genes under basal conditions [137]. Experimental evidence suggests that Nrf1 and Nrf2 exert cooperative and inter-regulatory functions in fine-tuning the expression of phase II detoxifying enzymes and ROS-scavenging systems. For instance, Nrf2 activation can promote Nrf1 transcription, while loss of either factor alters antioxidant gene expression and exacerbates oxidative stress. In addition, Nrf1 influences proteostasis and mitochondrial function, which are critical for cellular resilience against oxidative insults [138,139].

Beyond enzymatic systems, several non-enzymatic antioxidants contribute to myocardial redox control. Reduced glutathione (GSH) and coenzyme Q10 (CoQ10) directly scavenge free radicals and support enzymatic antioxidant recycling [140]. In diabetes, intracellular GSH pools are frequently depleted due to increased oxidative consumption and impaired synthesis or regeneration, thereby sustaining oxidative stress and inflammation. CoQ10, a lipophilic antioxidant and key component of the mitochondrial electron transport chain, is similarly affected; reduced tissue or circulating levels and impaired redox cycling have been reported in individuals with diabetes. Randomized trials and meta-analyses suggest that CoQ10 supplementation may improve selected metabolic and oxidative markers, although heterogeneity among studies and limited bioavailability constrain definitive conclusions [141,142].

Additional modulators of antioxidant defense further link metabolic status to redox homeostasis. Irisin, a myokine highly expressed in cardiomyocytes, enhances SIRT1 expression and has been shown to restore reduced GSH levels typically observed in diabetic cardiomyopathy [143]. High-density lipoproteins (HDLs) also exert anti-inflammatory and anti-fibrotic effects in the myocardium [144]: apolipoprotein A-1 inhibits the production of IL-1β and TNF-α [145] while promoting expression of the anti-inflammatory cytokine IL-10 [146]. Moreover, HDL downregulates angiotensin II type 1 receptor expression, leading to reduced NAD(P)H oxidase activity and lower ROS generation [147]. However, hyperglycemia-induced lipid peroxidation promotes the formation of oxidized and dysfunctional HDL, thereby limiting these protective effects [148]. Finally, cellular metabolic state can influence antioxidant capacity. Moderate elevations in circulating D-β-hydroxybutyrate have been shown to enhance resistance to oxidative stress by restoring thioredoxin and GPx4 expression [149]. These observations suggest that metabolic substrate shifts, such as partial reliance on ketone bodies, may help reinforce antioxidant defenses and mitigate oxidative injury in diabetic cardiomyopathy.

7. Emerging Therapeutic Strategies

Growing evidence indicates that the limited clinical success of conventional antioxidant therapies in diabetic cardiomyopathy does not undermine the pathogenic role of oxidative stress per se but rather reflects the inadequacy of non-specific reactive oxygen species scavenging strategies. Oxidative stress in the diabetic heart is not merely the result of excessive ROS accumulation but arises from a complex dysregulation of spatially and temporally controlled redox signaling pathways that are essential for cellular adaptation and survival. Accordingly, contemporary therapeutic approaches have shifted from indiscriminate antioxidant supplementation toward mechanism-based strategies aimed at

https://doi.org/10.3390/biom16030470

selectively modulating redox homeostasis at its primary sources and signaling nodes. These emerging interventions target mitochondrial ROS production, NADPH oxidase activity, redox-sensitive transcriptional programs, and organelle-specific oxidative signaling, with the goal of restoring physiological redox balance while preserving adaptive redox signaling [17,19]. Within this evolving framework, redox-targeted therapies are increasingly viewed as precision interventions capable of modifying disease progression rather than simply neutralizing oxidative by-products.

7.1. Mitochondria- and Enzyme-Targeted Redox Modulation

Excessive production of ROS, particularly from mitochondria and NOX, represents a major contributor to oxidative injury in the diabetic heart [49,150]. Persistent oxidative stress promotes lipid, protein, and DNA damage, mitochondrial dysfunction, inflammation, cardiomyocyte apoptosis, and myocardial fibrosis, ultimately leading to adverse ventricular remodeling and functional impairment [6,49,151]. Importantly, the limited or inconsistent efficacy observed with classical antioxidant supplementation in clinical studies, including vitamins C and E, has highlighted fundamental limitations of non-targeted ROS scavenging strategies [151,152]. These approaches may fail to reach critical intracellular compartments, interfere with physiological redox signaling, or be administered at stages when irreversible myocardial damage has already occurred. Consequently, the lack of clinical benefit should not be interpreted as evidence against the pathogenic role of oxidative stress in diabetic cardiomyopathy but rather as an indication that oxidative stress must be targeted with greater mechanistic precision [153–156]. This realization has driven the development of contemporary redox-based therapies aimed at selectively modulating ROS production and redox-sensitive pathways at their primary cellular sources. Pharmacological antioxidants have therefore been investigated as potential strategies to counteract this redox imbalance. In contrast to conventional antioxidants that primarily neutralize ROS downstream of their formation, mitochondria-targeted redox therapies are designed to act directly at the primary intracellular source of pathological ROS generation [157]. By selectively accumulating within mitochondria, agents such as MitoQ, MitoTEMPO, and SS-31 (elamipretide) directly modulate mitochondrial redox balance, preserve electron transport chain function, and interrupt self-amplifying cycles of oxidative damage at their origin [158,159]. MitoQ, a coenzyme Q analog conjugated to a lipophilic cation, has shown efficacy in preclinical models of diabetes; thus, further studies are needed to confirm these findings in functional models of the heart [158,159]. Similarly, MitoTEMPO has been demonstrated to selectively scavenge mitochondrial superoxide, improving endothelial function and attenuating cardiomyocyte apoptosis and hypertrophy in experimental models of both type 1 and type 2 diabetes [160]. Tempol, a membrane-permeable superoxide dismutase mimetic, has been shown to reduce superoxide accumulation, preserve nitric oxide bioavailability, and improve myocardial redox balance, resulting in cardioprotective effects in diabetic cardiovascular dysfunction [161]. In addition, apocynin, a NOX inhibitor, has been extensively studied in diabetic animal models, where it reduced cardiac hypertrophy and fibrosis, improved mitochondrial function, restored antioxidant enzyme activity, and attenuated adverse ventricular remodeling by limiting NOX-derived ROS production [162–164].

Despite encouraging preclinical data, translation of pharmacological antioxidants into clinical practice remains limited. Variability in bioavailability, lack of tissue specificity, and the complexity of redox signaling may account for the modest or inconsistent benefits observed in clinical studies, underscoring the need for more targeted and mechanism-based approaches.

https://doi.org/10.3390/biom16030470

7.2. Natural Compounds and Nutraceuticals

Natural compounds and nutraceuticals have attracted interest not merely for their intrinsic antioxidant properties but primarily for their ability to modulate redox-sensitive signaling pathways involved in mitochondrial function, inflammation, and metabolic regulation [165,166]. Polyphenolic compounds such as resveratrol and quercetin enhance mitochondrial biogenesis through activation of SIRT1 and AMP-activated protein kinase (AMPK), reduce oxidative damage, and improve calcium handling in cardiomyocytes [130,167]. Curcumin, derived from *Curcuma longa*, exerts cardioprotective effects by inhibiting NF-κB signaling, reducing advanced glycation end-product formation, and attenuating myocardial fibrosis [168]. Alpha-lipoic acid (ALA) functions as a redox modulator capable of regenerating endogenous antioxidants such as glutathione and vitamin C, while also improving endothelial function and insulin sensitivity [169]. Although these compounds demonstrate promising biological effects, clinical evidence remains heterogeneous. Differences in formulation, dosage, bioavailability, and study design limit definitive conclusions, highlighting the need for standardized clinical trials specifically addressing diabetic cardiomyopathy.

7.3. Looking to the Future: Gene Therapy and Redox Enzyme Modulation

Targeting redox-sensitive transcription factors and enzymatic systems involved in ROS regulation represents a promising strategy to correct the underlying oxidative imbalance in DCM. These approaches aim to restore redox homeostasis and re-establish adaptive cellular responses disrupted in the diabetic heart [154,170]. Modulation of Nrf2, NOX enzymes, and SIRT1 should therefore be viewed in future as an approach to re-establish physiological redox homeostasis, rather than as a form of exogenous antioxidant supplementation [8,47,135]. Activation of Nrf2 induces expression of antioxidant enzymes such as heme oxygenase-1 and glutathione peroxidase, conferring protection against oxidative myocardial injury [171]. Conversely, inhibition of NOX isoforms, particularly NOX2 and NOX4, has been shown to prevent cardiomyocyte apoptosis and pathological remodeling [172]. We have already discussed the effects of sirtuin-1 (SIRT1), a NAD^+-dependent deacetylase, on GSH levels. The modulation of SIRT1 has also shown to attenuate oxidative stress and exert protective effects in the diabetic heart [173]. Gene-based strategies aimed at enhancing these protective pathways may offer sustained correction of redox imbalance, although safety and long-term effects remain important challenges.

7.4. Lifestyle Interventions

Lifestyle interventions remain a cornerstone in the management of diabetes and exert significant effects on myocardial redox homeostasis. Caloric restriction reduces systemic inflammation, lowers circulating Fetuin-A levels, and downregulates Toll-like receptor signaling, thereby improving cardiac metabolism and contractility [174]. Regular physical activity enhances myocardial antioxidant enzyme activity, improves calcium handling, and increases mitochondrial efficiency [175]. In addition, strict glycemic control attenuates ROS generation and limits advanced glycation end-product accumulation, slowing the progression of myocardial fibrosis [176]. These interventions provide systemic benefits that synergize with pharmacological therapies.

7.5. Advanced Therapeutic Approaches

Recent advances in nanotechnology and peptide-based therapies have opened up new avenues for targeted redox modulation. Nanocarriers enable precise delivery of antioxidants, peptides, or microRNAs to specific tissues or subcellular compartments, overcoming limitations of conventional antioxidants such as poor bioavailability and off-target

https://doi.org/10.3390/biom16030470

effects [177–179]. Various platforms, including liposomes, dendrimers, biodegradable polymers, and bioinspired calcium phosphate nanoparticles, have been explored for cardiac applications [180]. Mitochondria-targeted peptides such as SS-31 (elamipretide) bind cardiolipin within the inner mitochondrial membrane, stabilizing the electron transport chain, reducing ROS production, and improving ATP synthesis. In experimental models of diabetic cardiomyopathy, SS-31 has been shown to attenuate fibrosis, reduce cardiac remodeling, and improve ventricular function [179]. The combination of organellar targeting with multifunctional nanocarriers may further enhance therapeutic efficacy while minimizing systemic toxicity, representing a promising direction for future precision medicine approaches [158,179,181]. Although these advanced strategies remain largely preclinical, they offer compelling prospects for selectively targeting the molecular drivers of oxidative stress in diabetic cardiomyopathy.

Collectively, these emerging strategies reflect a conceptual shift from non-specific antioxidant supplementation toward precision redox medicine in diabetic cardiomyopathy. Increasing evidence indicates that therapeutic efficacy depends on targeting the appropriate redox source, intracellular compartment, and disease stage, rather than indiscriminate ROS neutralization [8,154]. Moreover, interindividual heterogeneity in metabolic status, disease duration, and myocardial phenotype suggests that redox-targeted interventions may require patient stratification and biomarker-guided timing to achieve meaningful clinical benefit [17,150,182]. This evolving framework underscores the need to integrate mechanistic specificity, temporal intervention, and patient phenotyping when translating redox-based therapies into clinical practice. The main redox-targeted therapeutic strategies, their molecular targets, and the current level of experimental and clinical evidence are summarized in Table 1.

Table 1. Redox-targeted therapeutic strategies in diabetic cardiomyopathy: molecular targets and level of evidence.

Therapeutic Strategy	Representative Agents/Approaches	Primary Molecular Targets	Mechanism of Redox Modulation	Evidence	Refs.
Mitochondria-targeted redox therapies	MitoQ, MitoTEMPO, SS-31 (elamipretide), Tempol	Mitochondrial electron transport chain, cardiolipin	Direct modulation of mitochondrial ROS production or catalytic detoxification of superoxide and related radicals; preservation of electron transport chain function and interruption of self-amplifying oxidative damage cycles	Preclinical (robust); early clinical (SS-31)	[159–162,179]
NOX inhibition	Apocynin, NOX2/NOX4 inhibitors	NADPH oxidase isoforms	Source-specific reduction in ROS generation and attenuation of redox-driven inflammatory and fibrotic signaling	Preclinical	[162–165,172]
Redox-sensitive transcriptional reprogramming	Nrf2 activation strategies	Keap1–Nrf2 axis	Restoration of endogenous antioxidant and cytoprotective gene expression; re-establishment of redox homeostasis	Preclinical	[8,171]
Metabolic–redox regulators	SIRT1 activation (e.g., resveratrol), AMPK modulation	SIRT1, AMPK pathways	Recalibration of redox-sensitive metabolic signaling, mitochondrial biogenesis, and stress resistance	Preclinical; limited clinical	[130,167,173]
Nutraceutical redox modulators	Resveratrol, curcumin, α-lipoic acid	NF-κB, SIRT1, AMPK, endogenous antioxidant systems	Modulation of redox-sensitive inflammatory and metabolic pathways rather than direct ROS scavenging	Preclinical; heterogeneous clinical	[130,167–169]
Advanced delivery systems	Nanocarriers, mitochondria-targeted peptides	Subcellular redox compartments	Improved tissue and organelle specificity of redox-targeted interventions	Preclinical	[177–181]

Abbreviations: reactive oxygen species (ROS), mitochondrial electron transport chain (ETC), NADPH oxidase (NOX), nuclear factor erythroid 2-related factor 2 (Nrf2), sirtuin-1 (SIRT1), AMP-activated protein kinase (AMPK), nuclear factor-κB (NF-κB), mitochondria-targeted ubiquinone (MitoQ), mitochondria-targeted superoxide dismutase mimetic (MitoTEMPO), mitochondria-targeted peptide SS-31 (elamipretide).

8. Challenges and Future Directions

Despite significant advances in elucidating the molecular mechanisms underlying DCM, several critical challenges continue to limit the effective translation of experimental findings into clinical practice. One of the major obstacles lies in the substantial gap between

https://doi.org/10.3390/biom16030470

preclinical models and human disease. Most mechanistic insights into oxidative stress-driven myocardial injury derive from animal models or in vitro systems that only partially recapitulate the metabolic complexity, disease duration, and comorbidity burden observed in patients with diabetes [183,184]. As a result, therapeutic strategies, particularly those targeting redox imbalance, that demonstrate robust efficacy in experimental settings often fail to show consistent clinical benefits.

Another important challenge is the marked heterogeneity of DCM. In fact, DCM encompasses a broad spectrum of phenotypes, ranging from early diastolic dysfunction with preserved ejection fraction to advanced systolic failure, influenced by diabetes duration, glycemic control, obesity, sex, genetic background, and concomitant cardiovascular risk factors [16,185]. This heterogeneity suggests that a "one-size-fits-all" approach to redox-targeted therapies is unlikely to be effective. Future strategies should therefore move toward personalized and phenotype-driven interventions, potentially guided by metabolic, inflammatory, and oxidative stress profiles.

The identification and validation of reliable biomarkers of oxidative stress represent an additional unmet need. Currently available circulating markers often lack specificity for myocardial redox imbalance and are influenced by systemic inflammation and comorbid conditions [17]. Integration of redox-related biomarkers with advanced imaging modalities, such as cardiac magnetic resonance and positron emission tomography, may improve early detection of subclinical myocardial involvement and enable monitoring of disease progression and therapeutic response.

Emerging omics-based approaches, including transcriptomics, proteomics, metabolomics, and epigenomics, offer promising tools to unravel the complex molecular networks linking metabolism, oxidative stress, inflammation, and fibrosis in DCM and potentially drive precision medicine treatments [186,187]. However, their clinical utility will depend on integration with redox-specific biomarkers and functional cardiac endpoints, rather than on isolated molecular signatures. Systems biology and machine-learning-based analyses may help identify key regulatory nodes and stratify patients according to dominant pathogenic mechanisms, thereby refining therapeutic targeting [188]. Finally, future research should prioritize well-designed clinical trials specifically addressing diabetic cardiomyopathy, rather than extrapolating data from broader heart failure populations. Such trials should incorporate mechanistic endpoints related to oxidative stress, biomarker-driven patient stratification, and long-term follow-up to determine whether redox-targeted interventions can meaningfully alter the natural history of DCM.

9. Conclusions

DCM is a complex myocardial disorder that develops independently of coronary artery disease and hypertension and represents a major contributor to heart failure in patients with diabetes. Accumulating evidence supports oxidative stress as a central integrative mechanism linking metabolic dysregulation, mitochondrial dysfunction, inflammation, and fibrotic remodeling in the diabetic heart.

Multiple cellular sources of reactive oxygen species, including dysfunctional mitochondria, NADPH oxidases, uncoupled nitric oxide synthase, advanced glycation end-product signaling, and endoplasmic reticulum stress, converge to disrupt myocardial redox homeostasis. This imbalance promotes oxidative damage to lipids, proteins, and DNA, activates maladaptive signaling pathways, and drives progressive structural and functional deterioration of the myocardium. In parallel, epicardial and visceral adipose tissue have emerged as active contributors to myocardial oxidative stress through paracrine and systemic mechanisms, reinforcing inflammatory and fibrotic crosstalk. Although effective disease-specific therapies for diabetic cardiomyopathy are still lacking, targeting oxidative

https://doi.org/10.3390/biom16030470

stress and redox-regulated pathways represents a promising strategy for disease modification. Advances in pharmacological antioxidants, nutraceuticals, gene-based approaches, lifestyle interventions, and targeted delivery systems provide new opportunities to restore redox balance and preserve myocardial function. However, successful clinical translation will require improved biomarkers, refined patient stratification, and well-designed clinical trials to determine whether redox-targeted interventions can meaningfully alter the course of diabetic cardiomyopathy.

Author Contributions: Conceptualization, A.C., U.C., D.N. and C.M.; writing—original draft preparation, A.C., U.C., D.N., C.M., R.N. and S.S.; writing—review and editing, M.C.S., M.A.P., L.B., C.A., V.R. and F.C.S.; supervision: A.C. All authors have read and agreed to the published version of the manuscript.

Funding: This research received no external funding.

Institutional Review Board Statement: Not applicable.

Informed Consent Statement: Not applicable.

Data Availability Statement: No dataset was generated for the publication of this article.

Conflicts of Interest: The authors declare no conflicts of interest.

Abbreviations

The following abbreviations are used in this manuscript:

AGEs	Advanced glycation end-products
ALA	Alpha-lipoic acid
AMPK	AMP-activated protein kinase
ARE	Antioxidant response element
ATP	Adenosine triphosphate
BH_4	Tetrahydrobiopterin
CoQ10	Coenzyme Q10
DCM	Diabetic cardiomyopathy
DRP1	Dynamin-related protein 1
EAT	Epicardial adipose tissue
eNOS	Endothelial nitric oxide synthase
ER	Endoplasmic reticulum
ERK	Extracellular signal-regulated kinase
$FADH_2$	Flavin adenine dinucleotide (reduced form)
GPx	Glutathione peroxidase
GSH	Reduced glutathione
HDL	High-density lipoprotein
HFpEF	Heart failure with preserved ejection fraction
IL	Interleukin
IRE1α	Inositol-requiring enzyme 1 alpha
JNK	c-Jun N-terminal kinase
Keap1	Kelch-like ECH-associated protein 1
MAPKs	Mitogen-activated protein kinases
MDA	Malondialdehyde
MFN	Mitofusin
mtDNA	Mitochondrial DNA
NADH	Nicotinamide adenine dinucleotide (reduced form)
NADPH	Nicotinamide adenine dinucleotide phosphate
NF-κB	Nuclear factor kappa-light-chain-enhancer of activated B cells

https://doi.org/10.3390/biom16030470

NO	Nitric oxide
NOX	NADPH oxidase
Nrf1	Nuclear factor erythroid 2-related factor 1
Nrf2	Nuclear factor erythroid 2-related factor 2
OPA1	Optic atrophy protein 1
PERK	Protein kinase RNA-like endoplasmic reticulum kinase
PKC	Protein kinase C
RAGE	Receptor for advanced glycation end-products
ROS	Reactive oxygen species
SIRT1	Sirtuin-1
SOD	Superoxide dismutase
TGF-β	Transforming growth factor beta
TNF-α	Tumor necrosis factor alpha
UPR	Unfolded protein response
4-HNE	4-hydroxy-2-nonenal
8-OH-dG	8-hydroxy-2′-deoxyguanosine

References

1. Chen, S.; Cao, Z.; Chen, W.; Zhao, J.; Jiao, L.; Prettner, K.; Kuhn, M.; Pan, A.; Bärnighausen, T.W.; Bloom, D.E. The Global Macroeconomic Burden of Diabetes Mellitus. *Nat. Med.* **2026**, *32*, 126–138. [CrossRef]
2. Zhao, N.; Yu, X.; Zhu, X.; Song, Y.; Gao, F.; Yu, B.; Qu, A. Diabetes Mellitus to Accelerated Atherosclerosis: Shared Cellular and Molecular Mechanisms in Glucose and Lipid Metabolism. *J. Cardiovasc. Trans. Res.* **2024**, *17*, 133–152. [CrossRef]
3. Chen, D.; Sindone, A.; Huang, M.L.H.; Peter, K.; Jenkins, A.J. Diabetic Cardiomyopathy: Insights into Pathophysiology, Diagnosis and Clinical Management. *J. Mol. Cell. Cardiol.* **2025**, *206*, 55–69. [CrossRef]
4. Salvatore, T.; Pafundi, P.C.; Galiero, R.; Albanese, G.; Di Martino, A.; Caturano, A.; Vetrano, E.; Rinaldi, L.; Sasso, F.C. The Diabetic Cardiomyopathy: The Contributing Pathophysiological Mechanisms. *Front. Med.* **2021**, *8*, 695792. [CrossRef] [PubMed]
5. Jin, S.; Kang, P.M. A Systematic Review on Advances in Management of Oxidative Stress-Associated Cardiovascular Diseases. *Antioxidants* **2024**, *13*, 923. [CrossRef] [PubMed]
6. Caturano, A.; Rocco, M.; Tagliaferri, G.; Piacevole, A.; Nilo, D.; Di Lorenzo, G.; Iadicicco, I.; Donnarumma, M.; Galiero, R.; Acierno, C.; et al. Oxidative Stress and Cardiovascular Complications in Type 2 Diabetes: From Pathophysiology to Lifestyle Modifications. *Antioxidants* **2025**, *14*, 72. [CrossRef]
7. Tang, Z.; Wang, P.; Dong, C.; Zhang, J.; Wang, X.; Pei, H. Oxidative Stress Signaling Mediated Pathogenesis of Diabetic Cardiomyopathy. *Oxid. Med. Cell. Longev.* **2022**, *2022*, 5913374. [CrossRef] [PubMed]
8. Li, B.; Ming, H.; Qin, S.; Nice, E.C.; Dong, J.; Du, Z.; Huang, C. Redox Regulation: Mechanisms, Biology and Therapeutic Targets in Diseases. *Signal Transduct. Target. Ther.* **2025**, *10*, 72. [CrossRef]
9. Shaito, A.; Aramouni, K.; Assaf, R.; Parenti, A.; Orekhov, A.; Yazbi, A.E.; Pintus, G.; Eid, A.H. Oxidative Stress-Induced Endothelial Dysfunction in Cardiovascular Diseases. *Front. Biosci.* **2022**, *27*, 105. [CrossRef]
10. D'Oria, R.; Schipani, R.; Leonardini, A.; Natalicchio, A.; Perrini, S.; Cignarelli, A.; Laviola, L.; Giorgino, F. The Role of Oxidative Stress in Cardiac Disease: From Physiological Response to Injury Factor. *Oxid. Med. Cell Longev.* **2020**, *2020*, 5732956. [CrossRef]
11. García-Sánchez, A.; Gámez-Nava, J.I.; Díaz-de La Cruz, E.N.; Cardona-Muñoz, E.G.; Becerra-Alvarado, I.N.; Aceves-Aceves, J.A.; Sánchez-Rodríguez, E.N.; Miranda-Díaz, A.G. The Effect of Visceral Abdominal Fat Volume on Oxidative Stress and Proinflammatory Cytokines in Subjects with Normal Weight, Overweight and Obesity. *Diabetes Metab. Syndr. Obes.* **2020**, *13*, 1077–1087. [CrossRef] [PubMed]
12. Fakih, W.; Alhelou, C.; Mroueh, A.; Kindo, M.; Mommerot, A.; Mazzucotelli, J.P.; Pieper, M.P.; Ohlmann, P.; Morel, O.; Schini-Kerth, V.; et al. Human Epicardial Adipose Tissue-Derived Factors Promote Atrial Endothelial Dysfunction: Role of pro-Inflammatory Cytokines and AT1R/NADPH Oxidases/SGLT2 pro-Oxidant Pathway. *Eur. Heart J.* **2024**, *45*, ehae666.3728. [CrossRef]
13. Galis, P.; Bartosova, L.; Farkasova, V.; Bartekova, M.; Ferenczyova, K.; Rajtik, T. Update on Clinical and Experimental Management of Diabetic Cardiomyopathy: Addressing Current and Future Therapy. *Front. Endocrinol.* **2024**, *15*, 1451100. [CrossRef] [PubMed]
14. Yang, H.-M. GLP-1 Agonists in Cardiovascular Diseases: Mechanisms, Clinical Evidence, and Emerging Therapies. *J. Clin. Med.* **2025**, *14*, 6758. [CrossRef]
15. Huang, K.; Luo, X.; Liao, B.; Li, G.; Feng, J. Insights into SGLT2 inhibitor treatment of diabetic cardiomyopathy: Focus on the mechanisms. *Cardiovasc. Diabetol.* **2023**, *22*, 86. [CrossRef]

https://doi.org/10.3390/biom16030470

16. Abdullah, A.R.; Seliem, M.A.; Khidr, E.G.; Sobhy, A.M.; El-Shiekh, R.A.; Hafeez, M.S.A.E.; El-Husseiny, A.A. A Comprehensive Review on Diabetic Cardiomyopathy (DCM): Histological Spectrum, Diagnosis, Pathogenesis, and Management with Conventional Treatments and Natural Compounds. *Naunyn-Schmiedeberg's Arch. Pharmacol.* **2025**, *398*, 9929–9969. [CrossRef]
17. Huo, J.-L.; Feng, Q.; Pan, S.; Fu, W.-J.; Liu, Z.; Liu, Z. Diabetic Cardiomyopathy: Early Diagnostic Biomarkers, Pathogenetic Mechanisms, and Therapeutic Interventions. *Cell Death Discov.* **2023**, *9*, 256. [CrossRef]
18. Ridwan, M.; Dimiati, H.; Syukri, M.; Lesmana, R. Potential Molecular Mechanism Underlying Cardiac Fibrosis in Diabetes Mellitus: A Narrative Review. *Egypt. Heart J.* **2023**, *75*, 46. [CrossRef]
19. Grubić Rotkvić, P.; Planinić, Z.; Liberati Pršo, A.-M.; Šikić, J.; Galić, E.; Rotkvić, L. The Mystery of Diabetic Cardiomyopathy: From Early Concepts and Underlying Mechanisms to Novel Therapeutic Possibilities. *Int. J. Mol. Sci.* **2021**, *22*, 5973. [CrossRef]
20. Spartalis, M.; Pantelidis, P.; Kontogiannis, C.; Paschou, S.A.; Spartalis, E.; Iliopoulos, D.C.; Siasos, G. The Complex Relationship Between Diabetes and Cardiac Arrhythmias: Pathophysiologyand Mechanisms. *Curr. Pharm. Des.* **2022**, *28*, 2129–2130. [CrossRef]
21. Caturano, A.; Galiero, R.; Vetrano, E.; Sardu, C.; Rinaldi, L.; Russo, V.; Monda, M.; Marfella, R.; Sasso, F.C. Insulin–Heart Axis: Bridging Physiology to Insulin Resistance. *Int. J. Mol. Sci.* **2024**, *25*, 8369. [CrossRef]
22. Sivri, D.; Akdevelioğlu, Y. Effect of Fatty Acids on Glucose Metabolism and Type 2 Diabetes. *Nutr. Rev.* **2025**, *83*, 897–907. [CrossRef] [PubMed]
23. Lee, H.K. Fatty Acid Overload to Compromised Oxidative Phosphorylation Activates Inflammation in Type 2 Diabetes: Hidden Beasts and How to Find Them. *J. Diabetes Investig.* **2020**, *11*, 290–293. [CrossRef] [PubMed]
24. Dikalov, S.; Panov, A.; Dikalova, A. Critical Role of Mitochondrial Fatty Acid Metabolism in Normal Cell Function and Pathological Conditions. *Int. J. Mol. Sci.* **2024**, *25*, 6498. [CrossRef] [PubMed]
25. Yang, Y.; Wu, Y.; Sun, X.-D.; Zhang, Y. Reactive Oxygen Species, Glucose Metabolism, and Lipid Metabolism. In *Oxidative Stress*; Huang, C., Zhang, Y., Eds.; Springer: Singapore, 2021; pp. 213–235.
26. Dabravolski, S.A.; Sadykhov, N.K.; Kartuesov, A.G.; Borisov, E.E.; Sukhorukov, V.N.; Orekhov, A.N. The Role of Mitochondrial Abnormalities in Diabetic Cardiomyopathy. *Int. J. Mol. Sci.* **2022**, *23*, 7863. [CrossRef]
27. Rottenberg, H.; Hoek, J.B. The Path from Mitochondrial ROS to Aging Runs through the Mitochondrial Permeability Transition Pore. *Aging Cell* **2017**, *16*, 943–955. [CrossRef]
28. Palma, F.R.; Gantner, B.N.; Sakiyama, M.J.; Kayzuka, C.; Shukla, S.; Lacchini, R.; Cunniff, B.; Bonini, M.G. ROS Production by Mitochondria: Function or Dysfunction? *Oncogene* **2024**, *43*, 295–303. [CrossRef]
29. Al Kury, L.T. Calcium Homeostasis in Ventricular Myocytes of Diabetic Cardiomyopathy. *J. Diabetes Res.* **2020**, *2020*, 1942086. [CrossRef]
30. Wani, Z.A.; Sharma, A.K.; Muzamil, S.; Mir, M.N.; Malik, I.A.; Naik, R.A.; Malik, S.S.; Badroo, I.A.; Hurra, A.R.; Lone, Y. Molecular Pathways of Oxidative Stress in Diabetes: Redox Imbalance and Insulin Pathway Dysregulation. *Mol. Biol. Rep.* **2025**, *53*, 222. [CrossRef]
31. Liu, S.; Liu, J.; Wang, Y.; Deng, F.; Deng, Z. Oxidative Stress: Signaling Pathways, Biological Functions, and Disease. *MedComm* **2025**, *6*, e70268. [CrossRef]
32. Tashkandi, A.J.; Gorman, A.; McGoldrick Mathers, E.; Carney, G.; Yacoub, A.; Setyaningsih, W.A.W.; Kuburas, R.; Margariti, A. Metabolic and Mitochondrial Dysregulations in Diabetic Cardiac Complications. *Int. J. Mol. Sci.* **2025**, *26*, 3016. [CrossRef] [PubMed]
33. Srivastava, S.P.; Kopasz-Gemmen, O.; Thurman, A.; Rajendran, B.K.; Selvam, M.M.; Kumar, S.; Srivastava, R.; Suresh, M.X.; Kumari, R.; Goodwin, J.E.; et al. The Molecular Determinants Regulating Redox Signaling in Diabetic Endothelial Cells. *Front. Pharmacol.* **2025**, *16*, 1563047. [CrossRef] [PubMed]
34. Salvatore, T.; Galiero, R.; Caturano, A.; Vetrano, E.; Rinaldi, L.; Coviello, F.; Di Martino, A.; Albanese, G.; Colantuoni, S.; Medicamento, G.; et al. Dysregulated Epicardial Adipose Tissue as a Risk Factor and Potential Therapeutic Target of Heart Failure with Preserved Ejection Fraction in Diabetes. *Biomolecules* **2022**, *12*, 176. [CrossRef] [PubMed]
35. Tian, Y.; Wang, P.; Dong, Z. Heart-Guarding or Heart-Harming? The Dual Role of Epicardial Adipose Tissue in Cardiovascular Health and Disease. *Int. J. Obes.* **2025**, *49*, 2156–2167. [CrossRef]
36. Akoumianakis, I.; Akawi, N.; Antoniades, C. Exploring the Crosstalk between Adipose Tissue and the Cardiovascular System. *Korean Circ. J.* **2017**, *47*, 670. [CrossRef]
37. Nan, J.; Li, J.; Lin, Y.; Saif Ur Rahman, M.; Li, Z.; Zhu, L. The interplay between mitochondria and store-operated Ca^{2+} entry: Emerging insights into cardiac diseases. *J. Cell Mol. Med.* **2021**, *25*, 9496–9512. [CrossRef]
38. Ahola, S.; Pazurek, L.A.; Mayer, F.; Lampe, P.; Hermans, S.; Becker, L.; Amarie, O.V.; Fuchs, H.; Gailus-Durner, V.; de Angelis, M.H.; et al. Opa1 processing is dispensable in mouse development but is protective in mitochondrial cardiomyopathy. *Sci. Adv.* **2024**, *10*, eadp0443. [CrossRef] [PubMed] [PubMed Central]
39. Peoples, J.N.; Saraf, A.; Ghazal, N.; Pham, T.T.; Kwong, J.Q. Mitochondrial Dysfunction and Oxidative Stress in Heart Disease. *Exp. Mol. Med.* **2019**, *51*, 1–13. [CrossRef]

https://doi.org/10.3390/biom16030470

40. Liu, Q.; Wang, S.; Cai, L. Diabetic Cardiomyopathy and Its Mechanisms: Role of Oxidative Stress and Damage. *J. Diabetes Investig.* **2014**, *5*, 623–634. [CrossRef]
41. Giacco, F.; Brownlee, M. Oxidative Stress and Diabetic Complications. *Circ. Res.* **2010**, *107*, 1058–1070. [CrossRef]
42. Zhou, Z.; Arroum, T.; Luo, X.; Kang, R.; Lee, Y.J.; Tang, D.; Hüttemann, M.; Song, X. Diverse functions of cytochrome c in cell death and disease. *Cell Death Differ.* **2024**, *31*, 387–404. [CrossRef] [PubMed]
43. Grivennikovam, V.G.; Vinogradov, A.D. Partitioning of superoxide and hydrogen peroxide production by mitochondrial respiratory complex I. *Biochim. Biophys. Acta—Bioenerg.* **2013**, *1827*, 446–454. [CrossRef] [PubMed]
44. Wu, J.; Jin, Z.; Zheng, H.; Yan, L.J. Sources and implications of NADH/NAD(+) redox imbalance in diabetes and its complications. *Diabetes Metab. Syndr. Obes.* **2016**, *9*, 145–153. [CrossRef] [PubMed]
45. Morse, P.T.; Arroum, T.; Wan, J.; Pham, L.; Vaishnav, A.; Bell, J.; Pavelich, L.; Malek, M.H.; Sanderson, T.H.; Edwards, B.F.P.; et al. Phosphorylations and Acetylations of Cytochrome c Control Mitochondrial Respiration, Mitochondrial Membrane Potential, Energy, ROS, and Apoptosis. *Cells* **2024**, *13*, 493. [CrossRef]
46. An, Y.; Xu, B.; Wan, S.; Ma, X.; Long, Y.; Xu, Y.; Jiang, Z. The Role of Oxidative Stress in Diabetes Mellitus-Induced Vascular Endothelial Dysfunction. *Cardiovasc. Diabetol.* **2023**, *22*, 237. [CrossRef]
47. Hansen, S.S.; Aasum, E.; Hafstad, A.D. The Role of NADPH Oxidases in Diabetic Cardiomyopathy. *Biochim. Biophys. Acta—Mol. Basis Dis.* **2018**, *1864*, 1908–1913. [CrossRef]
48. Ren, X.; Ren, L.; Wei, Q.; Shao, H.; Chen, L.; Liu, N. Advanced Glycation End-Products Decreases Expression of Endothelial Nitric Oxide Synthase through Oxidative Stress in Human Coronary Artery Endothelial Cells. *Cardiovasc. Diabetol.* **2017**, *16*, 52. [CrossRef]
49. Ye, Z.; Wang, S.; Wan, Z.; Huang, B.; Guo, J. Targeting NADPH oxidase-driven oxidative stress in diabetic cardiomyopathy: Mechanisms and therapeutic perspectives. *Front. Pharmacol.* **2025**, *16*, 1610429. [CrossRef]
50. Okuyama, T.; Shirakawa, J.; Nakamura, T.; Murase, T.; Miyashita, D.; Inoue, R.; Kyohara, M.; Togashi, Y.; Terauchi, Y. Association of the Plasma Xanthine Oxidoreductase Activity with the Metabolic Parameters and Vascular Compli-cations in Patients with Type 2 Diabetes. *Sci. Rep.* **2021**, *11*, 3768. [CrossRef]
51. Kotozaki, Y.; Satoh, M.; Nasu, T.; Tanno, K.; Tanaka, F.; Sasaki, M. Human Plasma Xanthine Oxidoreductase Activity in Cardiovascular Disease: Evidence from a Population-Based Study. *Biomedicines* **2023**, *11*, 754. [CrossRef]
52. Polito, L.; Bortolotti, M.; Battelli, M.G.; Bolognesi, A. Xanthine oxidoreductase: A leading actor in cardiovascular disease drama. *Redox Biol.* **2021**, *48*, 102195. [CrossRef]
53. Kundu, T.K.; Velayutham, M.; Zweier, J.L. Aldehyde oxidase functions as a superoxide generating NADH oxidase: An important redox regulated pathway of cellular oxygen radical formation. *Biochemistry* **2012**, *51*, 2930–2939. [CrossRef] [PubMed]
54. Chen, C.H.; Sun, L.; Mochly-Rosen, D. Mitochondrial aldehyde dehydrogenase and cardiac diseases. *Cardiovasc. Res.* **2010**, *88*, 51–57. [CrossRef] [PubMed]
55. Saghir, S.A.; Ansari, R.A.; Munir, S.T. Fate of chemicals following exposure III: Metabolism (biotransformation). In *Encyclopedia of Toxicology*, 4th ed.; Wexler, P., Ed.; Academic Press: Cambridge, MA, USA, 2024; pp. 635–668. [CrossRef]
56. Harkin, C.; Cobice, D.; Watt, J.; Kurth, M.J.; Brockbank, S.; Bolton, S.; Johnston, F.; Strzelecka, A.; Lamont, J.V.; Moore, T.; et al. Analysis of reactive aldehydes in urine and plasma of type-2 diabetes mellitus patients through liquid chromatography-mass spectrometry: Reactive aldehydes as potential markers of diabetic nephropathy. *Front. Nutr.* **2023**, *9*, 997015. [CrossRef] [PubMed]
57. Youn, J.Y.; Gao, L.; Cai, H. The p47phox- and NADPH oxidase organiser 1 (NOXO1)-dependent activation of NADPH oxidase 1 (NOX1) mediates endothelial nitric oxide synthase (eNOS) uncoupling and endothelial dysfunction in a streptozotocin-induced murine model of diabetes. *Diabetologia* **2012**, *55*, 2069–2079. [CrossRef]
58. Thum, T.; Fraccarollo, D.; Schultheiss, M.; Froese, S.; Galuppo, P.; Widder, J.D.; Tsikas, D.; Ertl, G.; Bauersachs, J. Endothelial nitric oxide synthase uncoupling impairs endothelial progenitor cell mobilization and function in diabetes. *Diabetes* **2007**, *56*, 666–674. [CrossRef]
59. Wautier, M.P.; Chappey, O.; Corda, S.; Stern, D.M.; Schmidt, A.M.; Wautier, J.L. Activation of NADPH oxidase by AGE links oxidant stress to altered gene expression via RAGE. *Am. J. Physiol. Endocrinol. Metab.* **2001**, *280*, E685–E694. [CrossRef]
60. Adamopoulos, C.; Piperi, C.; Gargalionis, A.N.; Dalagiorgou, G.; Spilioti, E.; Korkolopoulou, P.; Diamanti-Kandarakis, E.; Papavassiliou, A.G. Advanced glycation end products upregulate lysyl oxidase and endothelin-1 in human aortic endothelial cells via parallel activation of ERK1/2-NF-κB and JNK-AP-1 signaling pathways. *Cell Mol. Life Sci.* **2016**, *73*, 1685–1698. [CrossRef]
61. Koyama, Y.; Takeishi, Y.; Arimoto, T.; Niizeki, T.; Shishido, T.; Takahashi, H.; Nozaki, N.; Hirono, O.; Tsunoda, Y.; Nitobe, J.; et al. High serum level of pentosidine, an advanced glycation end product (AGE), is a risk factor of patients with heart failure. *J. Card. Fail.* **2007**, *13*, 199–206. [CrossRef]
62. Kang, Z.; Chen, F.; Wu, W.; Liu, R.; Chen, T.; Xu, F. UPRmt and coordinated UPRER in type 2 diabetes. *Front. Cell Dev. Biol.* **2022**, *10*, 974083. [CrossRef]
63. Congur, I.; Mingrone, G.; Guan, K. Targeting Endoplasmic Reticulum Stress as a Potential Therapeutic Strategy for Diabetic Cardiomyopathy. *Metabolism* **2025**, *162*, 156062. [CrossRef] [PubMed]

https://doi.org/10.3390/biom16030470

64. Gianazza, E.; Brioschi, M.; Fernandez, A.M.; Banfi, C. Lipoxidation in Cardiovascular Diseases. *Redox Biol.* **2019**, *23*, 101119. [CrossRef] [PubMed]
65. Huynh, K.; Kiriazis, H.; Du, X.J.; Love, J.E.; Gray, S.P.; Jandeleit-Dahm, K.A.; McMullen, J.R.; Ritchie, R.H. Targeting the upregulation of reactive oxygen species subsequent to hyperglycemia prevents type 1 diabetic cardiomyopathy in mice. *Free Radic. Biol. Med.* **2013**, *60*, 307–317. [CrossRef] [PubMed]
66. Dham, D.; Roy, B.; Gowda, A.; Pan, G.; Sridhar, A.; Zeng, X.; Thandavarayan, R.A.; Palaniyandi, S.S. 4-Hydroxy-2-Nonenal, a Lipid Peroxidation Product, as a Biomarker in Diabetes and Its Complications: Challenges and Opportunities. *Free Radic. Res.* **2021**, *55*, 547–561. [CrossRef]
67. Dalle-Donne, I.; Aldini, G.; Carini, M.; Colombo, R.; Rossi, R.; Milzani, A. Protein Carbonylation, Cellular Dysfunction, and Disease Progression. *J. Cell. Mol. Med.* **2006**, *10*, 389–406. [CrossRef]
68. Mihm, M.J.; Yu, F.; Weinstein, D.M.; Reiser, P.J.; Bauer, J.A. Intracellular Distribution of Peroxynitrite during Doxorubicin Cardiomyopathy: Evidence for Selective Impairment of Myofibrillar Creatine Kinase. *Br. J. Pharmacol.* **2002**, *135*, 581–588. [CrossRef]
69. Ide, T.; Tsutsui, H.; Hayashidani, S.; Kang, D.; Suematsu, N.; Nakamura, K.; Utsumi, H.; Hamasaki, N.; Takeshita, A. Mitochondrial DNA Damage and Dysfunction Associated with Oxidative Stress in Failing Hearts After Myocardial Infarction. *Circ. Res.* **2001**, *88*, 529–535. [CrossRef]
70. Suri, S.; Mitra, P.; Abhilasha, A.; Saxena, I.; Garg, M.K.; Bohra, G.K.; Sharma, P. Altered expression of specific antioxidant (SOD1 and SOD2) and DNA repair (XRCC1 and OGG1) genes in patients with newly diagnosed type-2 diabetes mellitus. *Minerva Endocrinol.* **2024**, *49*, 398–405. [CrossRef]
71. Golbidi, S.; Botta, A.; Gottfred, S.; Nusrat, A.; Laher, I.; Ghosh, S. Glutathione administration reduces mitochondrial damage and shifts cell death from necrosis to apoptosis in ageing diabetic mice hearts during exercise. *Br. J. Pharmacol.* **2014**, *171*, 5345–5360. [CrossRef]
72. Rudokas, M.W.; McKay, M.; Toksoy, Z.; Eisen, J.N.; Bögner, M.; Young, L.H.; Akar, F.G. Mitochondrial Network Remodeling of the Diabetic Heart: Implications to Ischemia Related Cardiac Dysfunction. *Cardiovasc. Diabetol.* **2024**, *23*, 261. [CrossRef]
73. Wang, H.; Yu, W.; Wang, Y.; Wu, R.; Dai, Y.; Deng, Y.; Wang, S.; Yuan, J.; Tan, R. P53 Contributes to Cardiovascular Diseases via Mitochondria Dysfunction: A New Paradigm. *Free Radic. Biol. Med.* **2023**, *208*, 846–858. [CrossRef] [PubMed]
74. Wang, X.; Liu, R.; Liu, D. The Role of the MAPK Signaling Pathway in Cardiovascular Disease: Pathophysiological Mechanisms and Clinical Therapy. *Int. J. Mol. Sci.* **2025**, *26*, 2667. [CrossRef] [PubMed]
75. Xu, Z.; Sun, J.; Tong, Q.; Lin, Q.; Qian, L.; Park, Y.; Zheng, Y. The Role of ERK1/2 in the Development of Diabetic Cardiomyopathy. *Int. J. Mol. Sci.* **2016**, *17*, 2001. [CrossRef] [PubMed]
76. Guo, Q.; Jin, Y.; Chen, X.; Ye, X.; Shen, X.; Lin, M.; Zeng, C.; Zhou, T.; Zhang, J. NF-κB in Biology and Targeted Therapy: New Insights and Translational Implications. *Signal Transduct. Target. Ther.* **2024**, *9*, 53. [CrossRef]
77. Liu, T.; Zhang, L.; Joo, D.; Sun, S.-C. NF-κB Signaling in Inflammation. *Signal Transduct. Target. Ther.* **2017**, *2*, 17023. [CrossRef]
78. He, X.; Ma, Q. Disruption of Nrf2 Synergizes with High Glucose to Cause Heightened Myocardial Oxidative Stress and Severe Cardiomyopathy in Diabetic Mice. *J. Diabetes Metab.* **2012**, *7*, 2–9. [CrossRef]
79. Ma, Q. Role of Nrf2 in Oxidative Stress and Toxicity. *Annu. Rev. Pharmacol. Toxicol.* **2013**, *53*, 401–426. [CrossRef]
80. Bugger, H.; Abel, E.D. Molecular Mechanisms of Diabetic Cardiomyopathy. *Diabetologia* **2014**, *57*, 660–671. [CrossRef]
81. Iacobellis, G. Epicardial Adipose Tissue in Contemporary Cardiology. *Nat. Rev. Cardiol.* **2022**, *19*, 593–606. [CrossRef]
82. Iacobellis, G.; Corradi, D.; Sharma, A.M. Epicardial Adipose Tissue: Anatomic, Biomolecular and Clinical Relationships with the Heart. *Nat. Rev. Cardiol.* **2005**, *2*, 536–543. [CrossRef]
83. Cho, E.J. Epicardial Adipose Tissue (Pericoronary Adipose Tissue) in Cardiovascular Diseases. *Int. J. Heart Fail.* **2025**, *7*, 240. [CrossRef] [PubMed]
84. Marchington, J.M.; Mattacks, C.A.; Pond, C.M. Adipose Tissue in the Mammalian Heart and Pericardium: Structure, Foetal Development and Biochemical Properties. *Comp. Biochem. Physiol. Part. B Comp. Biochem.* **1989**, *94*, 225–232. [CrossRef] [PubMed]
85. Sacks, H.S.; Fain, J.N.; Bahouth, S.W.; Ojha, S.; Frontini, A.; Budge, H.; Cinti, S.; Symonds, M.E. Adult Epicardial Fat Exhibits Beige Features. *J. Clin. Endocrinol. Metab.* **2013**, *98*, E1448–E1455. [CrossRef] [PubMed]
86. Cinti, F.; Leccisotti, L.; Sorice, G.P.; Capece, U.; D'Amario, D.; Lorusso, M.; Gugliandolo, S.; Morciano, C.; Guarneri, A.; Guzzardi, M.A.; et al. Dapagliflozin Treatment Is Associated with a Reduction of Epicardial Adipose Tissue Thickness and Epicardial Glucose Uptake in Human Type 2 Diabetes. *Cardiovasc. Diabetol.* **2023**, *22*, 349. [CrossRef]
87. Morciano, C.; Gugliandolo, S.; Capece, U.; Di Giuseppe, G.; Mezza, T.; Ciccarelli, G.; Soldovieri, L.; Brunetti, M.; Avolio, A.; Splendore, A.; et al. SGLT2 Inhibition and Adipose Tissue Metabolism: Current Outlook and Perspectives. *Cardiovasc. Diabetol.* **2024**, *23*, 449. [CrossRef]
88. Gruzdeva, O.V.; Dyleva, Y.A.; Belik, E.V.; Sinitsky, M.Y.; Stasev, A.N.; Kokov, A.N.; Brel, N.K.; Krivkina, E.O.; Bychkova, E.E.; Tarasov, R.S.; et al. Relationship between Epicardial and Coronary Adipose Tissue and the Expression of Adiponectin, Leptin, and Interleukin 6 in Patients with Coronary Artery Disease. *J. Pers. Med.* **2022**, *12*, 129. [CrossRef]

89. Zhao, Y.X.; Zhu, H.J.; Pan, H.; Liu, X.M.; Wang, L.J.; Yang, H.B.; Li, N.S.; Gong, F.Y.; Sun, W.; Zeng, Y. Comparative Proteome Analysis of Epicardial and Subcutaneous Adipose Tissues from Patients with or without Coronary Artery Disease. *Int. J. Endocrinol.* **2019**, *2019*, 6976712. [CrossRef]
90. Bodenstab, M.L.; Varghese, R.T.; Iacobellis, G. Cardio-Lipotoxicity of Epicardial Adipose Tissue. *Biomolecules* **2024**, *14*, 1465. [CrossRef]
91. Goldman, S.A.; Requena-Ibanez, J.A.; Devesa, A.; Santos-Gallego, C.G.; Badimon, J.J.; Fuster, V. Uncovering the Role of Epicardial Adipose Tissue in Heart Failure with Preserved Ejection Fraction. *JACC Adv.* **2023**, *2*, 100657. [CrossRef]
92. Zhu, R.; Wang, W.; Gao, Y.; Wang, J.; Li, B.; Cheng, Z.; Ji, C.; Gu, H.; Yuan, X.; Yang, S.; et al. Insulin Resistance Aggravates Myocardial Fibrosis in Non-Diabetic Hypertensive Patients by Altering the Function of Epicardial Adipose Tissue: A Cardiac Magnetic Resonance Study. *Diabetol. Metab. Syndr.* **2025**, *17*, 133. [CrossRef]
93. Janssen-Telders, C.; Eringa, E.C.; De Groot, J.R.; De Man, F.S.; Handoko, M.L. The Role of Epicardial Adipose Tissue Remodelling in Heart Failure with Preserved Ejection Fraction. *Cardiovasc. Res.* **2025**, *121*, 860–870. [CrossRef]
94. Cui, C.; Qin, H.; Zhu, X.; Lu, X.; Wang, B.; Wang, X.; Wang, J.; Jiao, J.; Chu, M.; Wang, C.; et al. Unsupervised machine learning reveals epicardial adipose tissue subtypes with distinct atrial fibrosis profiles in patients with persistent atrial fibrillation: A prospective 2-center cohort study. *Heart Rhythm* **2022**, *19*, 2033–2041. [CrossRef]
95. Thanassoulis, G.; Massaro, J.M.; O'Donnell, C.J.; Hoffmann, U.; Levy, D.; Ellinor, P.T.; Wang, T.J.; Schnabel, R.B.; Vasan, R.S.; Fox, C.S.; et al. Pericardial Fat ss Associated with Prevalent Atrial Fibrillation: The Framingham Heart Study. *Circ. Arrhythmia Electrophysiol.* **2010**, *3*, 345–350. [CrossRef] [PubMed]
96. Caturano, A.; Galiero, R.; Pafundi, P.C. Atrial Fibrillation and Stroke. A Review on the Use of Vitamin K Antagonists and Novel Oral Anticoagulants. *Medicina* **2019**, *55*, 617. [CrossRef] [PubMed]
97. Zhao, L.; Zhu, Q.; Li, M.; Yang, H.; Zhao, Y. Revisiting Brown Adipose Tissue Whitening: Physiological Mechanisms and Pathophysiological Consequences. *J. Physiol.* **2025**, *603*, 5299–5326. [CrossRef] [PubMed]
98. Da Eira, D.; Jani, S.; Ceddia, R.B. An Obesogenic Diet Impairs Uncoupled Substrate Oxidation and Promotes Whitening of the Brown Adipose Tissue in Rats. *J. Physiol.* **2023**, *601*, 69–82. [CrossRef]
99. Hill, B.G. Insights into an Adipocyte Whitening Program. *Adipocyte* **2015**, *4*, 75–80. [CrossRef]
100. Kotzbeck, P.; Giordano, A.; Mondini, E.; Murano, I.; Severi, I.; Venema, W.; Cecchini, M.P.; Kershaw, E.E.; Barbatelli, G.; Haemmerle, G.; et al. Brown Adipose Tissue Whitening Leads to Brown Adipocyte Death and Adipose Tissue Inflammation. *J. Lipid Res.* **2018**, *59*, 784–794. [CrossRef]
101. Murano, I.; Barbatelli, G.; Parisani, V.; Latini, C.; Muzzonigro, G.; Castellucci, M.; Cinti, S. Dead Adipocytes, Detected as Crown-like Structures, Are Prevalent in Visceral Fat Depots of Genetically Obese Mice. *J. Lipid Res.* **2008**, *49*, 1562–1568. [CrossRef]
102. Cinti, S.; Mitchell, G.; Barbatelli, G.; Murano, I.; Ceresi, E.; Faloia, E.; Wang, S.; Fortier, M.; Greenberg, A.S.; Obin, M.S. Adipocyte Death Defines Macrophage Localization and Function in Adipose Tissue of Obese Mice and Humans. *J. Lipid Res.* **2005**, *46*, 2347–2355. [CrossRef]
103. Kralova Lesna, I.; Kralova, A.; Cejkova, S.; Fronek, J.; Petras, M.; Sekerkova, A.; Thieme, F.; Janousek, L.; Poledne, R. Characterisation and comparison of adipose tissue macrophages from human subcutaneous, visceral and perivascular adipose tissue. *J. Transl. Med.* **2016**, *14*, 208. [CrossRef]
104. Menghoum, N.; Badii, M.C.; Leroy, M.; Parra, M.; Roy, C.; Lejeune, S.; Vancraeynest, D.; Pasquet, A.; Brito, D.; Casadei, B.; et al. Exploring the impact of metabolic comorbidities on epicardial adipose tissue in heart failure with preserved ejection fraction. *Cardiovasc. Diabetol.* **2025**, *24*, 134. [CrossRef] [PubMed]
105. Yang, H.; Xiong, B.; Xiong, T.; Wang, D.; Yu, W.; Liu, B.; She, Q. Identification of Key Genes and Mechanisms of Epicardial Adipose Tissue in Patients with Diabetes through Bioinformatic Analysis. *Front. Cardiovasc. Med.* **2022**, *9*, 927397. [CrossRef] [PubMed]
106. Nevola, R.; Alfano, M.; Pafundi, P.C.; Brin, C.; Gragnano, F.; Calabrò, P.; Adinolfi, L.E.; Rinaldi, L.; Sasso, F.C.; Caturano, A. Cardiorenal Impact of SGLT-2 Inhibitors: A Conceptual Revolution in The Management of Type 2 Diabetes, Heart Failure and Chronic Kidney Disease. *Rev. Cardiovasc. Med.* **2022**, *23*, 106. [CrossRef] [PubMed]
107. Manubolu, V.S.; Lakshmanan, S.; Kinninger, A.; Ahmad, K.; Susarla, S.; Seok, H.J.; Hamal, S.; Dahal, S.; Roy, S.K.; Budoff, M.J. Effect of Semaglutide on Epicardial Adipose Tissue in Type 2 Diabetes: Insights From the STOP (Semaglutide Treatment effect On coronary atherosclerosis Progression) Randomized Trial. *J. Am. Coll. Cardiol.* **2024**, *84*, 865–867. [CrossRef]
108. Cinti, F.; Morciano, C.; Guarneri, A.; Cappannoli, L.; Sorice, G.; Gugliandolo, S.; Capece, U.; Splendore, A.; Avolio, A.; Mezza, T.; et al. Coronary Flow Reserve Increase after 4-Year Dapagliflozin Treatment in Patients with Type 2 Diabetes: The DAPAHEART Follow-up Study. *Cardiovasc. Diabetol.* **2025**, *24*, 351. [CrossRef]
109. Cersosimo, A.; Salerno, N.; Sabatino, J.; Scatteia, A.; Bisaccia, G.; De Rosa, S.; Dellegrottaglie, S.; Bucciarelli-Ducci, C.; Torella, D.; Leo, I. Underlying Mechanisms and Cardioprotective Effects of SGLT2i and GLP-1Ra: Insights from Cardiovascular Magnetic Resonance. *Cardiovasc. Diabetol.* **2024**, *23*, 94. [CrossRef]

110. Ziegler, A.K.; Scheele, C. Human Adipose Depots' Diverse Functions and Dysregulations during Cardiometabolic Disease. *npj Metab. Health Dis.* **2024**, *2*, 34. [CrossRef]
111. Cinti, S. Adipose Organ Development and Remodeling. *Compr. Physiol.* **2018**, *8*, 1357–1431. [CrossRef]
112. Galeone, A.; Annicchiarico, A.; Buccoliero, C.; Barile, B.; Luciani, G.B.; Onorati, F.; Nicchia, G.P.; Brunetti, G. Diabetic Cardiomyopathy: Role of Cell Death, Exosomes, Fibrosis and Epicardial Adipose Tissue. *Int. J. Mol. Sci.* **2024**, *25*, 9481. [CrossRef]
113. Lieb, D.C.; Parson, H.K.; Mamikunian, G.; Vinik, A.I. Cardiac autonomic imbalance in newly diagnosed and established diabetes is associated with markers of adipose tissue inflammation. *Exp. Diabetes Res.* **2012**, *2012*, 878760. [CrossRef] [PubMed]
114. Abad-Jiménez, Z.; López-Domènech, S.; Díaz-Rúa, R.; Iannantuoni, F.; Gómez-Abril, S.Á.; Periañez-Gómez, D.; Morillas, C.; Víctor, V.M.; Rocha, M. Systemic Oxidative Stress and Visceral Adipose Tissue Mediators of NLRP3 Inflammasome and Autophagy Are Reduced in Obese Type 2 Diabetic Patients Treated with Metformin. *Antioxidants* **2020**, *9*, 892. [CrossRef] [PubMed]
115. Perri, A.; Lofaro, D.; La Russa, A.; Lupinacci, S.; Toteda, G.; Curti, A.; Urso, A.; Bonofiglio, R.; La Russa, D.; Pellegrino, D.; et al. Proinflammatory Profile of Visceral Adipose Tissue and Oxidative Stress in Severe Obese Patients Carrying the Variant rs4612666 C of *NLRP3* Gene. *Minerva Endocrinol.* **2021**, *46*, 309–316. [CrossRef] [PubMed]
116. Luo, B.; Li, B.; Wang, W.; Liu, X.; Xia, Y.; Zhang, C.; Zhang, M.; Zhang, Y.; An, F. NLRP3 gene silencing ameliorates diabetic cardiomyopathy in a type 2 diabetes rat model. *PLoS ONE* **2014**, *9*, e104771. [CrossRef]
117. Wen, H.; Gris, D.; Lei, Y.; Jha, S.; Zhang, L.; Huang, M.T.; Brickey, W.J.; Ting, J.P. Fatty acid-induced NLRP3-ASC inflammasome activation interferes with insulin signaling. *Nat. Immunol.* **2011**, *12*, 408–415. [CrossRef]
118. Liu, W.; Zhang, X.; Zhao, M.; Zhang, X.; Chi, J.; Liu, Y.; Lin, F.; Fu, Y.; Ma, D.; Yin, X. Activation in M1 but Not M2 Macrophages Contributes to Cardiac Remodeling after Myocardial Infarction in Rats: A Critical Role of the Calcium Sensing Receptor/NRLP3 Inflammasome. *Cell Physiol. Biochem.* **2015**, *35*, 2483–2500. [CrossRef]
119. Ding, K.; Song, C.; Hu, H.; Yin, K.; Huang, H.; Tang, H. The Role of NLRP3 Inflammasome in Diabetic Cardiomyopathy and Its Therapeutic Implications. *Oxid. Med. Cell. Longev.* **2022**, *2022*, 3790721. [CrossRef]
120. Turner, N.A. Inflammatory and fibrotic responses of cardiac fibroblasts to myocardial damage associated molecular patterns (DAMPs). *J. Mol. Cell. Cardiol.* **2016**, *94*, 189–200. [CrossRef]
121. Saxena, A.; Chen, W.; Su, Y.; Rai, V.; Uche, O.U.; Li, N.; Frangogiannis, N.G. IL-1 Induces Proinflammatory Leukocyte Infiltration and Regulates Fibroblast Phenotype in the Infarcted Myocardium. *J. Immunol.* **2013**, *191*, 4838–4848. [CrossRef]
122. Khalil, H.; Kanisicak, O.; Prasad, V.; Correll, R.N.; Fu, X.; Schips, T.; Vagnozzi, R.J.; Liu, R.; Huynh, T.; Lee, S.-J.; et al. Fibroblast-Specific TGF-β–Smad2/3 Signaling Underlies Cardiac Fibrosis. *J. Clin. Investig.* **2017**, *127*, 3770–3783. [CrossRef]
123. Wu, H.; Li, G.N.; Xie, J.; Li, R.; Chen, Q.H.; Chen, J.Z.; Wei, Z.H.; Kang, L.N.; Xu, B. Resveratrol ameliorates myocardial fibrosis by inhibiting ROS/ERK/TGF-β/periostin pathway in STZ-induced diabetic mice. *BMC Cardiovasc. Disord.* **2016**, *16*, 5. [CrossRef] [PubMed]
124. Maleszewska, M.; Moonen, J.-R.A.J.; Huijkman, N.; Van De Sluis, B.; Krenning, G.; Harmsen, M.C. IL-1β and TGFβ2 Synergistically Induce Endothelial to Mesenchymal Transition in an NFκB-Dependent Manner. *Immunobiology* **2013**, *218*, 443–454. [CrossRef] [PubMed]
125. Kaludercic, N.; Di Lisa, F. Mitochondrial ROS Formation in the Pathogenesis of Diabetic Cardiomyopathy. *Front. Cardiovasc. Med.* **2020**, *7*, 12. [CrossRef] [PubMed]
126. Zhao, J. Molecular Mechanisms of AGE/RAGE-Mediated Fibrosis in the Diabetic Heart. *World J. Diabetes* **2014**, *5*, 860. [CrossRef]
127. Chen, C.-C.; Kuo, C.-H.; Leu, Y.-L.; Wang, S.-H. Corylin Reduces Obesity and Insulin Resistance and Promotes Adipose Tissue Browning through SIRT-1 and B3-AR Activation. *Pharmacol. Res.* **2021**, *164*, 105291. [CrossRef]
128. Bindu, S.; Pillai, V.B.; Gupta, M.P. Role of Sirtuins in Regulating Pathophysiology of the Heart. *Trends Endocrinol. Metab.* **2016**, *27*, 563–573. [CrossRef]
129. Bagul, P.K.; Dinda, A.K.; Banerjee, S.K. Effect of resveratrol on sirtuins expression and cardiac complications in diabetes. *Biochem. Biophys. Res. Commun.* **2015**, *468*, 221–227. [CrossRef]
130. Ayissi, V.B.O.; Ebrahimi, A.; Schluesenner, H. Epigenetic Effects of Natural Polyphenols: A Focus on SIRT1-mediated Mechanisms. *Mol. Nutr. Food Res.* **2014**, *58*, 22–32. [CrossRef]
131. Ceriello, A. New Insights on Oxidative Stress and Diabetic Complications May Lead to a "Causal" Antioxidant Therapy. *Diabetes Care* **2003**, *26*, 1589–1596. [CrossRef]
132. Rains, J.L.; Jain, S.K. Oxidative Stress, Insulin Signaling, and Diabetes. *Free Radic. Biol. Med.* **2011**, *50*, 567–575. [CrossRef]
133. Baynes, J.W.; Thorpe, S.R. Role of Oxidative Stress in Diabetic Complications: A New Perspective on an Old Paradigm. *Diabetes* **1999**, *48*, 1–9. [CrossRef] [PubMed]
134. Jaiswal, A.K. Nrf2 signaling in coordinated activation of antioxidant gene expression. *Free Radic. Biol. Med.* **2004**, *36*, 1199–1207. [CrossRef] [PubMed]
135. Uruno, A.; Yagishita, Y.; Yamamoto, M. The Keap1–Nrf2 System and Diabetes Mellitus. *Arch. Biochem. Biophys.* **2015**, *566*, 76–84. [CrossRef] [PubMed]

136. Qin, S.; Chen, Z.; Wen, Y.; Yi, Y.; Lv, C.; Zeng, C.; Chen, L.; Shi, M. Phytochemical Activators of Nrf2: A Review of Therapeutic Strategies in Diabetes. *Acta Biochim. Biophys. Sin.* **2022**, *55*, 11–22. [CrossRef]
137. Patti, M.E.; Butte, A.J.; Crunkhorn, S.; Cusi, K.; Berria, R.; Kashyap, S.; Miyazaki, Y.; Kohane, I.; Costello, M.; Saccone, R.; et al. Coordinated Reduction of Genes of Oxidative Metabolism in Humans with Insulin Resistance and Diabetes: Potential Role of *PGC1* and *NRF1*. *Proc. Natl. Acad. Sci. USA* **2003**, *100*, 8466–8471. [CrossRef]
138. Morrish, F.; Gingras, H.; Noonan, J.; Huang, L.; Sweet, I.R.; Kuok, I.T.; Knoblaugh, S.E.; Hockenbery, D.M. Mitochondrial Diabetes in Mice Expressing a Dominant-Negative Allele of Nuclear Respiratory Factor-1 (Nrf1) in Pancreatic β-Cells. *Biochem. Biophys. Res. Commun.* **2024**, *737*, 150478. [CrossRef]
139. Li, J.; Yan, J.; Tu, G.; Jiang, W.; Qiu, Y.; Su, Y.; Miao, C.; Luo, Z.; Horng, T. NRF1 coordinates mitochondrial adaptations to dampen intracellular ROS and inflammatory responses during ischemia reperfusion. *Cell Death Discov.* **2025**, *11*, 236. [CrossRef]
140. Młynarska, E.; Hajdys, J.; Czarnik, W.; Fularski, P.; Leszto, K.; Majchrowicz, G.; Lisińska, W.; Rysz, J.; Franczyk, B. The Role of Antioxidants in the Therapy of Cardiovascular Diseases—A Literature Review. *Nutrients* **2024**, *16*, 2587. [CrossRef]
141. Li, Y.; El-Sehrawy, A.A.M.A.; Shankar, A.; Srivastava, M.; Mohammed, J.S.; Hjazi, A.; Singh, M.; Sapaev, I.B.; Mustafa, Y.F.; Abosaoda, M.K. Effects of Coenzyme Q10 Supplementation on Metabolic Indicators in Patients with Type 2 Diabetes: A Systematic Review and Meta-Analysis. *Clin. Ther.* **2025**, *47*, 235–243. [CrossRef]
142. Sharma, A.; Fonarow, G.C.; Butler, J.; Ezekowitz, J.A.; Felker, G.M. Coenzyme Q10 and Heart Failure: A State-of-the-Art Review. *Circ. Heart Fail.* **2016**, *9*, e002639. [CrossRef]
143. Tang, Y.; Zhang, Z.; Yan, T.; Chen, K.; Xu, G.; Xiong, S.; Wu, D.; Chen, J.; Jose, P.A.; Zeng, C.; et al. Irisin Attenuates Type 1 Diabetic Cardiomyopathy by Anti-Ferroptosis via SIRT1-Mediated Deacetylation of P53. *Cardiovasc. Diabetol.* **2024**, *23*, 116. [CrossRef]
144. Spillmann, F.; Van Linthout, S.; Tschope, C. Cardiac Effects of HDL and Its Components on Diabetic Cardiomyopathy. *Endocr. Metab. Immune Disord. Drug Targets* **2012**, *12*, 132–147. [CrossRef] [PubMed]
145. Carpintero, R.; Gruaz, L.; Brandt, K.J.; Scanu, A.; Faille, D.; Combes, V.; Grau, G.E.; Burger, D. HDL Interfere with the Binding of T Cell Microparticles to Human Monocytes to Inhibit Pro-Inflammatory Cytokine Production. *PLoS ONE* **2010**, *5*, e11869. [CrossRef] [PubMed]
146. Coronado, M.T.; Pozzi, A.O.; Punchard, M.A.; González, P.; Fantidis, P. Inflammation as a Modulator of the HDL Cholesterol-Induced Inteleukin-10 Production by Human Circulating Mononuclear Cells. *Atherosclerosis* **2009**, *202*, 183–184. [CrossRef] [PubMed]
147. Van Linthout, S.; Spillmann, F.; Lorenz, M.; Meloni, M.; Jacobs, F.; Egorova, M.; Stangl, V.; De Geest, B.; Schultheiss, H.-P.; Tschöpe, C. Vascular-Protective Effects of High-Density Lipoprotein Include the Downregulation of the Angiotensin II Type 1 Receptor. *Hypertension* **2009**, *53*, 682–687. [CrossRef]
148. Perségol, L.; Vergès, B.; Foissac, M.; Gambert, P.; Duvillard, L. Inability of HDL from Type 2 Diabetic Patients to Counteract the Inhibitory Effect of Oxidised LDL on Endothelium-Dependent Vasorelaxation. *Diabetologia* **2006**, *49*, 1380–1386. [CrossRef]
149. Thai, P.N.; Miller, C.V.; King, M.T.; Schaefer, S.; Veech, R.L.; Chiamvimonvat, N.; Bers, D.M.; Dedkova, E.N. Ketone Ester D-β-Hydroxybutyrate-(R)-1,3 Butanediol Prevents Decline in Cardiac Function in Type 2 Diabetic Mice. *J. Am. Heart Assoc.* **2021**, *10*, e020729. [CrossRef]
150. Byrne, N.J.; Rajasekaran, N.S.; Abel, E.D.; Bugger, H. Therapeutic Potential of Targeting Oxidative Stress in Diabetic Cardiomyopathy. *Free Radic. Biol. Med.* **2021**, *169*, 317–342. [CrossRef]
151. Ashor, A.W.; Siervo, M.; Lara, J.; Oggioni, C.; Afshar, S.; Mathers, J.C. Effect of Vitamin C and Vitamin E Supplementation on Endothelial Function: A Systematic Review and Meta-Analysis of Randomised Controlled Trials. *Br. J. Nutr.* **2015**, *113*, 1182–1194. [CrossRef]
152. Biswas, S.K. Does the Interdependence between Oxidative Stress and Inflammation Explain the Antioxidant Paradox? *Oxid. Med. Cell Longev.* **2016**, *2016*, 5698931. [CrossRef]
153. Murphy, M.P. Targeting Lipophilic Cations to Mitochondria. *Biochim. Biophys. Acta—Bioenerg.* **2008**, *1777*, 1028–1031. [CrossRef] [PubMed]
154. Sies, H. Oxidative Stress: A Concept in Redox Biology and Medicine. *Redox Biol.* **2015**, *4*, 180–183. [CrossRef] [PubMed]
155. Tsutsui, H.; Kinugawa, S.; Matsushima, S. Oxidative Stress and Heart Failure. *Am. J. Physiol. Heart Circ. Physiol.* **2011**, *301*, H2181–H2190. [CrossRef] [PubMed]
156. Chen, X.; Xie, N.; Feng, L.; Huang, Y.; Wu, Y.; Zhu, H.; Tang, J.; Zhang, Y. Oxidative Stress in Diabetes Mellitus and Its Complications: From Pathophysiology to Therapeutic Strategies. *Chin. Med. J.* **2025**, *138*, 15–27. [CrossRef]
157. Nahar, N.; Sohag, M.d.S.U. Advancements in Mitochondrial-Targeted Antioxidants: Organelle-Specific Drug Delivery for Disease Management. *Adv. Redox Res.* **2025**, *17*, 100142. [CrossRef]
158. Smith, R.A.J.; Porteous, C.M.; Gane, A.M.; Murphy, M.P. Delivery of Bioactive Molecules to Mitochondria In Vivo. *Proc. Natl. Acad. Sci. USA* **2003**, *100*, 5407–5412. [CrossRef]

https://doi.org/10.3390/biom16030470

159. Parker, A.M.; Lees, J.G.; Tate, M.; Phang, R.J.; Velagic, A.; Deo, M.; Bishop, T.; Krieg, T.; Murphy, M.P.; Lim, S.Y.; et al. MitoQ Protects Against Oxidative Stress-Induced Mitochondrial Dysregulation in Human Cardiomyocytes. *J. Mol. Cell. Cardiol. Plus* **2025**, *13*, 100469. [CrossRef]
160. Ni, R.; Cao, T.; Xiong, S.; Ma, J.; Fan, G.-C.; Lacefield, J.C.; Lu, Y.; Tissier, S.L.; Peng, T. Therapeutic Inhibition of Mitochondrial Reactive Oxygen Species with Mito-TEMPO Reduces Diabetic Cardiomyopathy. *Free Radic. Biol. Med.* **2016**, *90*, 12–23. [CrossRef]
161. Wilcox, C.S. Effects of Tempol and Redox-Cycling Nitroxides in Models of Oxidative Stress. *Pharmacol. Ther.* **2010**, *126*, 119–145. [CrossRef]
162. Heumüller, S.; Wind, S.; Barbosa-Sicard, E.; Schmidt, H.H.H.W.; Busse, R.; Schröder, K.; Brandes, R.P. Apocynin Is Not an Inhibitor of Vascular NADPH Oxidases but an Antioxidant. *Hypertension* **2008**, *51*, 211–217. [CrossRef]
163. Qiu, J.; Liu, D.; Li, P.; Zhou, L.; Zhou, L.; Liu, X.; Zhang, Y.; Yuan, M.; Tse, G.; Li, G.; et al. NADPH Oxidase Mediates Oxidative Stress and Ventricular Remodeling through SIRT3/FOXO3a Pathway in Diabetic Mice. *Antioxidants* **2022**, *11*, 1745. [CrossRef] [PubMed]
164. Bravo-Sánchez, E.; Peña-Montes, D.; Sánchez-Duarte, S.; Saavedra-Molina, A.; Sánchez-Duarte, E.; Montoya-Pérez, R. Effects of Apocynin on Heart Muscle Oxidative Stress of Rats with Experimental Diabetes: Implications for Mitochondria. *Antioxidants* **2021**, *10*, 335. [CrossRef] [PubMed]
165. Dama, A.; Shpati, K.; Daliu, P.; Dumur, S.; Gorica, E.; Santini, A. Targeting Metabolic Diseases: The Role of Nutraceuticals in Modulating Oxidative Stress and Inflammation. *Nutrients* **2024**, *16*, 507. [CrossRef] [PubMed]
166. Amin, A.; Akhtar, M.S.; Khalil, A.A.K.; Ali, S.; Zaman, W. Natural Products in Medicinal Chemistry: Targeting Inflammatory Pathways with Plant-Derived Compounds. *Med. Chem. Res.* **2026**, *35*, 61–84. [CrossRef]
167. Huang, S.H.; Xu, M.; Wu, H.M.; Wan, C.X.; Wang, H.B.; Wu, Q.Q.; Liao, H.H.; Deng, W.; Tang, Q.Z. Isoquercitrin Attenuated Cardiac Dysfunction Via AMPKα-Dependent Pathways in LPS-Treated Mice. *Mol. Nutr. Food Res.* **2018**, *62*, e1800955. [CrossRef]
168. Fuloria, S.; Mehta, J.; Chandel, A.; Sekar, M.; Rani, N.N.I.M.; Begum, M.Y.; Subramaniyan, V.; Chidambaram, K.; Thangavelu, L.; Nordin, R.; et al. A Comprehensive Review on the Therapeutic Potential of *Curcuma longa* Linn. in Relation to its Major Active Constituent Curcumin. *Front. Pharmacol.* **2022**, *13*, 820806. [CrossRef]
169. Shay, K.P.; Moreau, R.F.; Smith, E.J.; Smith, A.R.; Hagen, T.M. Alpha-Lipoic Acid as a Dietary Supplement: Molecular Mechanisms and Therapeutic Potential. *Biochim. Biophys. Acta—Gen. Subj.* **2009**, *1790*, 1149–1160. [CrossRef]
170. Ritchie, R.H.; Abel, E.D. Basic Mechanisms of Diabetic Heart Disease. *Circ. Res.* **2020**, *126*, 1501–1525. [CrossRef]
171. Tan, Y.; Ichikawa, T.; Li, J.; Si, Q.; Yang, H.; Chen, X.; Goldblatt, C.S.; Meyer, C.J.; Li, X.; Cai, L.; et al. Diabetic Downregulation of Nrf2 Activity via ERK Contributes to Oxidative Stress-Induced Insulin Resistance in Cardiac Cells In Vitro and In Vivo. *Diabetes* **2011**, *60*, 625–633. [CrossRef]
172. Zhao, Q.D.; Viswanadhapalli, S.; Williams, P.; Shi, Q.; Tan, C.; Yi, X.; Bhandari, B.; Abboud, H.E. NADPH Oxidase 4 Induces Cardiac Fibrosis and Hypertrophy Through Activating Akt/mTOR and NFκB Signaling Pathways. *Circulation* **2015**, *131*, 643–655. [CrossRef]
173. Mihanfar, A.; Akbarzadeh, M.; Ghazizadeh Darband, S.; Sadighparvar, S.; Majidinia, M. SIRT1: A promising therapeutic target in type 2 diabetes mellitus. *Arch. Physiol. Biochem.* **2024**, *130*, 13–28. [CrossRef] [PubMed]
174. Cohen, K.; Waldman, M.; Abraham, N.G.; Laniado-Schwartzman, M.; Gurfield, D.; Aravot, D.; Arad, M.; Hochhauser, E. Caloric Restriction Ameliorates Cardiomyopathy in Animal Model of Diabetes. *Exp. Cell Res.* **2017**, *350*, 147–153. [CrossRef] [PubMed]
175. Le Douairon Lahaye, S.; Bekono, F.R.; Broderick, T. Physical activity and diabetic cardiomyopathy: Myocardial adaptation depending on exercise load. *Curr. Diabetes Rev.* **2014**, *10*, 371–390. [CrossRef] [PubMed]
176. Faria, A.; Persaud, S.J. Cardiac Oxidative Stress in Diabetes: Mechanisms and Therapeutic Potential. *Pharmacol. Ther.* **2017**, *172*, 50–62. [CrossRef]
177. Di Mauro, V.; Iafisco, M.; Salvarani, N.; Vacchiano, M.; Carullo, P.; Ramírez-Rodríguez, G.B.; Patrício, T.; Tampieri, A.; Miragoli, M.; Catalucci, D. Bioinspired Negatively Charged Calcium Phosphate Nanocarriers for Cardiac Delivery of MicroRNAs. *Nanomedicine* **2016**, *11*, 891–906. [CrossRef]
178. Caturano, A.; Nilo, R.; Nilo, D.; Russo, V.; Santonastaso, E.; Galiero, R.; Rinaldi, L.; Monda, M.; Sardu, C.; Marfella, R.; et al. Advances in Nanomedicine for Precision Insulin Delivery. *Pharmaceuticals* **2024**, *17*, 945. [CrossRef]
179. Du, X.; Zeng, Q.; Luo, Y.; He, L.; Zhao, Y.; Li, N.; Han, C.; Zhang, G.; Liu, W. Application Research of Novel Peptide Mitochondrial-Targeted Antioxidant SS-31 in Mitigating Mitochondrial Dysfunction. *Mitochondrion* **2024**, *75*, 101846. [CrossRef]
180. Chenxi, Z.; Hemmat, A.; Thi, N.; Afrand, M. Nanoparticle-Enhanced Drug Delivery Systems: An up-to-Date Review. *J. Mol. Liq.* **2025**, *424*, 126999. [CrossRef]
181. Krako Jakovljevic, N.; Pavlovic, K.; Jotic, A.; Lalic, K.; Stoiljkovic, M.; Lukic, L.; Milicic, T.; Macesic, M.; Stanarcic Gajovic, J.; Lalic, N.M. Targeting Mitochondria in Diabetes. *Int. J. Mol. Sci.* **2021**, *22*, 6642. [CrossRef]
182. Murphy, M.P.; Hartley, R.C. Mitochondria as a Therapeutic Target for Common Pathologies. *Nat. Rev. Drug Discov.* **2018**, *17*, 865–886. [CrossRef]

183. Xiong, Z.; Liao, Y.; Zhang, Z.; Wan, Z.; Liang, S.; Guo, J. Molecular Insights into Oxidative-Stress-Mediated Cardiomyopathy and Potential Therapeutic Strategies. *Biomolecules* **2025**, *15*, 670. [CrossRef]
184. Jia, G.; Hill, M.A.; Sowers, J.R. Diabetic Cardiomyopathy: An Update of Mechanisms Contributing to This Clinical Entity. *Circ. Res.* **2018**, *122*, 624–638. [CrossRef]
185. Choudhury, P.; Kosuru, R.; Cai, Y. Editorial: The Complex Phenotype of Diabetic Cardiomyopathy: Clinical Indicators and Novel Treatment Targets. *Front. Endocrinol.* **2024**, *15*, 1497352. [CrossRef]
186. Zhou, L.; Mei, S.; Ma, X.; Wuyun, Q.; Cai, Z.; Chen, C.; Ding, H.; Yan, J. Multi-Omics Insights into the Pathogenesis of Diabetic Cardiomyopathy: Epigenetic and Metabolic Profiles. *Epigenomics* **2025**, *17*, 33–48. [CrossRef]
187. Galiero, R.; Caturano, A.; Vetrano, E.; Monda, M.; Marfella, R.; Sardu, C.; Salvatore, T.; Rinaldi, L.; Sasso, F.C. Precision Medicine in Type 2 Diabetes Mellitus: Utility and Limitations. *Diabetes Metab. Syndr. Obes.* **2023**, *16*, 3669–3689. [CrossRef]
188. Oikonomou, E.K.; Khera, R. Machine Learning in Precision Diabetes Care and CardiovascularRisk Prediction. *Cardiovasc. Diabetol.* **2023**, *22*, 259. [CrossRef]

https://doi.org/10.3390/biom16030470

biomolecules

MDPI

Review

Mitochondrial Dysfunction in Cardiomyopathy and Heart Failure: From Energetic Collapse to Therapeutic Opportunity

Nikola Pavlović [1,2], Petar Todorović [3], Mirko Maglica [4,5], Marko Kumrić [1,2], Katarina Vukojević [3,6,7], Zenon Pogorelić [8,9,*] and Joško Božić [1,2]

1 Department of Pathophysiology, University of Split School of Medicine, 21000 Split, Croatia
2 Laboratory for Cardiometabolic Research, University of Split School of Medicine, 21000 Split, Croatia
3 Department of Anatomy, Histology and Embryology, University of Split School of Medicine, 21000 Split, Croatia
4 Department of Anatomy, School of Medicine, University of Mostar, 88000 Mostar, Bosnia and Herzegovina
5 Department of Otorhinolaryngology, University Hospital of Split, 21000 Split, Croatia
6 Mediterranean Institute for Life Sciences, University of Split, 21000 Split, Croatia
7 Center for Translational Research in Biomedicine, University of Split School of Medicine, 21000 Split, Croatia
8 Department of Pediatric Surgery, University Hospital of Split, 21000 Split, Croatia
9 Department of Surgery, School of Medicine, University of Split, 21000 Split, Croatia
* Correspondence: zenon.pogorelic@mefst.hr

Abstract

The heart's relentless contractile activity depends critically on mitochondrial function to meet its extraordinary bioenergetic demands. Mitochondria, through oxidative phosphorylation, not only supply ATP but also regulate metabolism, calcium homeostasis, and apoptotic signaling, ensuring cardiomyocyte viability and cardiac function. Mitochondrial dysfunction is a hallmark of cardiomyopathies and heart failure, characterized by impaired oxidative phosphorylation, excessive production of reactive oxygen species (ROS), dysregulated calcium handling, and disturbances in mitochondrial dynamics and mitophagy. These defects culminate in energetic insufficiency, cellular injury, and cardiomyocyte death, driving heart disease progression. Diverse cardiomyopathy phenotypes exhibit distinct mitochondrial pathologies, from acute ischemia-induced mitochondrial collapse to chronic remodeling seen in dilated, hypertrophic, restrictive, and primary mitochondrial cardiomyopathies. Mitochondria also orchestrate cell death and inflammatory pathways that worsen cardiac dysfunction. Therapeutic strategies targeting mitochondrial dysfunction, including antioxidants, modulators of mitochondrial biogenesis, metabolic therapies, and innovative approaches such as mitochondrial transplantation, show promise but face challenges in clinical translation. Advances in biomarker discovery and personalized medicine approaches hold promise for optimizing mitochondrial-targeted therapies. Unlike previous reviews that examined these pathways or interventions individually, this work summarizes insights into mechanisms with emerging therapeutic strategies, such as SGLT2 inhibition in HFpEF, NAD^+ repletion, mitochondrial transplantation, and biomarker-driven precision medicine, into a unified synthesis. This framework underscores the novel contribution of linking basic mitochondrial biology to translational and clinical opportunities in cardiomyopathy and heart failure. This review synthesizes the current understanding of mitochondrial biology in cardiac health and disease, delineates the molecular mechanisms underpinning mitochondrial dysfunction in cardiomyopathy and heart failure, and explores emerging therapeutic avenues aimed at restoring mitochondrial integrity and improving clinical outcomes in cardiac patients.

 https://doi.org/10.3390/biom15111572

Keywords: mitochondrial dysfunction; cardiomyopathy; heart failure; bioenergetics; mitochondrial dynamics

1. Introduction

The heart is an extraordinary organ that works incessantly throughout life, and is dependent on a continuous supply of energy to maintain its contractile activity and ensure haemodynamic stability [1–5]. This immense energy demand is primarily met by the mitochondria, the cell organelles known as the powerhouse of the cell [6–8]. In healthy heart muscle, the mitochondria are organised in such a way that they meet the high bioenergetic demands by producing adenosine triphosphate (ATP) through oxidative phosphorylation. These organelles not only generate energy but also play a crucial role in regulating cellular metabolism, calcium homeostasis, and apoptotic signalling, thus maintaining cardiomyocyte viability and overall cardiac function [1,9–12].

The fact that mitochondria make up over 30% of the volume of cardiomyocytes highlights how essential they are to heart physiology [13]. Under normal circumstances, fatty acids are the primary substrate for the adult heart's nearly entirely aerobic metabolism, which is augmented by the oxidation of glucose and lactate. The heart can effectively adjust to changing energy needs thanks to its metabolic flexibility [14–16]. As electrons go through several complexes connected to proton pumping and chemiosmotic ATP synthesis, the electron transport chain (ETC), which is enmeshed in the inner mitochondrial membrane, propels ATP production [17,18]. The heart's ability to maintain contractile action can be significantly impacted by any disruption in these precisely regulated mechanisms [8,19].

Beyond energy generation, mitochondrial dynamics, including the processes of biogenesis, fission, fusion, and mitophagy, are crucial for maintaining a healthy mitochondrial population [20,21]. Mitochondrial biogenesis is governed by transcription factors and coactivators such as PGC-1α, which coordinate the expression of nuclear and mitochondrial genes responsible for mitochondrial replication and function [22]. Mitochondrial fission and fusion enable the organelles to adapt to metabolic needs and facilitate the removal or repair of damaged regions [21]. Mitophagy, a selective form of autophagy, ensures the quality control of mitochondria by degrading dysfunctional mitochondria to prevent the accumulation of oxidative damage and maintain cellular homeostasis. Collectively, these mitochondrial quality control mechanisms are vital for the preservation of cardiac function, particularly under physiological stress [23,24].

The heart's reliance on mitochondrial integrity becomes glaringly evident in pathological states such as cardiomyopathy and heart failure, where mitochondrial dysfunction emerges as a central contributor to disease progression [25]. Cardiomyopathies encompass a heterogeneous group of myocardial disorders that impair cardiac structure and function, often culminating in heart failure, a leading cause of morbidity and mortality worldwide [26,27]. In these conditions, mitochondrial defects manifest as impaired oxidative phosphorylation, excessive reactive oxygen species (ROS) production, disrupted calcium handling, and imbalanced mitochondrial dynamics, which collectively drive energetic insufficiency and cellular damage (Figure 1) [28,29].

Energy deprivation in the myocardium initiates a downward spiral of contractile failure and adverse remodeling [3,30]. Moreover, mitochondria-derived signals can trigger cell death pathways and inflammatory cascades that exacerbate cardiac injury and promote disease progression. Thus, understanding mitochondrial biology in the context of cardiac health and disease is essential for the development of novel therapeutic strategies aimed at preserving or restoring mitochondrial function [31–33].

Although numerous reviews have examined individual aspects of mitochondrial biology in cardiovascular disease [7,8], few have provided an integrated summary that spans mechanistic insights, diverse cardiomyopathy phenotypes, and emerging therapeutic approaches. Prior work has typically focused on oxidative stress, substrate metabolism, or single therapeutic strategies in isolation. By contrast, the present review unifies recent advances, including SGLT2 inhibition in HFpEF, NAD^+ repletion, mitochondrial transplantation, and biomarker-driven precision medicine, within the broader framework of mitochondrial dysfunction across cardiomyopathy subtypes. In doing so, it highlights both mechanistic commonalities and disease-specific vulnerabilities, offering a novel perspective that bridges basic mitochondrial biology with translational and clinical opportunities in heart failure. This review aims to provide a comprehensive overview of mitochondrial dysfunction in cardiomyopathy and heart failure, outlining the underlying molecular mechanisms and emphasizing emerging therapeutic opportunities. We begin by thoroughly examining mitochondrial function in a healthy heart, laying the groundwork for later sections that explore pathophysiological changes, various cardiomyopathy phenotypes, mitochondria-driven cell death, and the potential for clinical applications of mitochondrial-targeted interventions.

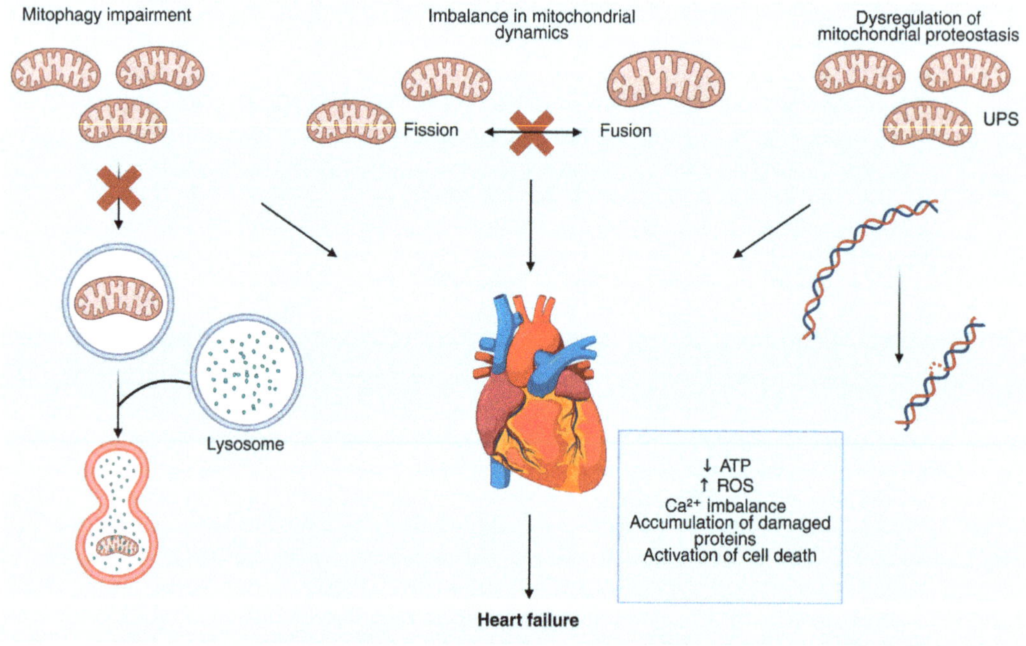

Figure 1. Mitochondrial quality control defects contribute to the development of heart failure. Mitochondrial dysfunction in cardiomyocytes arises from defects in multiple quality control pathways. Impaired mitophagy leads to the accumulation of dysfunctional mitochondria due to the inability of lysosomal degradation to occur. Imbalance in mitochondrial dynamics disrupts the fission–fusion cycle required for mitochondrial remodeling and adaptation, leading to altered morphology and bioenergetic failure. Dysregulation of mitochondrial proteostasis, including defective ubiquitin–proteasome system (UPS) activity, promotes accumulation of damaged proteins and mitochondrial DNA (mtDNA) instability. Collectively, these defects result in decreased ATP production, excessive generation of reactive oxygen species (ROS), calcium (Ca^{2+}) dysregulation, proteotoxic stress, and activation of cell death pathways, ultimately contributing to the progression of heart failure.

2. Mechanisms of Mitochondrial Dysfunction in Heart Failure

Mitochondrial dysfunction is a hallmark of cardiomyopathies and heart failure. It results from a complex interplay of molecular mechanisms that impair the energy production of the heart muscle and cellular homeostasis [34,35]. An important factor that sets this pathological cascade in motion is oxidative stress, which leads to damage to the mitochondrial DNA (mtDNA). The failing heart is characterised by an increased production of reactive oxygen species (ROS), which the mitochondrial ETC mainly generates [36]. In contrast to antioxidant enzymes, whose activities are relatively preserved, excessive mitochondrial ROS increasingly damages the mtDNA, which codes for essential subunits of the respiratory complexes. This leads to a vicious cycle in which the damaged mtDNA impairs mitochondrial function, further increasing ROS production and perpetuating cellular damage. ROS also initiates maladaptive myocardial remodeling by activating matrix metalloproteinases responsible for extracellular matrix degradation, promoting myocyte hypertrophy, apoptosis, and interstitial fibrosis, key features of progressive heart failure [37–39].

Beyond oxidative damage, mitochondrial dysfunction in cardiomyopathy is closely linked to impaired oxidative phosphorylation (OXPHOS) and consequent ATP depletion. Defects in substrate oxidation and the electron transport chain, affecting complexes I through V, lead to decreased bioenergetic capacity in failing myocardium [40–42]. The defects interrupt the cardiac phosphocreatine/ATP energy shuttle, leading to elevated levels of ADP that hinder contractile processes. Research consistently shows reduced oxidative phosphorylation (OXPHOS) activity in both human and animal models of heart failure, highlighting the mitochondrion as a key site of energy deficiency that contributes to contractile dysfunction [43–45]. Mitochondrial calcium overload further exacerbates OXPHOS dysfunction. Calcium regulates key dehydrogenases within the tricarboxylic acid cycle; however, excessive mitochondrial calcium accumulation triggers opening of the mitochondrial permeability transition pore (mPTP), dissipating membrane potential and collapsing ATP synthesis capability [46–48].

The regulation of mitochondrial calcium homeostasis is integral to cardiac function, involving calcium uptake primarily through the mitochondrial calcium uniporter (MCU) and efflux via Na^{+}/Ca^{2+} exchangers. In cardiomyopathy and heart failure, dysregulated calcium flux disrupts this delicate balance, favoring the opening of mPTP and subsequent cell death pathways. The mPTP is a high-conductance channel in the inner mitochondrial membrane, whose pathological opening not only depolarizes mitochondria but also releases pro-apoptotic factors such as cytochrome c, intensifying cardiomyocyte loss and contractile impairment [49–51].

At the same time, abnormalities in mitochondrial quality control mechanisms exacerbate cardiac mitochondrial dysfunction. Mitochondrial fission and fusion, a dynamic process essential for adapting mitochondrial morphology and function, become dysregulated in cardiomyopathic hearts. Impairments in fusion proteins (e.g., Mitofusins 1 and 2) and fission proteins (e.g., Drp1) disrupt mitochondrial network integrity and prevent segregation of damaged mitochondrial segments. This imbalance leads to the accumulation of dysfunctional mitochondria with reduced oxidative capacity and enhanced ROS release, further impairing cardiomyocyte function [52–54].

Mitophagy, the selective autophagic clearance of damaged mitochondria, is another critical facet of mitochondrial quality control negatively affected in heart failure. Deficient activation of mitophagy pathways, involving key regulators such as PINK1 and Parkin, results in the accumulation of structurally abnormal and bioenergetically compromised mitochondria [55–57]. This failure to clear damaged mitochondria intensifies oxidative stress and energetic deficits, accelerating cardiomyocyte dysfunction and death. Moreover, emerging evidence indicates that mitophagy and mitochondrial dynamics are tightly coordinated processes; disruption in one often impairs the other, creating a feedback loop that compromises mitochondrial homeostasis in failing hearts [58–60].

In summary, mitochondrial dysfunction in cardiomyopathy and heart failure arises from intertwined mechanisms: oxidative stress induces mtDNA damage and excessive ROS; impaired OXPHOS reduces ATP synthesis; disturbed calcium homeostasis leads to pathological mPTP opening; and dysregulated mitochondrial dynamics and deficient mitophagy degrade mitochondrial quality. These mechanisms are summarized schematically in Figure 2, illustrating their contribution to energetic deficits, cellular injury, and ultimately cardiac dysfunction leading to heart failure. Together, these molecular perturbations precipitate the energetic collapse and cellular damage characteristic of heart failure, highlighting the mitochondrion as a central nexus and promising therapeutic target in cardiac disease.

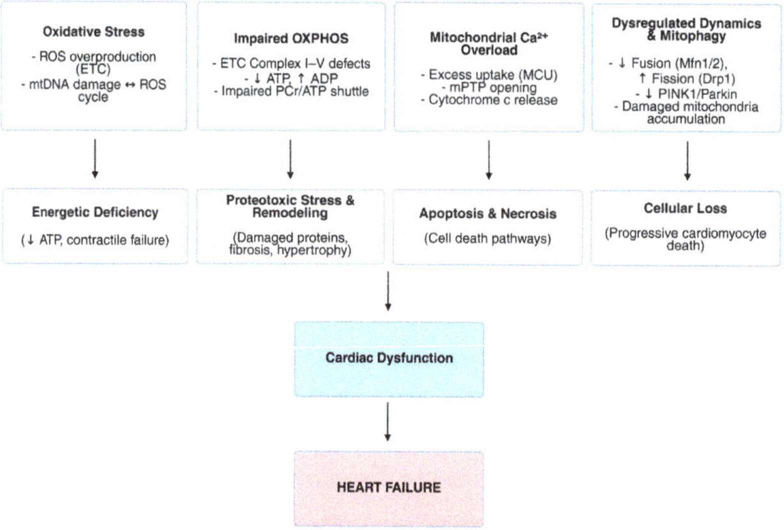

Figure 2. Mitochondrial dysfunction and its role in the pathogenesis of heart failure. This schematic illustrates key mitochondrial abnormalities that contribute to the progression of heart failure. The top row identifies four primary mitochondrial dysfunctions: oxidative stress characterized by reactive oxygen species (ROS) overproduction and mitochondrial DNA (mtDNA) damage; impaired oxidative phosphorylation (OXPHOS) involving defects in the electron transport chain (ETC) leading to reduced ATP production and disrupted energy shuttle; mitochondrial calcium overload resulting from excessive uptake (MCU) and mitochondrial permeability transition pore (mPTP) opening, causing cytochrome c release; and dysregulated mitochondrial dynamics and mitophagy, including decreased fusion, increased fission, downregulation of PINK1/Parkin pathways, and accumulation of damaged mitochondria. These disturbances trigger downstream pathological processes such as energetic deficiency, proteotoxic stress with remodeling, apoptosis and necrosis, and cellular loss. Together, these mechanisms culminate in cardiac dysfunction, ultimately leading to heart failure.

3. Mitochondrial Dysfunction Across Different Types of Cardiomyopathy

Mitochondrial dysfunction is a common axis in cardiomyopathies, but its mode and consequences vary by subtype, informing prognosis and therapy [61]. Mitochondrial energetic failure integrates deranged substrate utilization, impaired oxidative phosphorylation (OXPHOS), and redox stress, which cumulatively depresses ATP supply [61].

Although ischemic cardiomyopathy is not traditionally classified as a primary cardiomyopathy, since it is secondary to coronary artery disease, this heart failure phenotype exhibits a number of distinct mitochondrial characteristics. In ischemic cardiomyopathy, ischemia–reperfusion (I/R) triggers a cascade of mitochondrial calcium overload, reactive oxygen species (ROS) bursts, and the opening of the mPTP, ultimately driving necrosis and apoptosis [62]. During ischemia, fatty acid oxidation (FAO) is curtailed, and glucose oxidation becomes relatively favored, a shift that can be therapeutically exploited [63]. Pharmacologic FAO inhibition (e.g., malonyl-CoA decarboxylase inhibition) increases glucose oxidation and limits ischemic injury in preclinical models, highlighting metabolic control as a key component of cardioprotection [63]. Consistent with this, post-ischemic suppression of FAO promotes regeneration and functional recovery in mammalian hearts, linking substrate choice to repair capacity [64]. In contrast, non-ischemic etiologies typically impose chronic mitochondrial stress rather than an acute energetic collapse, as seen in viral myocarditis, where viral replication and innate immune signaling disrupt mitochondrial dynamics, bioenergetics, and antiviral response pathways, fostering progression to dilated phenotypes [65,66].

Human dilated cardiomyopathy (DCM) myocardium displays fragmented mitochondrial networks, cristae disorganization, and reduced respiratory chain capacity, indicating profound ultrastructural and functional remodeling [67]. Mitochondrial cristae defects in failing human hearts correlate with impaired OXPHOS efficiency and heightened susceptibility to cell death, placing inner-membrane architecture at the center of disease mechanisms [68]. In parallel, lipotoxicity from inefficient FAO and toxic lipid intermediates further impairs mitochondrial function, amplifies oxidative stress, and worsens ventricular dilation and failure [69]. Together, architectural remodeling and metabolic derangements converge on energetic failure in DCM [67–69].

In hypertrophic cardiomyopathy (HCM), emerging evidence underscores mitochondrial dysfunction as a key contributor to disease pathology. Mutations in sarcomeric proteins increase the ATP cost of contraction, termed "mechano-energetic uncoupling", creating an intrinsic energetic inefficiency even at rest [70–73]. Phosphorus magnetic resonance spectroscopy has consistently revealed reduced phosphocreatine/ATP ratios in HCM patients, which correlate with diastolic impairment and heightened arrhythmic risk [72,73]. Recent analyses of explanted HCM hearts have revealed structural disorganization of mitochondrial networks and uncoupling of respiration from NADH supply. Importantly, NADH-driven respiration could be rescued ex vivo, underscoring both the centrality and therapeutic tractability of mitochondrial dysfunction in HCM [74]. These observations are supported by studies, such as the feline HCM model by Christiansen et al., which directly demonstrated that mitochondrial OXPHOS capacity is impaired, accompanied by increased oxidative stress, establishing energy depletion as a fundamental aspect of HCM pathophysiology that likely contributes to diastolic dysfunction, arrhythmia risk, and progression to heart failure [75].

Restrictive cardiomyopathy (RCM) often arises from infiltrative disorders—most prominently cardiac amyloidosis and storage diseases such as Fabry, where diastolic dysfunction occurs with relatively preserved systolic function early on [76]. In amyloidosis, mitochondrial dysfunction is typically secondary to proteotoxic stress from amyloid species. Cardiotropic light chains and transthyretin (TTR) assemblies trigger oxidative injury and disrupt Ca^{2+} handling, impairing respiration and contractile performance [77–79]. Experimental work shows that amyloidogenic light chains directly provoke oxidant stress, contractile dysfunction, and apoptosis in cardiomyocytes via a non-canonical p38α MAPK pathway, supporting a causal link between proteotoxic aggregates and mitochondrial/energetic failure [78]. Complementary studies demonstrate metabolic (bioenergetic) dysfunction in human cardiomyocytes exposed to light-chain fibrils [80]. In TTR amyloidosis, cardiomyocytes exposed to TTR fibrils exhibit decreased force production and prolonged Ca^{2+} transients, indicating electrophysiologic–energetic coupling defects [77], while recent reviews synthesize evidence that oxidative stress, impaired mitochondrial function, and Ca^{2+} dysregulation are central to ATTR-CM pathogenesis [81]. Beyond amyloidosis, Fabry disease illustrates how lysosomal sphingolipid accumulation perturbs autophagy and secondarily injures mitochondria, contributing to myocardial remodeling and conduction/arrhythmic vulnerability [82,83].

Primary mitochondrial cardiomyopathies establish a direct causal link between mitochondrial dysfunction and cardiac disease [29]. The common mtDNA mutation m.3243A>G, classically associated with MELAS, can manifest as either hypertrophic or dilated cardiomyopathy depending on heteroplasmy levels and tissue distribution [84]. Functional studies using patient-derived iPSC-cardiomyocytes with high m.3243A>G loads demonstrate impaired mitochondrial respiration in all cases, but only cells from patients with clinical cardiomyopathy exhibited reduced ATP levels and disrupted calcium signaling, emphasizing respiratory chain dysfunction and metabolic heterogeneity as direct effects of the mutation [85]. Additional mtDNA variants continue to be identified; for instance, the m.4300A>G mutation in MT-TI (tRNA Ile) has been reported in hypertrophic cardiomyopathy [86]. Nuclear gene defects in mtDNA maintenance, such as mutations in TK2 and POLG, can result in mtDNA depletion or multiple deletions, undermining respiratory chain integrity and precipitating severe early-onset cardiomyopathy [87]. Barth syndrome, resulting from TAFAZZIN mutations and defective cardiolipin remodeling, destabilizes respiratory supercomplexes and disrupts cristae architecture, producing dilated or non-compaction cardiomyopathy with high risk of heart failure [88].

Collectively, ischemic cardiomyopathy represents an acute mitochondrial collapse driven by Ca^{2+} overload, ROS bursts, and mPTP opening [62–64], whereas DCM is dominated by chronic cristae remodeling and lipotoxic stress [67–69]. HCM is marked by mechano-energetic uncoupling, impaired PCr/ATP transfer, and OXPHOS inefficiency [70–73], while RCM largely reflects secondary mitochondrial injury imposed by proteotoxic amyloid species or lysosomal storage pathology [76–80]. Genetic mitochondrial cardiomyopathies firmly establish causality, as mtDNA point mutations (e.g., m.3243A > G), defects in mtDNA maintenance (TK2, POLG), and TAFAZZIN-related cardiolipin remodeling produce cardiomyopathy phenotypes, highlighting opportunities for precision therapies such as metabolic modulation, nucleoside supplementation, or cardiolipin stabilization [29,84–88].

In addition to the traditional energetic and structural changes observed in various cardiomyopathy subtypes, emerging research underscores the importance of cytoskeletal interactions and mitochondrial diversity in influencing cardiomyocyte performance and susceptibility. Incorporating these elements yields a fuller picture of mitochondrial disease mechanisms and could uncover novel avenues for precision therapies.

4. Mitochondrial-Driven Cell Death Pathways in Heart Failure Progression

Mitochondria play a central role in regulated cardiomyocyte death, orchestrating multiple cell-death modalities and therefore contributing to heart failure progression [1,41]. In failing myocardium, both apoptotic and non-apoptotic pathways converge on mitochondrial dysfunction, resulting in energy collapse and loss of viable cardiomyocytes [41,89]. Dysregulated mitochondria also act as initiators of inflammatory signaling via the release of damage-associated molecular patterns (DAMPs), amplifying cardiomyocyte injury [90,91].

Cardiomyocyte apoptosis in heart failure is predominantly executed through the intrinsic, mitochondria-dependent pathway. Pro-apoptotic BAX and BAK oligomerize in the outer mitochondrial membrane, inducing permeabilization and cytochrome c release [92–94]. In the cytosol, cytochrome c interacts with Apaf-1 and ATP to assemble the apoptosome, activating initiator caspase-9, followed by caspase-3, ultimately driving controlled cellular disintegration [95]. SMAC/DIABLO release further neutralizes IAPs (inhibitors of apoptosis proteins), amplifying caspase activation [92]. In experimental rodent models and human failing hearts, increased BAX expression, mitochondrial cristae remodeling, and elevated caspase activity strongly correlate with myocardial dysfunction [96,97]. Under acute stress (e.g., ischemia, reperfusion, or severe oxidative stress), opening of the mPTP collapses mitochondrial membrane potential, halting ATP synthesis and triggering osmotic swelling [98]. Outer membrane rupture releases pro-death-factors and ROS, leading to regulated necrotic cell death [99]. In parallel, necroptosis—mediated by RIPK1, RIPK3, and MLKL, has emerged as a genetically programmed necrotic pathway that frequently intersects with mitochondrial dysfunction [100]. In models of ischemia–reperfusion injury, inhibition of Cyclophilin D (an mPTP regulator) reduces necrosis, while blockade of RIPK3/MLKL attenuates infarct size and preserves cardiac function [101]. Thus, mitochondria-driven necrotic signaling critically contributes to cardiomyocyte loss. Importantly, apoptotic and necrotic pathways do not operate in isolation within cardiomyocytes. Emerging evidence suggests that they are interconnected through mitochondrial dysfunction, particularly via the sub-threshold opening of the mPTP. When the mPTP is low or partially open, it can trigger apoptotic signaling if ATP levels remain sufficient. However, if ATP levels fall severely or caspase activity is inhibited, the same stimulus may result in necrotic cell death instead [102]. This continuum between apoptosis and necrosis is described by the concept of "necroptosis", where the fate of the cell hinges on cellular energy status and caspase activation. Meanwhile, mitochondrial dysfunction leads to the release of DAMPs into the cytosol and extracellular space. These include mtDNA fragments, ROS, oxidized phospholipids such as cardiolipin, and mitochondrial peptides. Once released, these molecules engage innate immune sensors, such as TLR9, NLRP3 inflammasome, cGAS-STING, and TLR4, thus amplifying local inflammatory cascades [103,104]. Persistent mitochondrial-driven cell loss undermines myocardial integrity and contractility, establishing a vicious cycle that ultimately contributes to heart failure progression. Therapeutic strategies targeting components of these pathways hold promise: inhibiting BAX/BAK activation to prevent cytochrome c release, Cyclophilin D antagonists to block mPTP opening, and RIPK1/RIPK3/MLKL inhibitors to reduce necroptosis may preserve cardiomyocyte viability. Such approaches, by attenuating both energetic collapse and inflammatory amplification, could slow or reverse adverse remodeling in heart failure.

5. Therapeutic Strategies Targeting Mitochondrial Dysfunction

Mitochondrial dysfunction plays a central role in the pathogenesis of cardiomyopathy and heart failure, making it an attractive therapeutic target. Several strategies, ranging from pharmacologic interventions to metabolic and lifestyle modulators, have been investigated to restore mitochondrial integrity and function.

Oxidative stress is a hallmark of mitochondrial impairment in failing hearts. Antioxidant supplementation has been extensively tested, with coenzyme Q10 (CoQ10) receiving particular attention. In the randomized, double-blind Q-SYMBIO trial of 420 patients with chronic heart failure, long-term CoQ10 supplementation significantly reduced major adverse cardiovascular events and mortality compared with placebo, suggesting that improving electron transport and reducing oxidative stress can translate into clinical benefit [105].

Another pharmacological strategy involves inhibition of the mPTP, whose opening during ischemia–reperfusion contributes to cell death. Early preclinical studies suggested the protective effects of cyclosporine A [106]. However, large clinical trials failed to confirm these findings. The CIRCUS trial (n = 970 anterior STEMI patients undergoing PCI) showed no reduction in all-cause mortality, heart failure hospitalization, or adverse remodeling at one year [107]. Similarly, the CYCLE trial (n = 410 STEMI patients) found no benefit of cyclosporine on reperfusion success, infarct size, or left ventricular recovery [108]. Together, these trials underscore the challenges of translating mitochondrial-targeted cardioprotection into clinical efficacy.

Emerging therapeutic strategies aim to correct mitochondrial dysfunction directly. A central focus has been PGC-1α, the master regulator of mitochondrial biogenesis and oxidative metabolism. Knockout studies revealed that PGC-1α deficiency reduces oxidative phosphorylation gene expression, diminishes mitochondrial activity, and impairs the heart's ability to meet energetic demand [109]. Targeting the broader PGC-1 family (PGC-1α, PGC-1β, PRC) is also promising, as these coactivators coordinate OXPHOS, fatty acid oxidation, and antioxidant defense, and are increasingly recognized as candidate targets for cardiac metabolic therapy [110].

Another innovative approach is mitochondrial transplantation. Preclinical ischemia–reperfusion models show that delivery of healthy mitochondria into injured myocardium enhances ATP production, reduces ROS, and improves cardiomyocyte survival [111]. Systematic reviews confirm protective effects across animal models, though delivery methods and dosing vary [111]. Early human feasibility is encouraging: in pediatric patients with ischemic cardiomyopathy, autologous mitochondrial transplantation improved myocardial bioenergetics and enabled successful weaning from mechanical support in several cases [112]. Collectively, gene-based activation of mitochondrial biogenesis and mitochondrial transfer therapies provide promising avenues to restore cardiac energetics. While still in preclinical and early clinical stages, these strategies underscore the feasibility of directly targeting mitochondria to treat cardiomyopathy and heart failure.

Lifestyle and metabolic interventions are crucial strategies for combating mitochondrial dysfunction in heart failure. Exercise training is the most established non-pharmacological intervention. In a randomized trial, aerobic interval training was superior to moderate continuous training, producing larger improvements in VO2 peak, reverse remodeling, endothelial function, and skeletal muscle mitochondrial capacity in patients with post-infarction heart failure [113]. Caloric restriction mimetics such as spermidine also protect the heart through mitochondrial pathways. Eisenberg et al. demonstrated that oral spermidine reduced cardiac hypertrophy, preserved diastolic function, and enhanced mitophagy and respiration in mice. In humans, a higher dietary intake correlated with reduced blood pressure and cardiovascular events [114]. The metabolic modulation of substrate utilization has been clinically validated. In randomized trials, trimetazidine improved ejection fraction, NYHA class, and quality of life in systolic heart failure. Metabolic data showed reduced resting energy expenditure, consistent with a shift toward glucose oxidation [115]. The most transformative pharmacologic advances are the sodium–glucose cotransporter 2 (SGLT2) inhibitors. In the DAPA-HF trial, dapagliflozin reduced the risk of cardiovascular death or worsening heart failure in patients with HFrEF, independent of diabetes status [116]. The EMPEROR-Reduced trial confirmed that empagliflozin lowered this composite endpoint by 25% and slowed renal decline [117]. Benefits extended to preserved EF: the EMPEROR-Preserved [118] and DELIVER [119] trials both demonstrated significant reductions in cardiovascular death or heart failure hospitalization across HFpEF and HFmrEF populations. Mechanistically, SGLT2 inhibitors modulate myocardial energetics by promoting ketone body utilization, improving mitochondrial respiration, and reducing oxidative stress, thereby enhancing the efficiency of the failing heart [120,121]. A summary of pharmacological, genetic, lifestyle, and metabolic interventions targeting mitochondrial dysfunction is presented in Table 1.

Table 1. Summary of therapeutic strategies targeting mitochondrial dysfunction in heart failure.

Therapeutic Strategy	Mechanism/Rationale	Evidence/Key Findings	References
Antioxidant supplementation (Coenzyme Q10)	Improves electron transport, reduces oxidative stress	Q-SYMBIO trial (420 patients): reduced major CV events and mortality	[105]
mPTP inhibition (Cyclosporine A)	Prevents mitochondrial permeability transition pore opening during reperfusion	Preclinical benefit; CIRCUS and CYCLE trials: no clinical efficacy in STEMI patients	[106–108]
PGC-1 family activation (PGC-1α, PGC-1β, PRC)	Enhances mitochondrial biogenesis, OXPHOS, fatty acid oxidation, antioxidant defense	Knockout studies: PGC-1α deficiency reduces mitochondrial activity and cardiac energy supply	[109,110]
Mitochondrial transplantation	Delivery of healthy mitochondria to restore ATP production and reduce ROS	Preclinical: improved cardiomyocyte survival; Human feasibility: improved myocardial bioenergetics in pediatric ischemic cardiomyopathy	[111,112]
Exercise training	Enhances mitochondrial capacity, reverse remodeling, endothelial function	RCT: Aerobic interval training > moderate continuous training in post-infarction HF	[113]
Caloric restriction mimetic (Spermidine)	Promotes mitophagy, mitochondrial respiration, and cardioprotection	Animal studies: reduced hypertrophy, preserved diastolic function; Human data: higher intake → lower BP, fewer CV events	[114]
Metabolic modulation (Trimetazidine)	Shifts substrate utilization toward glucose oxidation	RCTs: improved EF, NYHA class, QoL; reduced resting energy expenditure	[115]
SGLT2 inhibitors (Dapagliflozin, Empagliflozin)	Enhance ketone utilization, mitochondrial respiration, and reduce oxidative stress	DAPA-HF, EMPEROR-Reduced, EMPEROR-Preserved, DELIVER trials: reduced CV death and HF hospitalization across HFrEF, HFpEF, HFmrEF	[116–121]

Taken together, exercise, caloric restriction mimetics, substrate modulators, and SGLT2 inhibitors converge on restoring mitochondrial health, positioning mitochondria as both a pathophysiologic driver and a therapeutic target in heart failure. Targeting mitochondrial dysfunction through pharmacologic, genetic, lifestyle, and metabolic interventions highlights mitochondria as both a central driver of cardiomyopathy and a promising therapeutic target in heart failure.

6. Conclusions and Future Perspectives

Despite the growing body of evidence linking mitochondrial dysfunction to the pathophysiology of cardiomyopathy and heart failure, translating this knowledge into effective clinical interventions remains a formidable challenge. Mitochondria are intricately integrated into nearly every facet of cardiomyocyte biology, making them both a promising and difficult therapeutic target. Their double-membrane structure, dynamic behavior, and complex interaction with nuclear signaling and cellular metabolism complicate drug delivery, specificity, and efficacy.

One of the main challenges in targeting mitochondria is achieving therapeutic precision without disrupting physiological mitochondrial functions [122]. Unlike conventional drug targets such as receptors or enzymes, mitochondria are not static structures; they constantly undergo structural changes, move within cells, and vary in quantity and function between tissues and even within cardiomyocytes. This plasticity demands tailored therapeutic approaches that can adapt to both temporal and spatial mitochondrial dynamics [123]. Furthermore, mitochondrial-targeted therapies must overcome pharmacokinetic obstacles such as crossing the plasma and mitochondrial membranes. Lipophilic cations (e.g., triphenylphosphonium-based compounds) have been used to deliver bioactive molecules directly into mitochondria, but these approaches may lead to off-target effects and accumulation-related toxicity [124]. Additionally, while numerous compounds (e.g., mPTP inhibitors, antioxidants, electron transport chain modulators) have shown promise in preclinical models, many have failed to demonstrate consistent efficacy in large-scale clinical trials due to differences in mitochondrial phenotype across patient populations and disease stages [125].

Another crucial barrier to clinical translation is the lack of reliable, validated biomarkers of mitochondrial health in cardiovascular disease. The ability to monitor mitochondrial function dynamically and non-invasively is essential for both diagnosis and therapeutic stratification [126]. Currently, most assessments rely on indirect markers such as circulating lactate levels, oxidative stress indicators, or biopsy-derived electron microscopy and respirometry data [127]. However, emerging candidates such as circulating mitochondrial DNA (mtDNA), cardiolipin oxidation products, and mitokines (e.g., FGF21, GDF15) offer new promise. Circulating mtDNA, released during mitochondrial damage or apoptosis, has been associated with inflammation, endothelial dysfunction, and adverse outcomes in heart failure [128]. Similarly, elevated levels of oxidized cardiolipin have been correlated with mitochondrial membrane destabilization and activation of the NLRP3 inflammasome [129]. Advanced imaging modalities, including PET tracers targeting mitochondrial membrane potential (e.g., 18F-TPP+) and hyperpolarized magnetic resonance spectroscopy [130,131], are also being explored as non-invasive techniques to assess mitochondrial metabolism and redox status in vivo. The future clinical utility of these tools lies not only in diagnosis but also in predicting therapeutic responsiveness and disease trajectory.

As the heterogeneity of mitochondrial dysfunction across different cardiomyopathies becomes more apparent, a one-size-fits-all treatment strategy is unlikely to be effective [30]. Instead, the future lies in personalized, phenotype-driven approaches [125]. Genomic, metabolomic, and proteomic profiling could help stratify patients based on specific mitochondrial signatures, such as respiratory chain defects, mitophagy insufficiency, or oxidative stress burden, and guide individualized treatment regimens [132]. Several ongoing clinical trials are exploring mitochondria-targeted therapies in cardiovascular diseases. For example, elamipretide, a peptide that targets cardiolipin to stabilize mitochondrial membranes and improve electron transport efficiency, is under investigation for heart failure with reduced ejection fraction [133]. Another promising agent, bendavia, aims to modulate mitochondrial bioenergetics and reduce ROS production [134]. Trials involving NAD^+ precursors (e.g., nicotinamideriboside) are also underway, based on their role in mitochondrial biogenesis and sirtuin inactivation [135]. In parallel, lifestyle and metabolic interventions, such as caloric restriction, intermittent fasting, and exercise, are being repurposed and studied for their mitochondria-enhancing properties. These strategies modulate mitochondrial dynamics and may offer low-cost adjuncts to pharmacological therapies [136,137].

As our understanding of mitochondrial biology in cardiac pathology deepens, the future of heart failure management may shift from conventional hemodynamic support to bioenergetic optimization. Yet, to realize this potential, several critical steps must be taken: refining delivery strategies, developing specific and safe modulators of mitochondrial pathways, and establishing robust biomarkers to monitor mitochondrial function longitudinally. Ultimately, mitochondria are not merely passive energy suppliers but central regulators of cardiomyocyte fate and immune–metabolic signaling. Their dysfunction is not a byproduct of disease; it is a driver. Therefore, targeting mitochondrial dysfunction represents not just a therapeutic opportunity, but a conceptual shift in how we understand and treat heart failure.

Continued integration of basic science, translational research, and precision medicine approaches will be essential to transform this opportunity into a clinical reality. If successful, this mitochondria-focused paradigm may unlock new avenues for cardioprotection, improve quality of life, and reduce mortality for millions of patients living with cardiomyopathy and heart failure.

Author Contributions: Conceptualization, N.P., P.T., M.M., M.K., K.V., Z.P. and J.B.; methodology, N.P., P.T., M.M., M.K., K.V., Z.P. and J.B.; software, N.P., P.T., M.M., M.K., K.V., Z.P. and J.B.; validation, N.P., P.T., M.M., M.K., K.V., Z.P. and J.B.; formal analysis, N.P., P.T., M.M., M.K., K.V., Z.P. and J.B.; investigation, N.P., P.T., M.M., M.K., K.V., Z.P. and J.B.; resources, N.P., P.T., M.M., M.K., K.V., Z.P. and J.B.; data curation, N.P., P.T., M.M., M.K., K.V., Z.P. and J.B.; writing—original draft preparation, N.P., P.T., M.M., M.K., K.V., Z.P. and J.B.; writing—review and editing, N.P., P.T., M.M., M.K., K.V., Z.P. and J.B.; visualization, N.P. and P.T.; supervision, K.V. and J.B.; project administration, J.B.; funding acquisition, J.B. All authors have read and agreed to the published version of the manuscript.

Funding: This research received no external funding.

Institutional Review Board Statement: Not applicable.

Informed Consent Statement: Not applicable.

Data Availability Statement: Not applicable.

Acknowledgments: The authors have reviewed and edited the output and take full responsibility for the content of this publication.

Conflicts of Interest: The authors declare no conflicts of interest.

References

1. Nguyen, B.Y.; Ruiz-Velasco, A.; Bui, T.; Collins, L.; Wang, X.; Liu, W. Mitochondrial Function in the Heart: The Insight into Mechanisms and Therapeutic Potentials. *Br. J. Pharmacol.* **2019**, *176*, 4302–4318. [CrossRef]
2. Lopaschuk, G.D.; Karwi, Q.G.; Tian, R.; Wende, A.R.; Abel, E.D. Cardiac Energy Metabolism in Heart Failure. *Circ. Res.* **2021**, *128*, 1487–1513. [CrossRef] [PubMed]
3. Doenst, T.; Nguyen, T.D.; Abel, E.D. Cardiac Metabolism in Heart Failure: Implications beyond Atp Production. *Circ. Res.* **2013**, *113*, 709–724. [CrossRef]
4. Sun, Q.; Karwi, Q.G.; Wong, N.; Lopaschuk, G.D. Advances in Myocardial Energy Metabolism: Metabolic Remodelling in Heart Failure and Beyond. *Cardiovasc. Res.* **2024**, *120*, 1996–2016. [CrossRef] [PubMed]
5. Vega, R.B.; Kelly, D.P. Cardiac Nuclear Receptors: Architects of Mitochondrial Structure and Function. *J. Clin. Investig.* **2017**, *127*, 1155–1164. [CrossRef]
6. Pavlović, N.; Križanac, M.; Kumrić, M.; Vukojević, K.; Božić, J. Mitochondrial Dysfunction: The Silent Catalyst of Kidney Disease Progression. *Cells* **2025**, *14*, 794. [CrossRef]
7. Bayeva, M.; Gheorghiade, M.; Ardehali, H. Mitochondria as a Therapeutic Target in Heart Failure. *J. Am. Coll. Cardiol.* **2013**, *61*, 599–610. [CrossRef]
8. Schwemmlein, J.; Maack, C.; Bertero, E. Mitochondria as Therapeutic Targets in Heart Failure. *Curr. Heart Fail. Rep.* **2022**, *19*, 27–37. [CrossRef] [PubMed]
9. Kumar, V.K.; Lackey, A.; Snyder, J.; Karhadkar, S.; Rao, A.D.; DiCarlo, A.; Sato, P.Y. Mitochondrial Membrane Intracellular Communication in Healthy and Diseased Myocardium. *Front. Cell Dev. Biol.* **2020**, *8*, 609241. [CrossRef]
10. Tokuyama, T.; Yanagi, S. Role of Mitochondrial Dynamics in Heart Diseases. *Genes* **2023**, *14*, 1876. [CrossRef]
11. Federico, M.; De la Fuente, S.; Palomeque, J.; Sheu, S.S. The Role of Mitochondria in Metabolic Disease: A Special Emphasis on Heart Dysfunction. *J. Physiol.* **2021**, *599*, 3477–3493. [CrossRef]
12. Nan, J.; Zhu, W.; Rahman, M.S.; Liu, M.; Li, D.; Su, S.; Zhang, N.; Hu, X.; Yu, H.; Gupta, M.P.; et al. Molecular Regulation of Mitochondrial Dynamics in Cardiac Disease. *Biochim. Biophys. Acta Mol. Cell Res.* **2017**, *1864*, 1260–1273. [CrossRef]
13. Morciano, G.; Vitto, V.A.M.; Bouhamida, E.; Giorgi, C.; Pinton, P. Mitochondrial Bioenergetics and Dynamism in the Failing Heart. *Life* **2021**, *11*, 436. [CrossRef] [PubMed]

14. Stanley, W.C.; Recchia, F.A.; Lopaschuk, G.D. Myocardial Substrate Metabolism in the Normal and Failing Heart. *Physiol. Rev.* **2005**, *85*, 1093–1129. [CrossRef] [PubMed]
15. Lopaschuk, G.D.; Ussher, J.R.; Folmes, C.D.L.; Jaswal, J.S.; Stanley, W.C. Myocardial Fatty Acid Metabolism in Health and Disease. *Physiol. Rev.* **2010**, *90*, 207–258. [CrossRef]
16. Tuomainen, T.; Tavi, P. The Role of Cardiac Energy Metabolism in Cardiac Hypertrophy and Failure. *Exp. Cell Res.* **2017**, *360*, 12–18. [CrossRef] [PubMed]
17. Brown, D.A.; O'Rourke, B. Cardiac Mitochondria and Arrhythmias. *Cardiovasc. Res.* **2010**, *88*, 241–249. [CrossRef]
18. Rich, P.R. The Molecular Machinery of Keilin's Respiratory Chain. *Proc. Biochem. Soc. Trans.* **2003**, *31*, 1095–1105. [CrossRef]
19. Murphy, E.; Ardehali, H.; Balaban, R.S.; DiLisa, F.; Dorn, G.W.; Kitsis, R.N.; Otsu, K.; Ping, P.; Rizzuto, R.; Sack, M.N.; et al. Mitochondrial Function, Biology, and Role in Disease. *Circ. Res.* **2016**, *118*, 1960–1991. [CrossRef]
20. Campos, J.C.; Bozi, L.H.M.; Bechara, L.R.G.; Lima, V.M.; Ferreira, J.C.B. Mitochondrial Quality Control in Cardiac Diseases. *Front. Physiol.* **2016**, *7*, 479. [CrossRef]
21. Youle, R.J.; Van Der Bliek, A.M. Mitochondrial Fission, Fusion, and Stress. *Science* **2012**, *337*, 1062–1065. [CrossRef]
22. Puigserver, P.; Spiegelman, B.M. Peroxisome Proliferator-Activated Receptor-γ Coactivator 1α (PGC-1α): Transcriptional Coactivator and Metabolic Regulator. *Endocr. Rev.* **2003**, *24*, 78–90. [CrossRef]
23. Ashrafi, G.; Schwarz, T.L. The Pathways of Mitophagy for Quality Control and Clearance of Mitochondria. *Cell Death Differ.* **2013**, *20*, 31–42. [CrossRef]
24. Kubli, D.A.; Gustafsson, Å.B. Mitochondria and Mitophagy: The Yin and Yang of Cell Death Control. *Circ. Res.* **2012**, *111*, 1208–1221. [CrossRef]
25. Dorn, G.W. Mitochondrial Dynamics in Heart Disease. *Biochim. Biophys. Acta Mol. Cell Res.* **2013**, *1833*, 233–241. [CrossRef] [PubMed]
26. Elliott, P.; Andersson, B.; Arbustini, E.; Bilinska, Z.; Cecchi, F.; Charron, P.; Dubourg, O.; Kühl, U.; Maisch, B.; McKenna, W.J.; et al. Classification of the Cardiomyopathies: A Position Statement from the European Society of Cardiology Working Group on Myocardial and Pericardial Diseases. *Eur. Heart J.* **2008**, *29*, 270–276. [CrossRef]
27. Maron, B.J.; Towbin, J.A.; Thiene, G.; Antzelevitch, C.; Corrado, D.; Arnett, D.; Moss, A.J.; Seidman, C.E.; Young, J.B. Contemporary Definitions and Classification of the Cardiomyopathies. *Circulation* **2006**, *113*, 1807–1816. [CrossRef] [PubMed]
28. St-Pierre, G.; Steinberg, C.; Dubois, M.; Sénéchal, M. What the Cardiologist Should Know About Mitochondrial Cardiomyopathy? *Can. J. Cardiol.* **2019**, *35*, 221–224. [CrossRef] [PubMed]
29. El-Hattab, A.W.; Scaglia, F. Mitochondrial Cardiomyopathies. *Front. Cardiovasc. Med.* **2016**, *3*, 25. [CrossRef]
30. Neubauer, S. The Failing Heart—An Engine Out of Fuel. *N. Engl. J. Med.* **2007**, *356*, 1140–1151. [CrossRef]
31. Paillard, M.; Abdellatif, M.; Andreadou, I.; Bär, C.; Bertrand, L.; Brundel, B.J.J.M.; Chiva-Blanch, G.; Davidson, S.M.; Dawson, D.; Di Lisa, F.; et al. Mitochondrial Targets in Ischaemic Heart Disease and Heart Failure, and Their Potential for a More Efficient Clinical Translation. A Scientific Statement of the ESC Working Group on Cellular Biology of the Heart and the ESC Working Group on Myocardial Function. *Eur. J. Heart Fail.* **2025**, *27*, 1720–1736. [CrossRef]
32. Luan, Y.; Luan, Y.; Feng, Q.; Chen, X.; Ren, K.D.; Yang, Y. Emerging Role of Mitophagy in the Heart: Therapeutic Potentials to Modulate Mitophagy in Cardiac Diseases. *Oxid. Med. Cell. Longev.* **2021**, *2021*, 3259963. [CrossRef]
33. Carinci, M.; Vezzani, B.; Patergnani, S.; Ludewig, P.; Lessmann, K.; Magnus, T.; Casetta, I.; Pugliatti, M.; Pinton, P.; Giorgi, C. Different Roles of Mitochondria in Cell Death and Inflammation: Focusing on Mitochondrial Quality Control in Ischemic Stroke and Reperfusion. *Biomedicines* **2021**, *9*, 169. [CrossRef]
34. Maack, C.; Dudek, J.; Bertero, E.; Tampakakis, E.; Vernon, H.J. Mitochondrial Cardiomyopathies: Pathogenesis, Diagnosis, and Treatment. *Eur. Heart J.* **2025**, *46*, 4060–4075. [CrossRef]
35. Yang, J.; Guo, Q.; Feng, X.; Liu, Y.; Zhou, Y. Mitochondrial Dysfunction in Cardiovascular Diseases: Potential Targets for Treatment. *Front. Cell Dev. Biol.* **2022**, *10*, 841523. [CrossRef]
36. Tsutsui, H.; Kinugawa, S.; Matsushima, S. Oxidative Stress and Mitochondrial DNA Damage in Heart Failure. *Circ. J.* **2008**, *72*, A31–A37. [CrossRef]
37. Dubois-Deruy, E.; Peugnet, V.; Turkieh, A.; Pinet, F. Oxidative Stress in Cardiovascular Diseases. *Antioxidants* **2020**, *9*, 864. [CrossRef]
38. Werbner, B.; Tavakoli-Rouzbehani, O.M.; Fatahian, A.N.; Boudina, S. The Dynamic Interplay between Cardiac Mitochondrial Health and Myocardial Structural Remodeling in Metabolic Heart Disease, Aging, and Heart Failure. *J. Cardiovasc. Aging* **2023**, *3*, 9. [CrossRef] [PubMed]
39. Marquez, J.; Lee, S.R.; Kim, N.; Han, J. Rescue of Heart Failure by Mitochondrial Recovery. *Int. Neurourol. J.* **2016**, *20*, 5–12. [CrossRef] [PubMed]
40. Peoples, J.N.; Saraf, A.; Ghazal, N.; Pham, T.T.; Kwong, J.Q. Mitochondrial Dysfunction and Oxidative Stress in Heart Disease. *Exp. Mol. Med.* **2019**, *51*, 1–13. [CrossRef] [PubMed]

41. Zhou, B.; Tian, R. Mitochondrial Dysfunction in Pathophysiology of Heart Failure. *J. Clin. Investig.* **2018**, *128*, 3716–3726. [CrossRef]
42. Bhullar, S.K.; Dhalla, N.S. Status of Mitochondrial Oxidative Phosphorylation during the Development of Heart Failure. *Antioxidants* **2023**, *12*, 1941. [CrossRef]
43. ten Hove, M.; Neubauer, S. MR Spectroscopy in Heart Failure—Clinical and Experimental Findings. *Heart Fail. Rev.* **2007**, *12*, 48–57. [CrossRef]
44. Hao, Y.D.; Zhao, Y.X.; Yang, S.W.; Zhou, Y.J. High-Energy Phosphates and Ischemic Heart Disease: From Bench to Bedside. *Front. Cardiovasc. Med.* **2021**, *8*, 675608. [CrossRef] [PubMed]
45. Hamilton, D.J. The Failing Heart: Energy Supply, Processing, and Transfer. *Methodist DeBakey Cardiovasc. J.* **2017**, *13*, 3. [CrossRef] [PubMed]
46. Dridi, H.; Santulli, G.; Bahlouli, L.; Miotto, M.C.; Weninger, G.; Marks, A.R. Mitochondrial Calcium Overload Plays a Causal Role in Oxidative Stress in the Failing Heart. *Biomolecules* **2023**, *13*, 1409. [CrossRef] [PubMed]
47. Johnson, E.; Albakri, J.S.; Allemailem, K.S.; Sultan, A.; Alwanian, W.M.; Alrumaihi, F.; Almansour, N.M.; Aldakheel, F.M.; Khalil, F.M.A.; Abduallah, A.M.; et al. Mitochondrial Dysfunction and Calcium Homeostasis in Heart Failure: Exploring the Interplay between Oxidative Stress and Cardiac Remodeling for Future Therapeutic Innovations. *Curr. Probl. Cardiol.* **2025**, *50*, 102968. [CrossRef]
48. Manolis, A.S.; Manolis, A.A.; Manolis, T.A.; Apostolaki, N.E.; Apostolopoulos, E.J.; Melita, H.; Katsiki, N. Mitochondrial Dysfunction in Cardiovascular Disease: Current Status of Translational Research/Clinical and Therapeutic Implications. *Med. Res. Rev.* **2021**, *41*, 275–313. [CrossRef]
49. Santulli, G.; Xie, W.; Reiken, S.R.; Marks, A.R. Mitochondrial Calcium Overload Is a Key Determinant in Heart Failure. *Proc. Natl. Acad. Sci. USA* **2015**, *112*, 11389–11394. [CrossRef]
50. Popoiu, T.-A.; Maack, C.; Bertero, E. Mitochondrial Calcium Signaling and Redox Homeostasis in Cardiac Health and Disease. *Front. Mol. Med.* **2023**, *3*, 1235188. [CrossRef]
51. Xu, H.X.; Cui, S.M.; Zhang, Y.M.; Ren, J. Mitochondrial Ca^{2+} Regulation in the Etiology of Heart Failure: Physiological and Pathophysiological Implications. *Acta Pharmacol. Sin.* **2020**, *41*, 1301–1309. [CrossRef]
52. Quiles, J.M.; Gustafsson, Å.B. The Role of Mitochondrial Fission in Cardiovascular Health and Disease. *Nat. Rev. Cardiol.* **2022**, *19*, 723–736. [CrossRef]
53. Svaguša, T.; Martinić, M.; Martinić, M.; Kovačević, L.; Šepac, A.; Miličić, D.; Bulum, J.; Starčević, B.; Sirotković-Skerlev, M.; Seiwerth, F.; et al. Mitochondrial Unfolded Protein Response, Mitophagy and Other Mitochondrial Quality Control Mechanisms in Heart Disease and Aged Heart. *Croat. Med. J.* **2020**, *61*, 126–138. [CrossRef] [PubMed]
54. Wang, Y.; Ren, T.; Li, C.; Wu, Q.; Liu, J.; Guan, X.; Chang, X.; Liu, Z.; Liu, R. Mechanisms Involved in the Regulation of Mitochondrial Quality Control by PGAM5 in Heart Failure. *Cell Stress. Chaperones* **2024**, *29*, 510–518. [CrossRef]
55. Javadifar, A.; Tahani, M.; Khayat, S.; Nasab, S.R.; Karav, S.; Kesharwani, P.; Sahebkar, A. Targeting Mitophagy in the Heart: Exploring the Therapeutic Potential of MicroRNAs. *Mech. Ageing Dev.* **2025**, *226*, 112082. [CrossRef]
56. Liu, Y.; Wang, Y.; Bi, Y.; Zhao, Z.; Wang, S.; Lin, S.; Yang, Z.; Wang, X.; Mao, J. Emerging Role of Mitophagy in Heart Failure: From Molecular Mechanism to Targeted Therapy. *Cell Cycle* **2023**, *22*, 906–918. [CrossRef] [PubMed]
57. Shires, S.E.; Gustafsson, Å.B. Mitophagy and Heart Failure. *J. Mol. Med.* **2015**, *93*, 253–262. [CrossRef] [PubMed]
58. Lin, J.; Duan, J.; Wang, Q.; Xu, S.; Zhou, S.; Yao, K. Mitochondrial Dynamics and Mitophagy in Cardiometabolic Disease. *Front. Cardiovasc. Med.* **2022**, *9*, 917135. [CrossRef]
59. Vásquez-Trincado, C.; García-Carvajal, I.; Pennanen, C.; Parra, V.; Hill, J.A.; Rothermel, B.A.; Lavandero, S. Mitochondrial Dynamics, Mitophagy and Cardiovascular Disease. *J. Physiol.* **2016**, *594*, 509–525. [CrossRef]
60. Qiu, Z.; Wei, Y.; Song, Q.; Du, B.; Wang, H.; Chu, Y.; Hu, Y. The Role of Myocardial Mitochondrial Quality Control in Heart Failure. *Front. Pharmacol.* **2019**, *10*, 1404. [CrossRef]
61. Ng, S.M.; Neubauer, S.; Rider, O.J. Myocardial Metabolism in Heart Failure. *Curr. Heart Fail. Rep.* **2023**, *20*, 63–75. [CrossRef]
62. Soares, R.O.S.; Losada, D.M.; Jordani, M.C.; Évora, P.; Castro-e-Silva, O. Ischemia/Reperfusion Injury Revisited: An Overview of the Latest Pharmacological Strategies. *Int. J. Mol. Sci.* **2019**, *20*, 5034. [CrossRef]
63. Dyck, J.R.B.; Cheng, J.F.; Stanley, W.C.; Barr, R.; Chandler, M.P.; Brown, S.; Wallace, D.; Arrhenius, T.; Harmon, C.; Yang, G.; et al. Malonyl Coenzyme a Decarboxylase Inhibition Protects the Ischemic Heart by Inhibiting Fatty Acid Oxidation and Stimulating Glucose Oxidation. *Circ. Res.* **2004**, *94*, e78–e84. [CrossRef]
64. Li, X.; Wu, F.; Günther, S.; Looso, M.; Kuenne, C.; Zhang, T.; Wiesnet, M.; Klatt, S.; Zukunft, S.; Fleming, I.; et al. Inhibition of Fatty Acid Oxidation Enables Heart Regeneration in Adult Mice. *Nature* **2023**, *622*, 619–626. [CrossRef]
65. Ying, C. Viral Myocarditis. *Yale J. Biol. Med.* **2024**, *97*, 515–520. [CrossRef]
66. Qi, Y.; Yin, J.; Xia, W.; Yang, S. Exploring the Role of Mitochondrial Antiviral Signaling Protein in Cardiac Diseases. *Front. Immunol.* **2025**, *16*, 1540774. [CrossRef]

67. Chen, J.; Shao, J.; Wang, Y.; Wu, K.; Huang, M. OPA1, a Molecular Regulator of Dilated Cardiomyopathy. *J. Cell. Mol. Med.* **2023**, *27*, 3017–3025. [CrossRef]
68. Hinton, A.; Claypool, S.M.; Neikirk, K.; Senoo, N.; Wanjalla, C.N.; Kirabo, A.; Williams, C.R. Mitochondrial Structure and Function in Human Heart Failure. *Circ. Res.* **2024**, *135*, 372–396. [CrossRef] [PubMed]
69. Da Dalt, L.; Cabodevilla, A.G.; Goldberg, I.J.; Norata, G.D. Cardiac Lipid Metabolism, Mitochondrial Function, and Heart Failure. *Cardiovasc. Res.* **2023**, *119*, 1905–1914. [CrossRef] [PubMed]
70. Sequeira, V.; Waddingham, M.T.; Tsuchimochi, H.; Maack, C.; Pearson, J.T. Mechano-Energetic Uncoupling in Hypertrophic Cardiomyopathy: Pathophysiological Mechanisms and Therapeutic Opportunities. *J. Mol. Cell. Cardiol. Plus* **2023**, *4*, 100036. [CrossRef] [PubMed]
71. Witjas-Paalberends, E.R.; Güclü, A.; Germans, T.; Knaapen, P.; Harms, H.J.; Vermeer, A.M.C.; Christiaans, I.; Wilde, A.A.M.; DosRemedios, C.; Lammertsma, A.A.; et al. Gene-Specific Increase in the Energetic Cost of Contraction in Hypertrophic Cardiomyopathy Caused by Thick Filament Mutations. *Cardiovasc. Res.* **2014**, *103*, 248–257. [CrossRef]
72. Tsampasian, V.; Cameron, D.; Sobhan, R.; Bazoukis, G.; Vassiliou, V.S. Phosphorus Magnetic Resonance Spectroscopy (31P MRS) and Cardiovascular Disease: The Importance of Energy. *Medicina* **2023**, *59*, 174. [CrossRef]
73. Crilley, J.G.; Boehm, E.A.; Blair, E.; Rajagopalan, B.; Blamire, A.M.; Styles, P.; McKenna, W.J.; Östman-Smith, I.; Clarke, K.; Watkins, H. Hypertrophic Cardiomyopathy Due to Sarcomeric Gene Mutations Is Characterized by Impaired Energy Metabolism Irrespective of the Degree of Hypertrophy. *J. Am. Coll. Cardiol.* **2003**, *41*, 1776–1782. [CrossRef]
74. Nollet, E.E.; Duursma, I.; Rozenbaum, A.; Eggelbusch, M.; Wüst, R.C.I.; Schoonvelde, S.A.C.; Michels, M.; Jansen, M.; Van Der Wel, N.N.; Bedi, K.C.; et al. Mitochondrial Dysfunction in Human Hypertrophic Cardiomyopathy Is Linked to Cardiomyocyte Architecture Disruption and Corrected by Improving NADH-Driven Mitochondrial Respiration. *Eur. Heart J.* **2023**, *44*, 1170–1185. [CrossRef]
75. Christiansen, L.B.; Dela, F.; Koch, J.; Hansen, C.N.; Leifsson, P.S.; Yokota, T. Impaired Cardiac Mitochondrial Oxidative Phosphorylation and Enhanced Mitochondrial Oxidative Stress in Feline Hypertrophic Cardiomyopathy. *Am. J. Physiol. Heart Circ. Physiol.* **2015**, *308*, H1237–H1247. [CrossRef]
76. Porcari, A.; Fontana, M.; Gillmore, J.D. Transthyretin Cardiac Amyloidosis. *Cardiovasc. Res.* **2022**, *118*, 3517–3535. [CrossRef] [PubMed]
77. Dittloff, K.T.; Spanghero, E.; Solís, C.; Banach, K.; Russell, B. Transthyretin Deposition Alters Cardiomyocyte Sarcomeric Architecture, Calcium Transients, and Contractile Force. *Physiol. Rep.* **2022**, *10*, e15207. [CrossRef] [PubMed]
78. Shi, J.; Guan, J.; Jiang, B.; Brenner, D.A.; Del Monte, F.; Ward, J.E.; Connors, L.H.; Sawyer, D.B.; Semigran, M.J.; Macgillivray, T.E.; et al. Amyloidogenic Light Chains Induce Cardiomyocyte Contractile Dysfunction and Apoptosis via a Non-Canonical P38α MAPK Pathway. *Proc. Natl. Acad. Sci. USA* **2010**, *107*, 4188–4193. [CrossRef]
79. Brenner, D.A.; Jain, M.; Pimentel, D.R.; Wang, B.; Connors, L.H.; Skinner, M.; Apstein, C.S.; Liao, R. Human Amyloidogenic Light Chains Directly Impair Cardiomyocyte Function Through an Increase in Cellular Oxidant Stress. *Circ. Res.* **2004**, *94*, 1008–1010. [CrossRef]
80. McWilliams-Koeppen, H.P.; Foster, J.S.; Hackenbrack, N.; Ramirez-Alvarado, M.; Donohoe, D.; Williams, A.; Macy, S.; Wooliver, C.; Wortham, D.; Morrell-Falvey, J.; et al. Light Chain Amyloid Fibrils Cause Metabolic Dysfunction in Human Cardiomyocytes. *PLoS ONE* **2015**, *10*, e0137716. [CrossRef] [PubMed]
81. Wu, D.; Chen, W. Molecular Mechanisms and Emerging Therapies in Wild-Type Transthyretin Amyloid Cardiomyopathy. *Heart Fail. Rev.* **2024**, *29*, 511–521. [CrossRef] [PubMed]
82. Pieroni, M.; Moon, J.C.; Arbustini, E.; Barriales-Villa, R.; Camporeale, A.; Vujkovac, A.C.; Elliott, P.M.; Hagege, A.; Kuusisto, J.; Linhart, A.; et al. Cardiac Involvement in Fabry Disease: JACC Review Topic of the Week. *J. Am. Coll. Cardiol.* **2021**, *77*, 922–936. [CrossRef]
83. Weissman, D.; Dudek, J.; Sequeira, V.; Maack, C. Fabry Disease: Cardiac Implications and Molecular Mechanisms. *Curr. Heart Fail. Rep.* **2024**, *21*, 81–100. [CrossRef]
84. Majamaa, K.; Moilanen, J.S.; Uimonen, S.; Remes, A.M.; Salmela, P.I.; Kärppä, M.; Majamaa-Voltti, K.A.M.; Rusanen, H.; Sorri, M.; Peuhkurinen, K.J.; et al. Epidemiology of A3243G, the Mutation for Mitochondrial Encephalomyopathy, Lactic Acidosis, and Strokelike Episodes: Prevalence of the Mutation in an Adult Population. *Am. J. Hum. Genet.* **1998**, *63*, 447–454. [CrossRef] [PubMed]
85. Ryytty, S.; Modi, S.R.; Naumenko, N.; Shakirzyanova, A.; Rahman, M.O.; Vaara, M.; Suomalainen, A.; Tavi, P.; Hämäläinen, R.H. Varied Responses to a High m.3243A>G Mutation Load and Respiratory Chain Dysfunction in Patient-Derived Cardiomyocytes. *Cells* **2022**, *11*, 2593. [CrossRef]
86. Lopes, L.R.; Macken, W.L.; Preez, S.D.; Kotwal, H.; Savvatis, K.; Sekhri, N.; Mohiddin, S.A.; Kabiljo, R.; Pitceathly, R.D.S. An Analysis of Mitochondrial Variation in Cardiomyopathy Patients from the 100,000 Genomes Cohort: M.4300A>G as a Cause of Genetically Elusive Hypertrophic Cardiomyopathy. *Hum. Genom.* **2024**, *18*, 136. [CrossRef] [PubMed]

87. Berardo, A.; Domínguez-González, C.; Engelstad, K.; Hirano, M. Advances in Thymidine Kinase 2 Deficiency: Clinical Aspects, Translational Progress, and Emerging Therapies. *J. Neuromuscul. Dis.* **2022**, *9*, 225–235. [CrossRef]
88. Dudek, J.; Maack, C. Barth Syndrome Cardiomyopathy. *Cardiovasc. Res.* **2017**, *113*, 399–410. [CrossRef]
89. Gallo, G.; Rubattu, S.; Volpe, M. Mitochondrial Dysfunction in Heart Failure: From Pathophysiological Mechanisms to Therapeutic Opportunities. *Int. J. Mol. Sci.* **2024**, *25*, 2667. [CrossRef]
90. Qi, X.; Zhu, Z.; Wang, Y.; Wen, Z.; Jiang, Z.; Zhang, L.; Pang, Y.; Lu, J. Research Progress on the Relationship between Mitochondrial Function and Heart Failure: A Bibliometric Study from 2002 to 2021. *Front. Mol. Biosci.* **2022**, *9*, 1036364. [CrossRef]
91. Watson, W.D.; Arvidsson, P.M.; Miller, J.J.J.; Lewis, A.J.; Rider, O.J. A Mitochondrial Basis for Heart Failure Progression. *Cardiovasc. Drugs Ther.* **2024**, *38*, 1161–1171. [CrossRef] [PubMed]
92. Jiang, B.; Xiao, W.; Shi, Y.; Liu, M.; Xiao, X. Heat Shock Pretreatment Inhibited the Release of Smac/DIABLO from Mitochondria and Apoptosis Induced by Hydrogen Peroxide in Cardiomyocytes and C2C12 Myogenic Cells. *Cell Stress. Chaperones* **2005**, *10*, 252–262. [CrossRef]
93. Kim, N.H.; Kang, P.M. Apoptosis in Cardiovascular Diseases: Mechanism and Clinical Implications. *Korean Circ. J.* **2010**, *40*, 299–305. [CrossRef] [PubMed]
94. Konstantinidis, K.; Whelan, R.S.; Kitsis, R.N. Mechanisms of Cell Death in Heart Disease. *Arterioscler. Thromb. Vasc. Biol.* **2012**, *32*, 1552–1562. [CrossRef]
95. Guerrero, A.D.; Schmitz, I. Promotion of Caspase Activation by Caspase-9-Mediated Feedback Amplification of Mitochondrial Damage. *J. Clin. Cell Immunol.* **2012**, *3*, 1000126. [CrossRef]
96. Chen, L.; Knowlton, A.A. Mitochondria and Heart Failure: New Insights into an Energetic Problem. *Minerva Cardioangiol.* **2010**, *58*, 213–229.
97. He, X.; Du, T.; Long, T.; Liao, X.; Dong, Y.; Huang, Z.P. Signaling Cascades in the Failing Heart and Emerging Therapeutic Strategies. *Signal Transduct. Target. Ther.* **2022**, *7*, 134. [CrossRef]
98. Javadov, S.; Jang, S.; Parodi-Rullán, R.; Khuchua, Z.; Kuznetsov, A.V. Mitochondrial Permeability Transition in Cardiac Ischemia–Reperfusion: Whether Cyclophilin D Is a Viable Target for Cardioprotection? *Cell. Mol. Life Sci.* **2017**, *74*, 2795–2813. [CrossRef]
99. Zhang, T.; Zhang, Y.; Cui, M.; Jin, L.; Wang, Y.; Lv, F.; Liu, Y.; Zheng, W.; Shang, H.; Zhang, J.; et al. CaMKII Is a RIP3 Substrate Mediating Ischemia- and Oxidative Stress-Induced Myocardial Necroptosis. *Nat. Med.* **2016**, *22*, 175–182. [CrossRef]
100. Ying, L.; Benjanuwattra, J.; Chattipakorn, S.C.; Chattipakorn, N. The Role of RIPK3-Regulated Cell Death Pathways and Necroptosis in the Pathogenesis of Cardiac Ischaemia-Reperfusion Injury. *Acta Physiol.* **2021**, *231*, e13541. [CrossRef] [PubMed]
101. Zhu, P.; Wan, K.; Yin, M.; Hu, P.; Que, Y.; Zhou, X.; Zhang, L.; Li, T.; Du, Y.; Xu, G.; et al. RIPK3 Induces Cardiomyocyte Necroptosis via Inhibition of AMPK-Parkin-Mitophagy in Cardiac Remodelling after Myocardial Infarction. *Oxid. Med. Cell Longev.* **2021**, *2021*, 6635955. [CrossRef]
102. Lemasters, J.J. V. Necrapoptosis and the Mitochondrial Permeability Transition: Shared Pathways to Necrosis and Apoptosis. *Am. J. Physiol.* **1999**, *276*, G1–G6. [CrossRef]
103. Nakayama, H.; Otsu, K. Mitochondrial DNA as an Inflammatory Mediator in Cardiovascular Diseases. *Biochem. J.* **2018**, *475*, 839–852. [CrossRef] [PubMed]
104. Silvis, M.J.M.; Kaffka genaamd Dengler, S.E.; Odille, C.A.; Mishra, M.; van der Kaaij, N.P.; Doevendans, P.A.; Sluijter, J.P.G.; de Kleijn, D.P.V.; de Jager, S.C.A.; Bosch, L.; et al. Damage-Associated Molecular Patterns in Myocardial Infarction and Heart Transplantation: The Road to Translational Success. *Front. Immunol.* **2020**, *11*, 599511. [CrossRef]
105. Mortensen, S.A.; Rosenfeldt, F.; Kumar, A.; Dolliner, P.; Filipiak, K.J.; Pella, D.; Alehagen, U.; Steurer, G.; Littarru, G.P. The Effect of Coenzyme Q10 on Morbidity and Mortality in Chronic Heart Failure: Results from Q-SYMBIO: A Randomized Double-Blind Trial. *JACC Heart Fail.* **2014**, *2*, 641–649. [CrossRef]
106. Leshnower, B.G.; Kanemoto, S.; Matsubara, M.; Sakamoto, H.; Hinmon, R.; Gorman, J.H.; Gorman, R.C. Cyclosporine Preserves Mitochondrial Morphology After Myocardial Ischemia/Reperfusion Independent of Calcineurin Inhibition. *Ann. Thorac. Surg.* **2008**, *86*, 1286–1292. [CrossRef]
107. Cung, T.-T.; Morel, O.; Cayla, G.; Rioufol, G.; Garcia-Dorado, D.; Angoulvant, D.; Bonnefoy-Cudraz, E.; Guérin, P.; Elbaz, M.; Delarche, N.; et al. Cyclosporine before PCI in Patients with Acute Myocardial Infarction. *N. Engl. J. Med.* **2015**, *373*, 1021–1031. [CrossRef]
108. Ottani, F.; Latini, R.; Staszewsky, L.; La Vecchia, L.; Locuratolo, N.; Sicuro, M.; Masson, S.; Barlera, S.; Milani, V.; Lombardi, M.; et al. Cyclosporine A in Reperfused Myocardial Infarction the Multicenter, Controlled, Open-Label CYCLE Trial. *J. Am. Coll. Cardiol.* **2016**, *67*, 365–374. [CrossRef]
109. Arany, Z.; He, H.; Lin, J.; Hoyer, K.; Handschin, C.; Toka, O.; Ahmad, F.; Matsui, T.; Chin, S.; Wu, P.H.; et al. Transcriptional Coactivator PGC-1α Controls the Energy State and Contractile Function of Cardiac Muscle. *Cell Metab.* **2005**, *1*, 259–271. [CrossRef] [PubMed]

110. Qian, L.; Zhu, Y.; Deng, C.; Liang, Z.; Chen, J.; Chen, Y.; Wang, X.; Liu, Y.; Tian, Y.; Yang, Y. Peroxisome Proliferator-Activated Receptor Gamma Coactivator-1 (PGC-1) Family in Physiological and Pathophysiological Process and Diseases. *Signal Transduct. Target. Ther.* **2024**, *9*, 50. [CrossRef] [PubMed]
111. Hayashida, K.; Takegawa, R.; Shoaib, M.; Aoki, T.; Choudhary, R.C.; Kuschner, C.E.; Nishikimi, M.; Miyara, S.J.; Rolston, D.M.; Guevara, S.; et al. Mitochondrial Transplantation Therapy for Ischemia Reperfusion Injury: A Systematic Review of Animal and Human Studies. *J. Transl. Med.* **2021**, *19*, 214. [CrossRef] [PubMed]
112. Hassanpour, P.; Sadeghsoltani, F.; Saghebasl, S.; Boroumand, S.; Khanicheragh, P.; Tafti, S.H.A.; Rahbarghazi, R.; Rahmati, M. Mitochondrial Transplantation for Cardioprotection and Induction of Angiogenesis in Ischemic Heart Disease. *Stem Cell Res. Ther.* **2025**, *16*, 54. [CrossRef]
113. Wisløff, U.; Støylen, A.; Loennechen, J.P.; Bruvold, M.; Rognmo, Ø.; Haram, P.M.; Tjønna, A.E.; Helgerud, J.; Slørdahl, S.A.; Lee, S.J.; et al. Superior Cardiovascular Effect of Aerobic Interval Training versus Moderate Continuous Training in Heart Failure Patients: A Randomized Study. *Circulation* **2007**, *115*, 3086–3094. [CrossRef]
114. Eisenberg, T.; Abdellatif, M.; Schroeder, S.; Primessnig, U.; Stekovic, S.; Pendl, T.; Harger, A.; Schipke, J.; Zimmermann, A.; Schmidt, A.; et al. Cardioprotection and Lifespan Extension by the Natural Polyamine Spermidine. *Nat. Med.* **2016**, *22*, 1428–1438. [CrossRef]
115. Fragasso, G.; Salerno, A.; Lattuada, G.; Cuko, A.; Calori, G.; Scollo, A.; Ragogna, F.; Arioli, F.; Bassanelli, G.; Spoladore, R.; et al. Effect of Partial Inhibition of Fatty Acid Oxidation by Trimetazidine on Whole Body Energy Metabolism in Patients with Chronic Heart Failure. *Heart* **2011**, *97*, 1495–1500. [CrossRef] [PubMed]
116. Ferry, A. Dapagliflozin for Patients with Heart Failure and Reduced Ejection Fraction. *J. Am. Acad. Physician Assist.* **2022**, *35*, 51–53. [CrossRef]
117. Packer, M.; Anker, S.D.; Butler, J.; Filippatos, G.; Pocock, S.J.; Carson, P.; Januzzi, J.; Verma, S.; Tsutsui, H.; Brueckmann, M.; et al. Cardiovascular and Renal Outcomes with Empagliflozin in Heart Failure. *N. Engl. J. Med.* **2020**, *383*, 1413–1424. [CrossRef] [PubMed]
118. Anker, S.D.; Butler, J.; Filippatos, G.; Ferreira, J.P.; Bocchi, E.; Böhm, M.; Brunner–La Rocca, H.-P.; Choi, D.-J.; Chopra, V.; Chuquiure-Valenzuela, E.; et al. Empagliflozin in Heart Failure with a Preserved Ejection Fraction. *N. Engl. J. Med.* **2021**, *385*, 1451–1461. [CrossRef]
119. Solomon, S.D.; McMurray, J.J.V.; Claggett, B.; de Boer, R.A.; DeMets, D.; Hernandez, A.F.; Inzucchi, S.E.; Kosiborod, M.N.; Lam, C.S.P.; Martinez, F.; et al. Dapagliflozin in Heart Failure with Mildly Reduced or Preserved Ejection Fraction. *N. Engl. J. Med.* **2022**, *387*, 1089–1098. [CrossRef]
120. Saucedo-Orozco, H.; Voorrips, S.N.; de Boer, R.A.; Westenbrink, B.D.; Yurista, S.R. SGLT2 Inhibitors and Ketone Metabolism in Heart Failure. *J. Lipid Atheroscler.* **2022**, *11*, 1. [CrossRef]
121. Lopaschuk, G.D.; Verma, S. Mechanisms of Cardiovascular Benefits of Sodium Glucose Co-Transporter 2 (SGLT2) Inhibitors: A State-of-the-Art Review. *JACC Basic Transl. Sci.* **2020**, *5*, 632–644. [CrossRef]
122. Bonora, M.; Wieckowski, M.R.; Sinclair, D.A.; Kroemer, G.; Pinton, P.; Galluzzi, L. Targeting Mitochondria for Cardiovascular Disorders: Therapeutic Potential and Obstacles. *Nat. Rev. Cardiol.* **2019**, *16*, 33–55. [CrossRef]
123. Kiyuna, L.A.; e Albuquerque, R.P.; Chen, C.H.; Mochly-Rosen, D.; Ferreira, J.C.B. Targeting Mitochondrial Dysfunction and Oxidative Stress in Heart Failure: Challenges and Opportunities. *Free Radic. Biol. Med.* **2018**, *129*, 155–168. [CrossRef]
124. Trnka, J.; Elkalaf, M.; Anděl, M. Lipophilic Triphenylphosphonium Cations Inhibit Mitochondrial Electron Transport Chain and Induce Mitochondrial Proton Leak. *PLoS ONE* **2015**, *10*, e0121837. [CrossRef] [PubMed]
125. Brown, D.A.; Perry, J.B.; Allen, M.E.; Sabbah, H.N.; Stauffer, B.L.; Shaikh, S.R.; Cleland, J.G.F.; Colucci, W.S.; Butler, J.; Voors, A.A.; et al. Expert Consensus Document: Mitochondrial Function as a Therapeutic Target in Heart Failure. *Nat. Rev. Cardiol.* **2017**, *14*, 238–250. [CrossRef]
126. Paraskevaidis, I.; Kourek, C.; Farmakis, D.; Tsougos, E. Mitochondrial Dysfunction in Cardiac Disease: The Fort Fell. *Biomolecules* **2024**, *14*, 1534. [CrossRef] [PubMed]
127. Shayota, B.J. Biomarkers of Mitochondrial Disorders. *Neurotherapeutics* **2024**, *21*, e00325. [CrossRef] [PubMed]
128. Zhang, Q.; Raoof, M.; Chen, Y.; Sumi, Y.; Sursal, T.; Junger, W.; Brohi, K.; Itagaki, K.; Hauser, C.J. Circulating Mitochondrial DAMPs Cause Inflammatory Responses to Injury. *Nature* **2010**, *464*, 104–107. [CrossRef]
129. Iyer, S.S.; He, Q.; Janczy, J.R.; Elliott, E.I.; Zhong, Z.; Olivier, A.K.; Sadler, J.J.; Knepper-Adrian, V.; Han, R.; Qiao, L.; et al. Mitochondrial Cardiolipin Is Required for Nlrp3 Inflammasome Activation. *Immunity* **2013**, *39*, 311–323. [CrossRef]
130. Pustylnikov, S.; Costabile, F.; Beghi, S.; Facciabene, A. Targeting Mitochondria in Cancer: Current Concepts and Immunotherapy Approaches. *Transl. Res.* **2018**, *202*, 35–51. [CrossRef]
131. Schroeder, M.A.; Clarke, K.; Neubauer, S.; Tyler, D.J. Hyperpolarized Magnetic Resonance: A Novel Technique for the in Vivo Assessment of Cardiovascular Disease. *Circulation* **2011**, *124*, 1580–1594. [CrossRef]
132. Murphy, M.P.; Hartley, R.C. Mitochondria as a Therapeutic Target for Common Pathologies. *Nat. Rev. Drug Discov.* **2018**, *17*, 865–886. [CrossRef]

133. Daubert, M.A.; Yow, E.; Dunn, G.; Marchev, S.; Barnhart, H.; Douglas, P.S.; O'Connor, C.; Goldstein, S.; Udelson, J.E.; Sabbah, H.N. Novel Mitochondria-Targeting Peptide in Heart Failure Treatment: A Randomized, Placebo-Controlled Trial of Elamipretide. *Circ. Heart Fail.* **2017**, *10*, e004389. [CrossRef]
134. Kloner, R.A.; Hale, S.L.; Dai, W.; Gorman, R.C.; Shuto, T.; Koomalsingh, K.J.; Gorman, J.H.; Sloan, R.C.; Frasier, C.R.; Watson, C.A.; et al. Reduction of Ischemia/Reperfusion Injury with Bendavia, a Mitochondria-Targeting Cytoprotective Peptide. *J. Am. Heart Assoc.* **2012**, *1*, e001644. [CrossRef]
135. Diguet, N.; Trammell, S.A.J.; Tannous, C.; Deloux, R.; Piquereau, J.; Mougenot, N.; Gouge, A.; Gressette, M.; Manoury, B.; Blanc, J.; et al. Nicotinamide Riboside Preserves Cardiac Function in a Mouse Model of Dilated Cardiomyopathy. *Circulation* **2018**, *137*, 2256–2273. [CrossRef] [PubMed]
136. Mattson, M.P.; Longo, V.D.; Harvie, M. Impact of Intermittent Fasting on Health and Disease Processes. *Ageing Res. Rev.* **2017**, *39*, 46–58. [CrossRef] [PubMed]
137. López-Lluch, G.; Navas, P. Calorie Restriction as an Intervention in Ageing. *J. Physiol.* **2016**, *594*, 2043–2060. [CrossRef] [PubMed]

biomolecules

MDPI

Review

Molecular Pathogenesis of Arrhythmogenic Cardiomyopathy: Mechanisms and Therapeutic Perspectives

Eliza Popa [1,2] and Sorin Hostiuc [1,2,*]

[1] National Institute of Legal Medicine, 042122 Bucharest, Romania; eliza.popa@inml-mm.ro
[2] Department of Legal Medicine and Bioethics, Carol Davila University of Medicine and Pharmacy, 020021 Bucharest, Romania
* Correspondence: soraer@gmail.com

Abstract

Arrhythmogenic cardiomyopathy (ACM) is a genetic cardiac disease characterized by a progressive loss of cardiomyocytes associated with fibrofatty myocardial replacement, resulting in a heightened risk of ventricular arrhythmias and sudden cardiac death. ACM is a common cause of sudden death in young individuals, and exercise has been proven to be a factor in disease progression. Current therapeutic strategies, including lifestyle modification, antiarrhythmic pharmacological therapy, catheter ablation, and the placement of implantable cardioverter-defibrillators, remain primarily palliative options rather than addressing the underlying molecular substrate. The pathogenesis of ACM includes complex molecular and cellular mechanisms, linking genetic mutations to structural and electrical anomalies of the ventricle. The lack of targeted therapies contributes to a challenging approach to the disease. It highlights the need for a better understanding of the mechanisms that lead to myocardial remodeling and arrhythmic predisposition. With the help of animal models (especially murine) and induced pluripotent stem cells, there have been advances in understanding the molecular pathogenesis of ACM. In this review, we summarized some of the pathogenic molecular pathways involved in the development of ACM and emerging therapies targeted towards disease modification, not just prevention.

Keywords: arrhythmogenic cardiomyopathy; molecular pathogenesis; gene therapy

1. Introduction

Arrhythmogenic cardiomyopathy (ACM) is a hereditary disease characterized by a progressive loss of cardiomyocytes and their replacement with fibrous and fatty tissue [1]. The replacement of the healthy heart muscle with fibrofatty tissue creates a "myocardial scar," leading to alteration of the typical architecture and electrophysiological properties of the heart, which results in regional or global ventricular dysfunction that predisposes patients to ventricular arrhythmias, irrespective of the systolic ventricular function [2].

ACM is a rare disease with a prevalence ranging from 1:2000 to 1:5000 individuals; however, it poses a significant risk factor for sudden cardiac death, especially among younger patients [3,4]. Even though it is considered a rare disease, the number of reported cases might be underestimated because the symptoms are often subtle or even concealed in their early phases [5]. The disease was initially described as arrhythmogenic right ventricular dysplasia (ARVD) before the involvement of the left ventricle was recognized, which led to the term ACM [6].

ACM is one of the main causes of sudden cardiac death (SCD) in young adults, especially in the under-35-years-old group. ACM holds significant importance in cases of exercise-related sudden deaths in athletes. SCD might be the first clinical presentation, suggesting the need for a better screening strategy oriented towards identifying high-risk groups [7,8]. James et al. demonstrated in their epidemiological study the link between exercise and the development of ACM. They suggested that intense physical activity and competitive sport participation can accelerate the progression of the disease and increase the risk of arrhythmia, while also increasing the likelihood of diagnosis [9]. The importance of physical activity as an environmental modifier for ACM has gained attention as it is one of the most common causes of sudden cardiac death in young individuals, especially in athletes [10]. There is no agreement regarding the effect of moderate physical activity or recreational sports, as some studies reported increased risk of arrhythmia, while others showed otherwise [11]. Studies on mice have shown that forced exercise protocols (running on a treadmill or swimming) can exacerbate ACM phenotypes, leading to increased fibrosis, reduced cardiac function, and an increased occurrence of arrhythmias or cardiomyocyte death [12,13]. At the same time, one study observed that, in some cases, low to moderate intensity voluntary exercise had a positive effect, suggesting the potential benefit of long-term light exercise [14]. Connexin-43 (Cx43) is a protein that is involved in cellular communication and cell-to-cell adhesion and is a component of gap junctions, allowing the passage between cells of ions and small molecules [15]. While Cx43 improves cardiac function, in ACM, there is a reduced expression and mislocalization of Cx43 at intercalated discs, contributing to impaired conduction and arrhythmogenesis [16]. Moderate exercise has been observed to increase Cx43 expression (in non-ACM models) [17,18], but high-intensity prolonged exercise tends to reduce Cx43 expression [19]. Taking this into account, the "Cx43 paradox" suggests that although enhancing Cx43 expression could be a therapeutic solution, excessive exercise can counteract the potential positive effects.

The clinical manifestations of ACM are most commonly caused by arrhythmias, with symptoms including palpitations, syncope, and cardiac arrest. Even though heart failure with its specific symptoms appears later in the course of the disease, for some patients, it may be the first sign [20]. The natural history for ACM usually consists of four phases: (i) the concealed phase in which there are no structural modifications, but the patients are still at risk, especially during strenous physical activity; (ii) the second phase is characterized by EKG abnormalities, like inverted T wave and arrhytmias with left bundle branch block; (iii) the third phase exhibiting the loss of healthy heart muscle that is being replaced by fibrous and fatty tissue that are usually detected by magnetic resonance; the end-phase is characterized by heart failure affecting both of the ventricles [21].

Conventional therapy strategies, such as implantable cardioverter-defibrillators (ICD), pharmacological options, or lifestyle changes, do not affect the molecular substrate that causes the disease to progress [22]. Recently, significant advances in molecular cardiology and genetics have provided insights into ACM pathogenesis. A consistent component of ACM pathogenesis involves genes that encode desmosomal proteins, resulting in the disruption of cell-to-cell adhesion and intracellular signaling [23]. By understanding the molecular mechanisms that define the disease, there is a better chance of an early diagnosis and an opportunity for novel therapeutic strategies.

This narrative review aims to integrate current knowledge of the molecular pathogenesis of ACM and explore new therapeutic perspectives that extend beyond traditional pharmacological and device-based treatments, moving towards approaches designed to alter the disease process itself.

2. Genetic Basis of ACM

The current classification of ACM consists of three subtypes, each defined by the preponderance of ventricular involvement [24]. The first and most common type is the right ventricle-dominant arrhythmogenic cardiomyopathy, formerly known as arrhythmogenic right ventricular dysplasia (ARVD) or arrhythmogenic right ventricular cardiomyopathy (ARVC) [25]. In this subtype, the right ventricle is dilated and exhibits focal wall motion anomalies. Commonly, the subtricuspid valve region and the right ventricular outflow tract are the most predisposed regions to remodeling, often leading to the formation of aneurysms [26]. The second type is the biventricular disease, with the involvement of both ventricles; this variant is associated with a greater risk of heart failure and earlier presentation of arrhythmia [27]. The third type refers to a more pronounced involvement of the left ventricle, known as arrhythmogenic left-ventricular cardiomyopathy (ALVC) or left-dominant arrhythmogenic cardiomyopathy, with no or minimal abnormalities of the right ventricle. Patients with ALVC have a higher incidence of ventricular arrhythmias, and episodes that can mimic myocarditis are common, with chest pain, troponin elevation, and ECG abnormalities [28]. The classification of ACM and the key clinical features are summarized in Table 1.

Table 1. Classification of ACM and key clinical features—based on [25,27,29].

ACM Variant	Predominant Ventricular Involvement	Key Clinical Features
ARVC	Right ventricle	RV fibro-fatty replacement, ventricular arrhythmias, typical onset in adolescence or early adulthood.
BiV-ACM	Both ventricles	Early LV involvement alongside RV, higher risk of heart failure.
ALVC	Left ventricle	LV arrhythmias, subepicardial fibrosis, and misdiagnosis as DCM are common.

ARVC = Arrhythmogenic Right Ventricular Cardiomyopathy; BiV-ACM = Biventricular Arrhythmogenic Cardiomyopathy; ALVC = Arrhythmogenic Left Ventricular Cardiomyopathy; RV = Right Ventricle; LV = Left Ventricle; DCM = Dilated Cardiomyopathy.

ACM is a familial disease in 30–50% of cases, usually with autosomal dominant transmission with variable penetrance [30]. It is now known to be a disease centered around the desmosome, with almost 90% of cases being linked to mutations in genes encoding desmosomal proteins [31]. Desmosomes are cell-to-cell junctions that anchor intermediate filaments to the plasma membrane [32]. Reports of familial clustering of ACM pointed to a genetic cause of the disease. By analyzing cases from the Naxos island in Greece, Protonotarios et al. observed that there were individuals who had ACM along with keratoderma and woolly hair [33]. Studies have pointed to two genetic causes for Naxos syndrome: mutations in genes that encode proteins involved in cell-to-cell adhesion—plakoglobin and desmoplakin [34]. Similarly to Naxos, Carvajal syndrome was reported in Ecuador, characterized by keratoderma, dry and blister-prone skin, woolly hair, and cardiomyopathy. Genetic studies also found a mutation in a gene that encodes desmoplakin [11]. While Naxos and Carvajal syndromes are autosomal recessive, most phenotypes of ACM have autosomal dominant transmission [21].

As stated previously, the known pathogenesis of ACM is usually centered around the desmosome. The most common desmosomal genes involved in ACM are PKP2 (encoding plakophilin-2), DSG2 (encoding desmoglein-2), DSC2 (encoding desmocollin-2), JUP (encoding plakoglobin), and DSP (encoding desmoplakin). Still, non-desmosomal protein involvement has also been reported, such as transmembrane protein 43 (TMEM43), desmin

(DES), phospholamban (PLB), N-cadherin (CDH2), sodium voltage-gated channel alpha subunit 5 (SCN5A), titin (TTN), and transforming growth factor 3 beta (TGF3β) [21]. The classic form of ACM involving the right ventricle is usually caused by the mutation of the PKP gene [35]. The main genes involved in ACM are listed in Table 2.

Table 2. Genes involved in ACM.

Genes	Protein Encoded	Evidence Level
PKP2	Plakophilin-2	Strong [36–39]
DSG2	Desmoglein-2	Strong [38,40]
DSC2	Desmocollin-2	Strong [38,40,41]
JUP	Plakoglobin	Strong [38,42]
DSP	Desmoplakin	Strong [38,43]
TMEM43	Transmembrane protein 43	Strong [38,44]
DES	Desmin	Moderate [38,45]
PLB(PLN)	Phospholamban	Moderate [38,46]
CDH2	N-cadherin	Limited [38,47]
SCN5A	Sodium voltage-gated channel alpha subunit 5	Limited [38,48]
TGF3β	Transforming growth factor 3 beta	Limited [38,49]
LMNA	Lamin A/C	Limited [38,50]
FLNC	Filamin C	Not defined yet [38,51]

3. Molecular and Cellular Pathogenesis

The pathogenesis of arrhythmogenic cardiomyopathy (ACM) consists of complex molecular and cellular mechanisms, linking genetic mutations to structural and electrical anomalies of the heart. These processes can be described as: desmosomal protein mutations with mechanical failure, dysregulation of signal transduction, programmed cell death, immune responses, and fibrofatty replacement of the cardiomyocytes. In this section, the processes that contribute to the pathogenic pathway leading to ACM are described.

3.1. Desmosomal Dysfunction

Desmosomes are specialized intercellular junctional structures, integrated into the intercalated discs of cardiomyocytes, with an essential role in maintaining the mechanical integrity of the cardiac muscle [52]. Three gene superfamilies contribute to the formation of desmosomes: cadherins, the armadillo family of nuclear and junctional proteins, and the plakins [23]. The cadherins are calcium-dependent adhesion molecules with multiple roles: embryogenesis, maintaining cell polarity, and signal transduction [53]. The desmosomal cadherins include two types of proteins encoded by separate genes: desmoglein (*DSG2)* and desmocollin (*DSC2*) [54]. The cardiac armadillo family consists of plakophilin-2 (*PKP2*) and junctional plakoglobin (*JUP*). They connect cadherins to desmoplakin and mediate the connection between desmoplakin and desmin, the intermediate filament [55]. The plakin family is represented by the desmoplakin (*DSP*), which connects the desmosome to the desmin [55].

Most genotype-positive cases have been linked to a genetic malfunction in at least one of the genes that encode the desmosomal proteins. Under mechanical strain, this compromised cell-to-cell adhesion causes cardiomyocytes to detach from one another, undergo apoptosis, and necrosis [56,57]. The repair mechanisms often trigger inflammatory pathways, fibrosis, and adipogenic pathways. In the end, these changes result in the hallmark of ACM: replacement of the ventricular myocardium with fibrous and fatty tissue [58,59].

3.2. *Signal Transduction Pathways*

3.2.1. The Wnt/β-Catenin Signaling Pathway

The Wnt pathway is a signaling pathway involved in both embryo development and adult tissue homeostasis [60]. The canonical Wnt pathway, dependent on β-catenin, is activated when Wnt ligands attach to the Frizzled receptor and its co-receptor (lipoprotein receptor-related protein-LRP) [61,62]. The absence of Wnt stimulation leads to the phosphorylation of β-catenin, which is mediated by glycogen synthase kinase-3β (GSK-3β) and induces its degradation [62]. When Wnt ligands bind to the receptor complex, glycogen synthase kinase-3β (GSK-3β) activity is inhibited, resulting in the accumulation of β-catenin in the cytoplasm [62,63]. The β-catenin will translocate into the nucleus, where it interacts with TCF (T-cell factor) transcription factors, initiating signaling cascades [63]. In ACM, the inhibition of canonical Wnt/β-catenin has been associated with adipogenesis, fibrogenesis, and apoptosis [63–65]. As a structural relative of β-catenin, plakoglobin can interfere with β-catenin's transcriptional activity by competing for binding, leading to degradation of β-catenin [11]. Desmosomal mutations increase the cytosolic plakoglobin (also known as γ-catenin), which can also be phosphorylated by GSK-3β and translocate into the nucleus, leading to the inhibition of β-catenin-mediated transcription and stimulating adipogenic signals [66]. Figure 1 illustrates a schematic representation of the Wnt/β-catenin pathway.

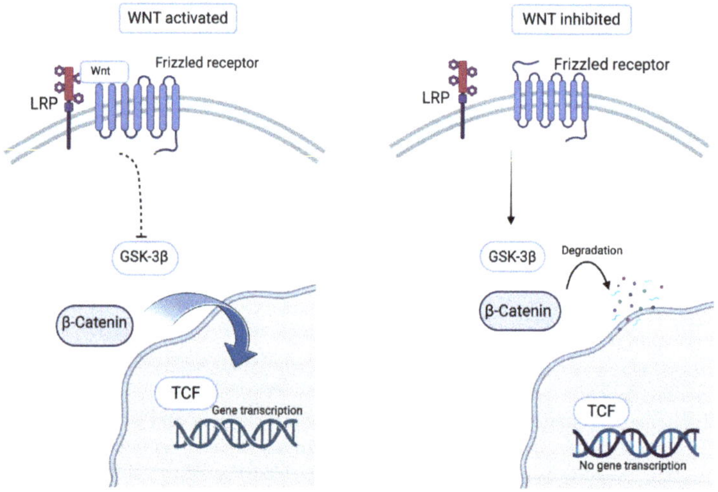

Figure 1. The Wnt/β-catenin pathway (Created in BioRender. Popa, E. (2025) https://BioRender.com/1vkur7t (accessed on 16 October 2025)).

3.2.2. Hippo Pathway Activation

The Hippo pathway is a signaling pathway that controls organ size and tumor growth by regulating cell proliferation and apoptosis [67]. It is primarily composed of the kinases STE20-like protein kinase (MST), large tumor suppressor kinase (LATS1/2), and the transcriptional coactivators Yes-associated protein (YAP) and transcriptional coactivator with PDZ-binding motif (TAZ) [68]. YAP and TAZ are regulated negatively by the activation of the Hippo pathway [69]. When YAP and TAZ are dephosphorylated by inactive Hippo signaling, they translocate into the nucleus and interact with TEA domain (TEAD) transcription factor, leading to proliferation, differentiation, and apoptosis [63,70]. The remodeling process at the intercalated discs in ACM leads to abnormal activation of the Hippo pathway, resulting in increased cytoplasmic retention of phosphorylated YAP and TAZ levels, and changes in gene expression that promote adipogenesis and apoptosis [71].

The activation of the Hippo pathway also suppresses canonical Wnt signaling, further inhibiting YAP/TAZ activity, contributing to cardiomyocyte proliferation, differentiation, and apoptosis [71,72]. Figure 2 shows the components of the Hippo pathway.

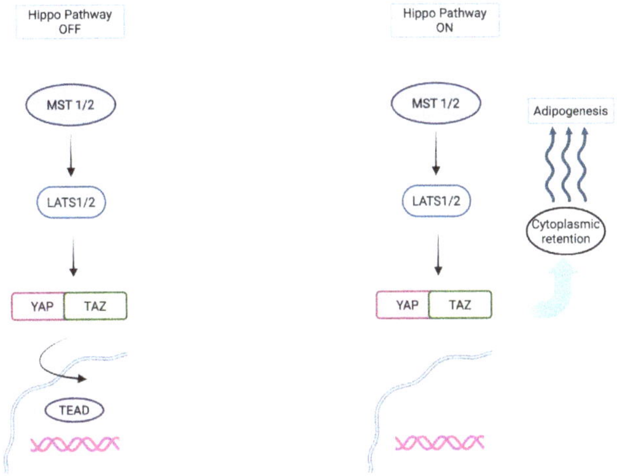

Figure 2. The Hippo Pathway (Created in BioRender. Popa, E. (2025) https://BioRender.com/ahfj5v4 (accessed on 16 October 2025)).

3.2.3. PPARγ Pathway

Peroxisome proliferator-activated receptor gamma (PPARγ) is a ligand-dependent transcription factor, part of the nuclear receptor family of peroxisome proliferator-activated receptors (PPARs) [73]. PPARs play an important role in adipogenesis, lipid metabolism, inflammation and overall metabolic homeostasis by forming a heterodimeric complex with retinoid X receptors (RXR), which can then bind to specific DNA sequences known as PPAR-responsive elements (PPREs) [74]. Overactivation of the PPARγ pathway has been involved as a downstream effector in ACM pathogenesis, as it contributes to the hallmark of fibro-fatty myocardial replacement [75,76]. Studies have observed in both human-induced pluripotent stem cell-derived cardiomyocytes and mouse models of ACM that an exaggerated activation of PPARγ leads to increased adipogenic differentiation and arrhythmogenic substrate formation [75,77]. Figure 3 illustrates the components mentioned above.

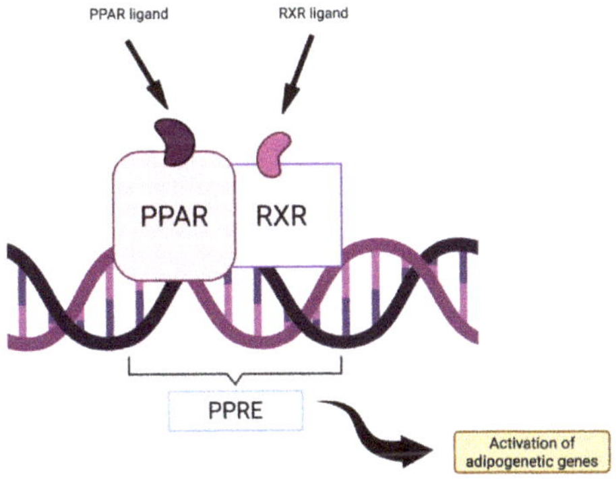

Figure 3. The PPARγ Pathway (Created in BioRender. Popa, E. (2025) https://BioRender.com/e3f7etr (accessed on 16 October 2025)).

3.3. Cardiomyocyte Loss

While cardiomyocyte death plays a central role in the development of ACM, the precise mechanisms of cell death have not been fully identified [78]. Loss of cardiomyocytes can result from various causes, including apoptosis in programmed cell death pathways, necrosis, or cell injury, either from structural damage or inflammation-induced damage [11]. It has been indicated that programmed cell death is involved in cardiomyocyte loss in ACM [79]. In some studies, apoptosis had been associated with unusual transduction of mechanical signals [80–82]. In a survey conducted on neonatal rat ventricular myocytes by Hariharan et al., it was observed that, even though the overexpression of mutant plakoglobin did not affect cell adhesion, it promoted apoptosis through shear stress or cyclic stretch [80]. Another study conducted on human induced pluripotent stem cells with plakophilin-2 (*PKP2*) mutations, the authors showed evidence of increased apoptosis in cardiomyocytes through elevated rates of caspase-3 activation and mitochondrial cytochrome-c release. Also, treatment with apoptosis inhibitors, such as caspase blockers, minimized myocyte loss in cardiomyocyte cultures [81]. Moreover, desmoplakin suppression via small interfering RNA (siRNA) in cultured mouse atrial cardiomyocytes (HL-1 cardiomyocytes) and cardiac-specific desmoplakin (*DSP*) deficient mice led to upregulation of apoptotic markers and increased myocyte apoptosis [82]. Collectively, these findings suggest that apoptosis is a key pathogenic event in arrhythmogenic cardiomyopathy.

3.4. Inflammation and Immune Activation

Autopsy samples from patients with ARVC revealed that inflammatory infiltrates associated with severe structural changes are often present in the affected hearts [83,84]. Multiple studies have reported that inflammation is common in ACM hearts, especially in so-called "hot phases" that mimic myocarditis [85–87]. However, it is unclear whether inflammation is a primary event that leads to arrhythmia and damage to the myocardium or if it is a response caused by cardiomyocyte death [57]. A continuous activation of cardiac inflammatory pathways leads to fibrosis, resulting in myocardium remodeling [88]. Fibrosis creates a substrate for electrical and mechanical disturbances and reduces nutrient supply, resulting in a cycle of fibrosis, cardiomyocyte death, and inflammation [88,89]. A study by Lubos et al. used two murine models, one with mutant desmoglein 2 and one with absent desmoglein, to examine the inflammatory response during different phases of ARVC progression [90]. The results of this study suggested that inflammation in desmoglein-2 mutant mice is not only secondary to cardiomyocyte death [90]. In early phases, cardiomyocyte necrosis was observed with consequent neutrophil inflammatory response and upregulated chemokine pathways, showing that innate immune activation precedes significant structural remodeling [90]. Even in chronic phases, when fibrosis is already present, macrophages and T cells were still observed, suggesting that there is an inflammatory process that sustains disease progression [90].

A study by Asimaki et al. used neonatal rat ventricular myocytes to explore the link between inflammation and the desmosomal disruptions in ARVC [91]. They observed that even low concentrations of IL-17, TNF-α, and IL-6 led to the translocation of plakoglobin from intercellular junctions to intracellular sites, suggesting that inflammation is indeed involved in the pathogenesis of ARVC and can also drive the disease [91].

3.5. Adipogenesis

As noted, the replacement of healthy cardiomyocytes with fibrofatty tissue is the hallmark of ACM. The accumulation of adipose tissue occurs through actual adipogenic differentiation, resulting in the formation of mature adipocytes, as opposed to being a cellular accumulation of lipids [92]. However, even if adipogenesis is the central element of ACM, the exact origin and mechanisms remain unclear [93].

The main pathways that have been proposed to be involved in adipogenesis are Wnt, Hippo, and PPARγ [63,94]. Regarding the Wnt/β-catenin pathway, in a study by Garcia-Gras et al., using HL-1 cardiac myocytes with suppressed desmoplakin expression, they observed that, by destabilizing the desmosome, the plakoglobin was freed from the junctions and translocated into the nucleus, where it competes with β-catenin for binding to specific transcription factors, suppressing the canonical Wnt/β-catenin signaling [82]. Normally, Wnt activity inhibits adipogenic differentiation, so when it is suppressed, adipogenic genes are no longer inhibited [82]. The study also showed upregulation of PPARγ (peroxisome proliferator-activated receptor gamma), which is a known regulator of adipogenesis, and C/EBP-α (CCAAT/enhancer-binding protein-alpha), which cooperates with PPARγ in adipocyte differentiation [82,95,96]. To test the potential involvement of the Hippo pathway in adipogenesis, Chen et al. used samples from ACM patient hearts, mouse models (desmoplakin-deficient and plakoglobin-altered), and HL-1 cardiac myocytes with PKP2 knockdown [71]. Decreased PKC-α (protein kinase C alpha) activity with consequent activation of neurofibromin (an upstream molecule of the pathway), along with phosphorylation of downstream molecules such as MST1/2, LATS1/2, and YAP, inhibited Wnt/β-catenin, leading to initiation of adipogenic transcriptional program [71].

Initially, it was believed that the adipocytes derive from cardiomyocyte transdifferentiation. In a study by d'Amati et al., it was observed that transitional cells between cardiomyocytes and adipocytes expressed both vimentin, which is specific for muscular tissue, and desmin, which is specific for adipose tissue [97]. Another study that analyzed an ACM patient's heart biopsy found an area in the ventricular myocardium, between a large amount of adipose tissue, where the cardiomyocytes exhibited abnormalities. They observed a change in the shape of the nucleus, perinuclear vacuolization, accumulation of granules resembling lipid droplets, and disruption of the plasma membrane, resulting in the discharge of intracellular contents into the interstitial space [98]. However, it is crucial to keep in mind that both of the studies were based on single cases. In an in vivo study, researchers found evidence that some adipocytes present in the hearts of desmoplakin knockout mouse models originate directly from cardiomyocytes. These adipocytes were located in the sub-epicardial region, except the midwall, and this conclusion was drawn by using the ventricular isoform of myosin light chain 2 (MLC2v) labeling on the cardiomyocytes [99]. Dorn et al. proposed the presence of shared progenitor cells in the normal heart that have the potential to differentiate into either cardiomyocytes or adipocytes. These cells were identified by the expression of Isl1 (ISL LIM Homeobox 1) and Wt1 (Wilms tumor 1). Depending on the stimuli, they could be pushed or biased towards an adipogenic fate, leading to the fatty replacement that characterizes ACM [100]. The epicardium, the outermost layer of the heart, is composed of mesothelial cells that are usually inactive in the adult heart, unless there is a response to a disease. In the context of development or disease, epicardial cells can transition to mesenchymal cells, allowing them to differentiate into various cardiac cell types [101]. Given this and the fact that fibro-fatty tissue is predominant in the sub-epicardial area in ACM hearts, the epicardial cells have been proposed as a potential source for the fibro-fatty tissue [94].

3.6. Arrhythmogenesis

As noted, the fibrofatty replacement of the normal ventricular tissue is the main characteristic of ACM, creating a substrate that is highly prone to life-threatening arrhythmias. Notably, even in earlier phases, when the ventricle appears to be structurally normal, the risk of arrhythmias is still present [102].

Along with desmosomes, other adhesion junctions, gap junctions, and sodium channels reside at the intercalated disk [103]. While desmosomes provide mechanical support for the intercalated disks, they also maintain the proper function of ion channels located in cardiac intercalated discs [104]. The cardiac sodium channel Na(v)1.5 is an essential component for the initiation and conduction of action potentials, therefore responsible for the regular rhythm in the heart [105]. It has been observed that there is a tight connection between the desmosomal protein complex and the cardiac sodium channel, which explains the early arrhythmias observed in patients with no or limited altered structure at a macroscopic level [102]. In their study, Rizzo et al. noted that this happens because of a reduction in action potential upstroke velocity consequent to a decrease in cardiac sodium current density [102]. The cardiac sodium channel had been previously associated with components of the desmosome, specifically plakophilin-2 [106]. A study has noted that loss of PKP2 expression was associated with reduced peak current density, negatively shifted voltage-dependence of inactivation, and prolonged recovery from inactivation of sodium currents [106]. A study also investigated the prevalence and functional implications of the sodium channel mutations in patients without known desmosomal mutations [48]. The authors performed whole-exome sequencing on six patients and found a rare missense variant (R1898H) [48]. The sodium channels and currents were studied in R1898H human induced pluripotent stem cell-derived cardiomyocytes (R1898H hiPSC-CM). Patch-clamping revealed a 36% reduction in peak sodium current compared to corrected controls, and fluorescence microscopy demonstrated decreased sodium channels and N-cadherin clusters at the intercalated disc [48]. This study also suggests that the sodium channel is part of a complex with adhesion molecules, like N-cadherin. Disrupting this complex may impair sodium current and intercalated disk integrity.

Connexin 43 (Cx43) is the predominant gap junction protein in ventricular myocardium, forming low-resistance electrical channels that allow rapid propagation of action potentials between cardiomyocytes [107]. Cx43 is localized at intercalated discs, where it forms hexameric connexons that attach to neighboring cells to form gap junctions [108]. The phosphorylation state and precise organization of Cx43 are essential for maintaining normal cardiac rhythm [109]. In ACM, the mutations in genes encoding desmosomal proteins cause secondary remodeling of gap junctions and reduced expression of Cx43, even in histologically unaffected regions of the miocardium [23]. Reduced expression and altered localization of Cx43 have been demonstrated both in human samples and experimental models. Oxford et al. reported a decreased expression and a redistribution of Cx43 in PKP2 knockdown rat cells, with Cx3 shifting from the plasma membrane contact zones to a more intracellular localization [110]. Asimaki et al. performed immunohistochemical analysis of human myocardial samples from patients with ACM. They observed that the biopsy samples from all subjects had a reduction in the amount of junctional signal for connexin 43 [111].

4. Current Clinical Management

ACM should be suspected particularly in younger individuals who present symptoms such as palpitations, syncope, or aborted sudden death [112]. The main goals of clinical management for patients with ACM are to improve symptoms by slowing the progression of structural remodeling and preventing complications such as ventricular tachycardia, development of heart failure, or sudden cardiac death [113].

Lifestyle and activity modifications are essential; it is strongly recommended that patients with ACM avoid intense and competitive sports, as prolonged and strenuous physical activity has been shown to increase the risk of ventricular arrhythmias and sudden cardiac death [114].

Pharmacological therapy is currently the cornerstone of disease management, with beta-blockers, amiodarone, sotalol, and flecainide being the preferred treatments [60]. Among these, beta-blockers are the first choice because they reduce adrenergic stimulation. This approach is especially beneficial during exercise. Titrating to the maximum tolerated dose of beta-blockers has been associated with increased survival from ventricular arrhythmias [115]. If beta-blockers are ineffective to control ventricular extrasystoles or tachycardia, additional antiarrhythmic agents such as amiodarone or flecainide can be added [116].

In ACM, the fibrofatty tissue replacing healthy myocardium creates a highly arrhythmogenic substrate, predisposing patients to sustained ventricular tachycardia and ventricular fibrillation. When pharmacological therapy fails to control recurrent arrhythmias adequately, catheter ablation can be an option [113]. Modern ablation strategies often use three-dimensional electroanatomical mapping to identify scar regions and abnormal conduction pathways, enabling a more complex endocardial and epicardial approach. Still, this method is associated with a high recurrence rate [113].

Another option that can be used if pharmacological therapy cannot control the ACM symptoms is the implantation of a cardioverter defibrillator. The implantable cardioverter defibrillator is usually recommended for secondary prevention of sudden cardiac death [60]. The indication for primary prophylactic implantation in patients with ACM remains debatable due to the lack of large randomized clinical trials that provide clear evidence of benefit [116]. Unlike secondary prevention, where a life-threatening arrhythmia has been established and then the placement of an ICD has been decided, the criteria for primary prevention rely on individual risk assessment models. The ARVC risk calculator has become a standard for predicting the 5-year risk of life-threatening arrhythmias and guiding the ICD placement decision [116,117]. According to the European Society of Cardiology (ESC) 2023 guidelines [60], ICD implantation is strongly recommended (Class I) for patients with ACM who have survived cardiac arrest or experienced sustained ventricular tachycardia associated with syncope, as these are clear markers of high risk for sudden death. Class IIa recommendation suggests that ICD should be implanted in patients who have sustained ventricular tachycardia without syncope but have other high-risk clinical features such as arrhythmic syncope, non-sustained ventricular tachycardia (NSVT), right ventricular ejection fraction (RVEF) < 40%, left ventricular ejection fraction (LVEF) < 45%, or sustained monomorphic ventricular tachycardia (SMVT) during programmed electrical stimulation (PES) [60].

5. Emerging Therapeutic Approaches

Actual treatment for ACM focuses on symptom control, arrhythmic risk reduction, and prevention of sudden cardiac death [60]. Unfortunately, current therapeutic options for ACM have seen little evolution and often have limited success. While lifestyle changes, along with the pharmacological and invasive approaches mentioned above, can improve patient outcomes, they remain largely palliative and frequently insufficient, especially in severe forms of the disease [118]. Recent studies have provided a somewhat enhanced understanding of the molecular pathogenesis of ACM, with opportunities for more targeted therapies. However, these emerging therapeutic options are still in early stages and lack evidence of their long-term safety.

5.1. Gene Therapy

Gene replacement therapy consists of the delivery of a healthy transgene to the target organ, usually using a viral vector [118]. Currently, the most commonly used and efficient viral vectors are derived from retroviruses, lentiviruses, adenoviruses, and adeno-associated viruses (AAV) [119]. The adeno-associated virus (AAV) is a small, non-pathogenic parvovirus with a non-enveloped protein capsid that carries single-stranded DNA [120]. Among these viral vectors, recombinant adeno-associated viruses (rAAV) are the leading option for delivering genes to the heart due to their high efficiency in gene transfer, low immune response, and unique capsid-dependent tissue tropism [121]. The capsid of the AAV has a vital role in determining vector tropism and influences both the efficiency of receptor-mediated entry and the level of transgene expression [120].

The *PKP2* gene has been the focus of most preclinical studies, mainly based on its high prevalence in clinical cohorts [121]. Specifically, a deficit of the *PKP2* gene leads to reduced contractility, disruption of the intercalated disc structure, and damaged desmosomal assembly [122].

Bradford et al. utilized the recombinant adeno-associated (rAAV) vector to deliver a functional copy of the plakophilin-2 (*PKP2*) gene to the ventricular cardiac tissue of a mouse model with ARVC [123]. They administered AAV9-PKP2 (serotype 9 of adeno-associated virus that carries a copy of the plakophilin-2 gene) to mice with established ACM disease and observed a notable decrease in arrhythmic events, improved cardiac function, and even partial reversal of fibrofatty infiltration. The authors used a knock-in mouse model that carries a *PKP2* splice-site mutation. Homozygous mutants exhibited sudden death starting at approximately 4 weeks, with a median survival time of 11 weeks, and no mice survived beyond 26 weeks. The homozygous mice displayed early onset of arrhythmias, a prolonged QRS complex, ventricular dysfunction, fibrofatty myocardial replacement, desmosomal protein loss, and inflammation. The administration of AAV9-PKP2 in young mice, before the onset of disease symptoms, led to a reduction in arrhythmic events, improved cardiac function, and preserved myocardial structure. In early therapeutic intervention, treated *PKP2* homozygous mice had 100% survival up to 6 months. By administering AAV-PKP2 when the disease is already established, there were still some improvements, such as the restoration of desmosomal proteins, reduced arrhythmias, and decreased fibrosis, fatty replacement, and inflammation. 100% survival was observed at 20 weeks in this case. Regarding the safety of AAV-PKP2 delivery, the mice showed no liver toxicity after long-term use, with normal liver enzyme levels [123].

Another study, by Kyriakopoulou et al., also used an AAV-mediated delivery of a functional *PKP2* gene [124]. They used induced pluripotent stem cell (iPSC)-derived cardiomyocytes with *PKP2* c.2013delC/WT haploinsufficiency to evaluate efficacy and corresponding in vivo mouse models. AAV6 vectors expressing wild-type human *PKP2* (AAV6-PKP2) managed to restore *PKP2* protein levels and normalize related junctional proteins like *JUP* and *DSP*, while *DSC2* and *DSG2* remained unchanged. Whole-cell patch-clamp recordings measured the sodium conduction to evaluate the peak transient sodium current density. These authors also reported an abnormal sodium conduction that, with treatment, was normalized at a level comparable with healthy controls. Since *PKP2* plays an essential role in myocyte adhesion and interaction, the authors created a three-dimensional engineered human myocardium containing 70% iPS-cell-derived cardiomyocytes and 30% human foreskin fibroblasts in a hydrogel made of type I collagen. The engineered human myocardium had reduced contractile force and kinetic anomalies (regarding beating frequency, contraction velocity, and relaxation velocity). With AAV6-PKP2 treatment, contraction amplitude was improved and kinetics were normalized (elongation of contraction and relaxation), showing excellent results by day 42 of treatment. Mice that carry the murine equivalent of the human *PKP2* mutation received the treatment intraperitoneally in 5-day-old pups and intravenously in adults. Improved desmosomal protein levels were observed, particularly in *JUP*, *DSP*, and *DSG2*, while *DSC2* remained unchanged, similar to the in vitro findings. Mice treated at 2 months with a single dose were followed up to 12 months, showing restored desmosomal integrity and prevention of diastolic dysfunction.

PKP2 gene replacement showed great promise, as it not only restored the primary protein but also other junctional proteins, producing long-lasting results without side effects in healthy models [124].

Wu et al. also determined the efficacy of adeno-associated virus 9 (AAV9) vector in restoring *PKP2* expression in mouse models [125]. The authors engineered a tamoxifen-inducible, cardiac-specific *PKP2* knockout mouse to induce ARVC. They also tested in vitro human induced pluripotent stem cell-derived cardiomyocytes (iPSC-CMs) with *PKP2* silencing. One dose was administered intravenously to the mice, either before disease onset or after the disease was already established. In vitro, *PKP2* silencing resulted in down-regulation of desmosome, sarcomere, and ion-channel gene expression with consequent impaired contraction and electrical signaling. In vivo, those mice developed arrhythmias, dilation of the right ventricle, reduced ejection fraction, remodeling, and rapid mortality (weeks). The results indicate that a single dose can prevent the development of cardiomyopathy before the onset of the disease and prolongs median survival by up to 58 weeks compared to untreated mice, where the median survival was approximately 4.7 weeks. When the cardiomyopathy has developed, it can slow its progression. There was also an extended median survival to >50 weeks. By delivering a functional *PKP2* gene, the reduction in the left ventricular ejection fraction was prevented, the frequency of ventricular arrhythmias was lower, and the dilation of the right ventricle was either prevented or even reversed. RNA sequencing showed a mass-normalization of gene expression across desmosome, sarcomere, ion channel, inflammatory, apoptotic, and fibrosis pathways. The U.S. Food and Drug Administration has approved the Investigational New Drug (IND) application, allowing the beginning of clinical trials for this new therapy [125].

Another study by Opbergen et al. investigated the effects of a gene therapy using an AAVrh.74 vector (AAVrh.74-PKP2a, also designated RP-A60) that delivers a functional human PKP2 transcript to mice with tamoxifen-induced, cardiomyocyte-specific *PKP2* knockout [126]. A single dose was administered via tail-vein injection, and after 28 days, tamoxifen was given to induce *PKP2* knockout. Some of the mice received the treatment dose after disease onset, at 7 or 14 days after the administration of tamoxifen. *PKP2* levels were increased after treatment, and immunofluorescence studies showed *PKP2* localization at the intercalated discs. Treated mice also showed a decrease in myocardial fibrosis, as well as smaller collagen deposits in the ventricles. While untreated mice had 100% mortality at about 50 days after tamoxifen induction, the ones that received the gene therapy survived beyond 5 months. This relatively late administration (at 7 and 14 days after receiving tamoxifen) in the setting of established disease, fielded promising results with increased survival and preserved cardiac function [126].

The results of these studies suggest that using a viral vector to deliver the *PKP2* gene holds great promise for the future treatment of patients with ACM, especially those with *PKP2* mutations.

5.2. Current Clinical Trials

At present, three human clinical trials are recruiting to evaluate this new possible therapy: HEROIC-PKP2 (identifier: NCT06109181), RIDGE-1 (identifier NCT06228924), and RP-A601-0323 (identifier NCT05885412) [118]. The participants will receive a single dose administered intravenously and will be monitored for the next five years. All three studies will evaluate modifications in *PKP2* expression through cardiac biopsy. Results from these trials could determine if gene therapy can become a real disease-modifying treatment option, potentially changing the course of a condition that currently benefits only from palliative approaches.

The clinical trial for HEROIC-PKP2 (identifier: NCT06109181) by Lexeo Therapeutics is still ongoing, but their most recent year-end 2024 report has already highlighted some promising results. Six participants enrolled in the trial, divided into two cohorts. Post-treatment cardiac biopsies were obtained from two of the participants in cohort 1, while the third participant decided not to undergo the biopsy procedure. The biopsies from the first two participants showed increased *PKP-2* protein expression, measured using Western blot assay. Increases of 71% and 115% were observed in *PKP-2* protein expression compared to the pre-treatment baseline. The first participant who reached six months post-treatment showed a 67% reduction in premature ventricular contractions from baseline and normalized QRS duration. The report mentions that the treatment has been well-tolerated with no reports of serious adverse events [127].

Thenaya Therapeutics has already enrolled the first three participants at a dose of 3×10^{13} vg/kg in the RIDGE-1 Phase 1b trial of TN-401 (identifier NCT06228924). In July 2025, the trial's Data Safety Monitoring Board (DSMB) reviewed the available data and approved dose escalation to 6×10^{13} vg/kg (Cohort2), along with the expansion of Cohort1 enrollment. At the moment, the first of three participants in the higher dose cohort has received the gene therapy. Currently, additional data and outcomes from the trial have not yet been reported [128].

Rocket Pharmaceuticals presented their preliminary data from Phase 1 of the clinical trial of RP-A601-0323 (identifier NCT05885412) at the 28th Annual Meeting of the American Society of Gene and Cell Therapy. Preliminary data show that in cardiac biopsies of three patients with ACM, there was increased PKP2 protein expression. Two of the participants with a low baseline level of PKP2 have an increase in the protein by 110% and 398% at the six-month follow-up. RP-A601 also increased protein expression and promoted desmosomal localization of Desmocollin-2 and Cadherin-2 in all three patients. There are also some notes on stabilization or improvement in cardiac function such as: maintained normal proper ventricular function in all patients, improved KCCQ-12 scores by 34–41 points and improved NYHA class from II to I in patients followed beyond 6 months, reduction in premature ventricular contractions by 9–63%, complete resolution of non-sustained ventricular tachycardias at 6 months, decrease from six to two leads of T-wave inversions in one patient at 6 months. At the moment, the tested therapy is well tolerated, with no dose-limiting toxicities observed in patients followed up to 12 months, while most adverse events were mild/moderate, with only one patient experiencing a more severe adverse reaction that was solved within two months, likely resulting from the immunomodulation regimen [129].

5.3. *Targeted Molecular Therapies*

As we mentioned above, the canonical Wnt/β-catenin pathway is crucial for myocyte differentiation and tissue integrity and is involved in ACM pathogenesis. Glycogen synthase kinase 3 beta (GSK-3β) is a serine/threonine kinase that is engaged in the regulation of the canonical Wnt signaling pathway [130]. Given GSK-3β's central role in the downregulation of Wnt signaling, the inhibition of the kinase has been proposed as a possible therapeutic option in ACM [31]. By investigating a GSK-3β inhibitor, SB2, researchers reported the involvement of dysregulated GSK-3β signaling. The SB2 occupies the ATP-binding site of the kinase, blocking it from phosphorylating downstream substrates [131,132]. Lithium is known for its inhibiting properties on the GSK-3β [133]. Accordingly, in a study by Hamstra et al., low doses of lithium were given to wild mice for 6 weeks to inhibit the kinase. The effects on the cardiac SERCA (sarco/endoplasmic reticulum calcium ATPase) function compared with control-fed mice [134]. SERCA uses the energy from ATP hydrolysis to move cytosolic calcium against its concentration gradient into the sarcoplasmic reticulum, resulting in muscle relaxation [135]. Lithium-fed mice showed reduced GSK3β activity compared to control-fed mice and an increased calcium affinity of SERCA. While therapeutic doses of lithium that are commonly used in psychiatry have significant toxicity, this study indicated that a low-dose lithium, which inhibits GSK3β, could improve SERCA pump activity. It thus may be a potential therapy option to improve heart failure and cardiomyopathy [134]. The work of Chelko et al. demonstrated that SB2 can protect against myocyte damage and cardiac dysfunction in vivo in two murine models of ACM [105]. In disease-free hearts, GSK3β is located diffusely in the cytoplasm, but the authors of the study observed that in the ACM mice, the kinase accumulates intracellularly near intercalated discs. In mice with desmosomal mutations, the kinase inhibitor reduced ventricular ectopy, fibrosis and inflammation compared with vehicle-treated controls [136]. SB2 improved left ventricular function and survival rate in mice with a homozygous mutation of desmoglein-2 (*DSG2*), both in sedentary conditions and in exercise. This kinase inhibitor also restored the normal localization of intercalated disc proteins, such as plakoglobin, connexin43 and SAP97 [136]. While these findings underline the central role of GSK3β signaling, they also reveal a promising therapeutic option with the use of SB2.

6. Future Directions

Current clinical management of arrhythmogenic cardiomyopathy remains palliative mainly, focusing on symptom control and prevention of life-threatening arrhythmia through lifestyle changes, anti-arrhythmics and implantable devices.

The future of ACM therapy lies in strategies capable of disease modification by directly addressing the underlying molecular substrate. While gene therapy targets the primary genetic defect, the dysregulated signaling pathways, specifically Wnt/β-catenin/GSK-3β, present potential targets for future treatment to stop the progression of myocardial damage.

The most promising avenue for a curative approach seems to be gene therapy, particularly for the most common genotype associated with PKP2 mutations. Preclinical studies utilizing recombinant adeno-associated viruses to deliver a functional copy of the PKP2 gene have shown remarkable results. They demonstrate the capacity to prevent disease onset when administered early or even partially reverse established disease by restoring desmosomal integrity [123–126]. These encouraging preclinical findings provided the rationale for advancing PKP2 gene therapy into clinical trials. As of late 2024/early 2025, gene therapy candidates are being evaluated in human clinical trials and overall, early-phase clinical trials of gene therapies targeting PKP-2-related arrhythmogenic cardiomyopathy show promising results [127–129]. From the information currently available regarding the ongoing studies, therapies have consistently increased PKP-2 protein expression in cardiac tissue, resulting in reduced arrhythmias and improved ventricular function. So far, the gene therapy has generally been well-tolerated with few adverse events and no dose-limiting toxicities observed. However, these clinical studies are in the early phases, and further follow-up and additional data are needed to confirm long-term safety and true efficacy, as well as to determine the optimal dosage.

Although gene therapies present a potential cure and significant progress has been made in their development, current trials are largely limited to the PKP-2 genotype. In contrast, other common ones (DSP, DSG2) remain unaddressed. Furthermore, strategies to treat advanced disease, where fibrofatty replacement is already established, remain unexplored. Given the efficacy of gene therapy in disease prevention, identifying at-risk individuals in the concealed phase is crucial. Future research should therefore be focused on improving screening methods to enable earlier detection and intervention.

7. Conclusions

ACM is a multifaceted cardiac disease linked to genetic and molecular disturbances that end in structural and electrophysiological abnormalities. While conventional treatment focuses on prevention, it does not stop disease progression. Advances in molecular cardiology have discovered plausible new therapeutic targets, paving the way for a disease-modifying approach. Gene therapy using viral vectors to deliver the normal *PKP2* gene and GSK-3β inhibitors has shown excellent results in preclinical studies on animals, with preservation of myocardial structure and function. Current clinical trials are still ongoing with promising but minimal data at the moment. A more proper understanding of ACM's molecular mechanisms could provide therapies that not only prevent the disease but also modify its trajectory.

Author Contributions: Conceptualization, E.P. and S.H.; methodology, E.P. and S.H.; software, E.P. and S.H.; investigation, E.P. and S.H.; writing—original draft preparation, E.P. and S.H.; writing—review and editing, E.P. and S.H.; supervision, E.P. and S.H. All authors have read and agreed to the published version of the manuscript.

Funding: This research received no external funding.

Institutional Review Board Statement: Not applicable.

Informed Consent Statement: Not applicable.

Data Availability Statement: Not applicable.

Acknowledgments: During the preparation of this manuscript/study, the authors used Chat GPT-4.5 and Grammarly (1.124.1.0) for the purposes of proofreading and summarizing. Figures were created with Biorender (https://www.biorender.com/). The authors have reviewed and edited the output and take full responsibility for the content of this publication.

Conflicts of Interest: The authors declare no conflicts of interest.

Abbreviations

The following abbreviations are used in this manuscript:

AAV	Adeno-Associated Virus
AAV6	Adeno-Associated Virus, Serotype 6
AAV9	Adeno-Associated Virus, Serotype 9
ACM	Arrhythmogenic Cardiomyopathy
ALVC	Arrythmogenic Left Ventricular Cardiomyopathy
ARVC	Arrhythmogenic Right Ventricular Cardiomyopathy
ARVD	Arrhythmogenic Right Ventricular Dysplasia
β-catenin	Beta-Catenin (Transcription Factor)
C/EBP-α	CCAAT/Enhancer-Binding Protein Alpha
Cx43	Connexin-43
CDH2	N-Cadherin
DCM	Dilated Cardiomyopathy
DES	Desmin
DSC2	Desmocollin-2
DSG2	Desmoglein-2
DSP	Desmoplakin
EKG	Electrocardiogram
ESC	European Society Of Cardiology
FDA	Food and Drug Administration
GSK-3β	Glycogen Synthase Kinase 3 Beta
ICD	Implantable Cardioverter-Defibrillator
IND	Investigational New Drug

IL-1β	Interleukin 1 Beta
IL-6	Interleukin 6
iPSC	Induced Pluripotent Stem Cells
iPSC-CM	Induced Pluripotent Stem Cell-Derived Cardiomyocyte
ISL1	ISL LIM Homeobox 1
JUP	Junctional plakoglobin
LATS1/2	Large Tumor Suppressor Kinase 1/2
LVEF	Left Ventricular Ejection Fraction
LV	Left Ventricle
MLC2v	Myosin Light Chain 2 (ventricular isoform)
MST1/2	Mammalian STE20-like Protein Kinase
NSVT	Non-Sustained Ventricular Tachycardia
PES	Programmed Electrical Stimulation
PKC-α	Protein Kinase C Alpha
PKP2	Plakophilin-2
PLB	Phospholamban
PPARγ	Peroxisome Proliferator-Activated Receptor Gamma
rAAV	Recombinant Adeno-Associated Virus
RVEF	Right Ventricular Ejection Fraction
RV	Right Ventricle
SAP97	Synapse-Associated Protein 97
SCN5A	Sodium-Voltage-Gated Channel Alpha Subunit 5
SB2	Selective GSK-3β Inhibitor
SCD	Sudden Cardiac Death
SERCA	Sarco/Endoplasmic Reticulum Calcium ATPase
siRNA	Small Interfering RNA
SMVT	Sustained Monomorphic Ventricular Tachycardia
TAZ	Transcriptional Coactivator with PDZ-binding Motif
TCF	T-cell Factor
TEAD	TEA Domain Family Member
TGF-β	Transforming Growth Factor Beta
TGF3β	Transforming Growth Factor 3 Beta
TMEM43	Transmembrane Protein 43
TNF-α	Tumor Necrosis Factor Alpha
TTN	Titin
WT1	Wilms Tumor 1
Wnt	Wingless-related integration site (signaling pathway)
YAP	Yess-associated Protein

References

1. Corrado, D.; Basso, C.; Judge, D.P. Arrhythmogenic cardiomyopathy. *Circ. Res.* **2017**, *121*, 784–802. [PubMed]
2. Corrado, D.; Perazzolo Marra, M.; Zorzi, A.; Beffagna, G.; Cipriani, A.; Lazzari, M.D.; Migliore, F.; Pilichou, K.; Rampazzo, A.; Rigato, I.; et al. Diagnosis of arrhythmogenic cardiomyopathy: The Padua criteria. *Int. J. Cardiol.* **2020**, *319*, 106–114. [CrossRef]
3. Thiene, G.; Nava, A.; Corrado, D.; Rossi, L.; Pennelli, N. Right ventricular cardiomyopathy and sudden death in young people. *N. Engl. J. Med.* **1988**, *318*, 129–133. [PubMed]
4. Vimalanathan, A.K.; Ehler, E.; Gehmlich, K. Genetics of and pathogenic mechanisms in arrhythmogenic right ventricular cardiomyopathy. *Biophys. Rev.* **2018**, *10*, 973–982. [CrossRef]
5. Basso, C.; Corrado, D.; Marcus, F.I.; Nava, A.; Thiene, G. Arrhythmogenic right ventricular cardiomyopathy. *Lancet* **2009**, *373*, 1289–1300. [CrossRef]
6. Engel, M.; Shiel, E.A.; Chelko, S.P. Basic and translational mechanisms in inflammatory arrhythmogenic cardiomyopathy. *Int. J. Cardiol.* **2024**, *397*, 131602.
7. Kochi, A.N.; Vettor, G.; Dessanai, M.A.; Pizzamiglio, F.; Tondo, C. Sudden cardiac death in athletes: From the basics to the practical work-up. *Medicina* **2021**, *57*, 168. [CrossRef]

8. Corrado, D.; Basso, C.; Thiene, G. Sudden cardiac death in young people with apparently normal heart. *Cardiovasc. Res.* **2001**, *50*, 399–408. [CrossRef] [PubMed]
9. James, C.A.; Bhonsale, A.; Tichnell, C.; Murray, B.; Russell, S.D.; Tandri, H.; Tedford, R.J.; Judge, D.P.; Calkins, H. Exercise increases age-related penetrance and arrhythmic risk in arrhythmogenic right ventricular dysplasia/cardiomyopathy-associated desmosomal mutation carriers. *J. Am. Coll. Cardiol.* **2013**, *62*, 1290–1297.
10. de la Guía-Galipienso, F.; Ugedo-Alzaga, K.; Grazioli, G.; Quesada-Ocete, F.J.; Feliu-Rey, E.; Perez, M.V.; Quesada-Dorador, A.; Sanchis-Gomar, F. Arrhythmogenic cardiomyopathy and athletes: A dangerous relationship. *Curr. Probl. Cardiol.* **2023**, *48*, 101799. [CrossRef]
11. Austin, K.M.; Trembley, M.A.; Chandler, S.F.; Sanders, S.P.; Saffitz, J.E.; Abrams, D.J.; Pu, W.T. Molecular mechanisms of arrhythmogenic cardiomyopathy. *Nat. Rev. Cardiol.* **2019**, *16*, 519–537. [CrossRef]
12. Morgan, J.A.; Corrigan, F.; Baune, B.T. Effects of physical exercise on central nervous system functions: A review of brain region specific adaptations. *J. Mol. Psychiatry* **2015**, *3*, 3. [CrossRef]
13. Chelko, S.P.; Keceli, G.; Carpi, A.; Doti, N.; Agrimi, J.; Asimaki, A.; Beti, C.B.; Miyamoto, M.; Amat-Codina, N.; Bedja, D.; et al. Exercise triggers CAPN1-mediated AIF truncation, inducing myocyte cell death in arrhythmogenic cardiomyopathy. *Sci. Transl. Med.* **2021**, *13*, eabf0891. [CrossRef]
14. Hammer, K.P.; Mustroph, J.; Stauber, T.; Birchmeier, W.; Wagner, S.; Maier, L.S. Beneficial effect of voluntary physical exercise in Plakophilin2 transgenic mice. *PLoS ONE* **2021**, *16*, e0252649. [CrossRef] [PubMed]
15. Pecoraro, M.; Verrilli, V.; Pinto, A.; Popolo, A. Role of connexin 43 in cardiovascular diseases. *Eur. J. Pharmacol.* **2015**, *768*, 71–76. [CrossRef]
16. Coscarella, I.L.; Landim-Vieira, M.; Pinto, J.R.; Chelko, S.P. Arrhythmogenic Cardiomyopathy: Exercise pitfalls, role of Connexin-43, and moving beyond antiarrhythmics. *Int. J. Mol. Sci.* **2022**, *23*, 8753. [CrossRef]
17. Veeranki, S.; Givvimani, S.; Kundu, S.; Metreveli, N.; Pushpakumar, S.; Tyagi, S.C. Moderate intensity exercise prevents diabetic cardiomyopathy associated contractile dysfunction through restoration of mitochondrial function and connexin 43 levels in db/db mice. *J. Mol. Cell Cardiol.* **2016**, *92*, 163–173. [CrossRef]
18. Sanchis-Gomar, F.; Malkani, K.; Perez-Quilis, C. Exercise training intensity and connexin 43 expression in hypertrophic cardiomyopathy. *J. Mol. Cell Cardiol.* **2017**, *109*, 60. [CrossRef] [PubMed]
19. Chang, Y.; Yu, T.; Yang, H.; Peng, Z. Exhaustive exercise-induced cardiac conduction system injury and changes of cTnT and Cx43. *Int. J. Sports Med.* **2015**, *36*, 1–8. [CrossRef] [PubMed]
20. Corrado, D.; Fontaine, G.; Marcus, F.I.; McKenna, W.J.; Nava, A.; Thiene, G.; Wichter, T. Arrhythmogenic right ventricular dysplasia/cardiomyopathy. *Circulation* **2000**, *101*, e101–e106. [CrossRef]
21. Costa, S.; Cerrone, M.; Saguner, A.M.; Brunckhorst, C.; Delmar, M.; Duru, F. Arrhythmogenic cardiomyopathy: An in-depth look at molecular mechanisms and clinical correlates. *Trends Cardiovasc. Med.* **2021**, *31*, 395–402. [CrossRef]
22. Gandjbakhch, E.; Redheuil, A.; Pousset, F.; Charron, P.; Frank, R. Clinical diagnosis, imaging, and genetics of arrhythmogenic right ventricular cardiomyopathy/dysplasia: JACC state-of-the-art review. *J. Am. Coll. Cardiol.* **2018**, *72*, 784–804. [CrossRef]
23. Delmar, M.; McKenna, W.J. The cardiac desmosome and arrhythmogenic cardiomyopathies: From gene to disease. *Circ. Res.* **2010**, *107*, 700–714. [CrossRef]
24. Corrado, D.; Anastasakis, A.; Basso, C.; Bauce, B.; Blomström-Lundqvist, C.; Bucciarelli-Ducci, C.; Cipriani, A.; De Asmundis, C.; Gandjbakhch, E.; Jiménez-Jáimez, J.; et al. Proposed diagnostic criteria for arrhythmogenic cardiomyopathy: European Task Force consensus report. *Int. J. Cardiol.* **2024**, *395*, 131447. [CrossRef] [PubMed]
25. Patel, V.; Asatryan, B.; Siripanthong, B.; Munroe, P.B.; Tiku-Owens, A.; Lopes, L.R.; Khanji, M.Y.; Protonotarios, A.; Santangeli, P.; Muser, D.; et al. State of the art review on genetics and precision medicine in arrhythmogenic cardiomyopathy. *Int. J. Mol. Sci.* **2020**, *21*, 6615. [CrossRef] [PubMed]
26. Akdis, D.; Brunckhorst, C.; Duru, F.; Saguner, A.M. Arrhythmogenic cardiomyopathy: Electrical and structural phenotypes. *Arrhythm. Electrophysiol. Rev.* **2016**, *5*, 90–101. [CrossRef] [PubMed]
27. Varrenti, M.; Preda, A.; Frontera, A.; Baroni, M.; Gigli, L.; Vargiu, S.; Colombo, G.; Carbonaro, M.; Paolucci, M.; Giordano, F.; et al. Arrhythmogenic cardiomyopathy: Definition, classification and arrhythmic risk stratification. *J. Clin. Med.* **2024**, *13*, 456. [CrossRef]
28. Bariani, R.; Rigato, I.; Cason, M.; Marinas, M.B.; Celeghin, R.; Pilichou, K.; Bauce, B. Genetic background and clinical features in arrhythmogenic left ventricular cardiomyopathy: A systematic review. *J. Clin. Med.* **2022**, *11*, 4313. [CrossRef]
29. Mattesi, G.; Cipriani, A.; Bauce, B.; Rigato, I.; Zorzi, A.; Corrado, D. Arrhythmogenic left ventricular cardiomyopathy: Genotype-phenotype correlations and new diagnostic criteria. *J. Clin. Med.* **2021**, *10*, 2212. [CrossRef]
30. Pinamonti, B.; Brun, F.; Mestroni, L.; Sinagra, G. Arrhythmogenic right ventricular cardiomyopathy: From genetics to diagnostic and therapeutic challenges. *World J. Cardiol.* **2014**, *6*, 1234–1244. [CrossRef]
31. Stevens, T.L.; Wallace, M.J.; Refaey, M.E.; Roberts, J.D.; Koenig, S.N.; Mohler, P.J. Arrhythmogenic cardiomyopathy: Molecular insights for improved therapeutic design. *J. Cardiovasc. Dev. Dis.* **2020**, *7*, 21. [CrossRef] [PubMed]

32. Delva, E.; Tucker, D.K.; Kowalczyk, A.P. The desmosome. *Cold Spring Harb. Perspect. Biol.* **2009**, *1*, a002543. [CrossRef] [PubMed]
33. Protonotarios, N.; Tsatsopoulou, A.; Patsourakos, P.; Alexopoulos, D.; Gezerlis, P.; Simitsis, S.; Scampardonis, G. Cardiac abnormalities in familial palmoplantar keratosis. *Br. Heart J.* **1986**, *56*, 321–326. [CrossRef] [PubMed]
34. Leopoulou, M.; Mattsson, G.; LeQuang, J.A.; Pergolizzi, J.V.; Varrassi, G.; Wallhagen, M.; Magnusson, P. Naxos disease—A narrative review. *Expert. Rev. Cardiovasc. Ther.* **2020**, *18*, 801–808. [CrossRef]
35. van Tintelen, J.P.; Entius, M.M.; Bhuiyan, Z.A.; Jongbloed, R.; Wiesfeld, A.C.P.; Wilde, A.A.M.; van der Smagt, J.; Boven, L.G.; Mannens, M.M.A.M.; van Langen, I.M.; et al. Plakophilin-2 mutations are the major determinant of familial arrhythmogenic right ventricular dysplasia/cardiomyopathy. *Circulation* **2006**, *113*, 1650–1658. [CrossRef]
36. Groeneweg, J.A.; Bhonsale, A.; James, C.A.; te Riele, A.S.; Dooijes, D.; Tichnell, C.; Murray, B.; Wiesfeld, A.C.P.; Sawant, A.C.; Kassamali, B.; et al. Clinical presentation, long-term follow-up, and outcomes of 1001 arrhythmogenic right ventricular dysplasia/cardiomyopathy patients and family members. *Circ. Cardiovasc. Genet.* **2015**, *8*, 437–446. [CrossRef]
37. Biernacka, E.K.; Borowiec, K.; Franaszczyk, M.; Szperl, M.; Rampazzo, A.; Woźniak, O.; Roszczynko, M.; Śmigielski, W.; Lutyńska, A.; Hoffman, P. Pathogenic variants in plakophilin-2 gene (PKP2) are associated with better survival in arrhythmogenic right ventricular cardiomyopathy. *J. Appl. Genet.* **2021**, *62*, 613–620. [CrossRef]
38. James, C.A.; Jongbloed, J.D.H.; Hershberger, R.E.; Morales, A.; Judge, D.P.; Syrris, P.; Pilichou, K.; Domingo, A.M.; Murray, B.; Cadrin-Tourigny, J.; et al. International evidence based reappraisal of genes associated with arrhythmogenic right ventricular cardiomyopathy using the Clinical Genome Resource framework. *Circ. Genom. Precis. Med.* **2021**, *14*, e003273. [CrossRef]
39. Alcalde, M.; Campuzano, O.; Berne, P.; García-Pavía, P.; Doltra, A.; Arbelo, E.; Sarquella-Brugada, G.; Iglesias, A.; Alonso-Pulpon, L.; Brugada, J.; et al. Stop-gain mutations in PKP2 are associated with a later age of onset of arrhythmogenic right ventricular cardiomyopathy. *PLoS ONE* **2014**, *9*, e100560. [CrossRef]
40. Xu, Z.; Zhu, W.; Wang, C.; Huang, L.; Zhou, Q.; Hu, J.; Cheng, X.; Hong, K. Genotype-phenotype relationship in patients with arrhythmogenic right ventricular cardiomyopathy caused by desmosomal gene mutations: A systematic review and meta-analysis. *Sci. Rep.* **2017**, *7*, 41387. [CrossRef]
41. Heuser, A.; Plovie, E.R.; Ellinor, P.T.; Grossmann, K.S.; Shin, J.T.; Wichter, T.; Basson, C.T.; Lerman, B.B.; Sasse-Klaassen, S.; Thierfelder, L.; et al. Mutant desmocollin-2 causes arrhythmogenic right ventricular cardiomyopathy. *Am. J. Hum. Genet.* **2006**, *79*, 1081–1088. [CrossRef]
42. Li, D.; Liu, Y.; Maruyama, M.; Zhu, W.; Chen, H.; Zhang, W.; Reuter, S.; Lin, S.-F.; Haneline, L.S.; Field, L.J.; et al. Restrictive loss of plakoglobin in cardiomyocytes leads to arrhythmogenic cardiomyopathy. *Hum. Mol. Genet.* **2011**, *20*, 4582–4596. [CrossRef]
43. Bauce, B.; Rampazzo, A.; Basso, C.; Mazzotti, E.; Rigato, I.; Steriotis, A.; Beffagna, G.; Lorenzon, A.; De Bortoli, M.; Pilichou, K.; et al. Clinical phenotype and diagnosis of arrhythmogenic right ventricular cardiomyopathy in pediatric patients carrying desmosomal gene mutations. *Heart Rhythm.* **2011**, *8*, 1686–1695. [CrossRef]
44. Merner, N.D.; Hodgkinson, K.A.; Haywood, A.F.M.; Connors, S.; French, V.M.; Drenckhahn, J.-D.; Kupprion, C.; Ramadanova, K.; Thierfelder, L.; McKenna, W.; et al. Arrhythmogenic right ventricular cardiomyopathy type 5 is a fully penetrant, lethal arrhythmic disorder caused by a missense mutation in the TMEM43 gene. *Am. J. Hum. Genet.* **2008**, *82*, 809–821. [CrossRef]
45. Bermúdez-Jiménez, F.J.; Carriel, V.; Brodehl, A.; Alaminos, M.; Campos, A.; Schirmer, I.; Milting, H.; Abril, B.Á.; Álvarez, M.; López-Fernández, S.; et al. Novel Desmin mutation p.Glu401Asp impairs filament formation, disrupts cell membrane integrity, and causes severe arrhythmogenic left ventricular cardiomyopathy/dysplasia. *Circulation* **2018**, *137*, 1595–1610. [CrossRef] [PubMed]
46. van der Heijden, J.F.; Hassink, R.J. The phospholamban p.Arg14del founder mutation in Dutch patients with arrhythmogenic cardiomyopathy. *Neth. Heart J.* **2013**, *21*, 284–285. [CrossRef]
47. Mayosi, B.M.; Fish, M.; Shaboodien, G.; Mastantuono, E.; Kraus, S.; Wieland, T.; Kotta, M.-C.; Chin, A.; Laing, N.; Ntusi, N.B.A.; et al. Identification of cadherin 2 (CDH2) mutations in arrhythmogenic right ventricular cardiomyopathy. *Circ. Cardiovasc. Genet.* **2017**, *10*, e001605. [CrossRef]
48. Te Riele, A.S.J.M.; Agullo-Pascual, E.; James, C.A.; Leo-Macias, A.; Cerrone, M.; Zhang, M.; Lin, X.; Lin, B.; Sobreira, N.L.; Amat-Alarcon, N.; et al. Multilevel analyses of SCN5A mutations in arrhythmogenic right ventricular dysplasia/cardiomyopathy suggest non-canonical mechanisms for disease pathogenesis. *Cardiovasc. Res.* **2017**, *113*, 102–111. [CrossRef]
49. Beffagna, G.; Occhi, G.; Nava, A.; Vitiello, L.; Ditadi, A.; Basso, C.; Bauce, B.; Carraro, G.; Thiene, G.; Towbin, J.A.; et al. Regulatory mutations in transforming growth factor-beta3 gene cause arrhythmogenic right ventricular cardiomyopathy type 1. *Cardiovasc. Res.* **2005**, *65*, 366–373. [CrossRef]
50. Quarta, G.; Syrris, P.; Ashworth, M.; Jenkins, S.; Zuborne Alapi, K.; Morgan, J.; Muir, A.; Pantazis, A.; McKenna, W.J.; Elliott, P.M. Mutations in the Lamin A/C gene mimic arrhythmogenic right ventricular cardiomyopathy. *Eur. Heart J.* **2012**, *33*, 1128–1136. [CrossRef]
51. Carruth, E.D.; Qureshi, M.; Alsaid, A.; Kelly, M.A.; Calkins, H.; Murray, B.; Tichnell, C.; Sturm, A.C.; Regeneron Genetics Center; Baras, A.; et al. Loss-of-function FLNC variants are associated with arrhythmogenic cardiomyopathy phenotypes when identified through exome sequencing of a general clinical population. *Circ. Genom. Precis. Med.* **2022**, *15*, e003645. [CrossRef]

52. Al-Amoudi, A.; Frangakis, A.S. Structural studies on desmosomes. *Biochem. Soc. Trans.* **2008**, *36*, 181–187. [CrossRef]
53. Dusek, R.L.; Godsel, L.M.; Green, K.J. Discriminating roles of desmosomal cadherins: Beyond desmosomal adhesion. *J. Dermatol. Sci.* **2007**, *45*, 7–21. [CrossRef]
54. Green, K.J.; Simpson, C.L. Desmosomes: New perspectives on a classic. *J. Investig. Dermatol.* **2007**, *127*, 2499–2515. [CrossRef]
55. Najor, N.A. Desmosomes in human disease. *Annu. Rev. Pathol.* **2018**, *13*, 51–70. [CrossRef] [PubMed]
56. Hoorntje, E.T.; Te Rijdt, W.P.; James, C.A.; Pilichou, K.; Basso, C.; Judge, D.P.; Bezzina, C.R.; van Tintelen, J.P. Arrhythmogenic cardiomyopathy: Pathology, genetics, and concepts in pathogenesis. *Cardiovasc. Res.* **2017**, *113*, 1521–1531. [CrossRef]
57. Meraviglia, V.; Alcalde, M.; Campuzano, O.; Bellin, M. Inflammation in the pathogenesis of arrhythmogenic cardiomyopathy: Secondary event or active driver? *Front. Cardiovasc. Med.* **2021**, *8*, 784715. [CrossRef] [PubMed]
58. McGregor, S.M.; Husain, A.N. A brief review and update of the clinicopathologic diagnosis of arrhythmogenic cardiomyopathy. *Arch. Pathol. Lab. Med.* **2015**, *139*, 1181–1186. [CrossRef]
59. van der Voorn, S.M.; Te Riele, A.S.J.M.; Basso, C.; Calkins, H.; Remme, C.A.; van Veen, T.A.B. Arrhythmogenic cardiomyopathy: Pathogenesis, pro-arrhythmic remodelling, and novel approaches for risk stratification and therapy. *Cardiovasc. Res.* **2020**, *116*, 1571–1584. [CrossRef]
60. Steinhart, Z.; Angers, S. Wnt signaling in development and tissue homeostasis. *Development* **2018**, *145*, dev146589. [CrossRef]
61. Lorenzon, A.; Calore, M.; Poloni, G.; De Windt, L.J.; Braghetta, P.; Rampazzo, A. Wnt/β-catenin pathway in arrhythmogenic cardiomyopathy. *Oncotarget* **2017**, *8*, 60640–60655. [CrossRef] [PubMed]
62. MacDonald, B.T.; Tamai, K.; He, X. Wnt/beta-catenin signaling: Components, mechanisms, and diseases. *Dev. Cell* **2009**, *17*, 9–26. [CrossRef]
63. Gao, S.; Puthenvedu, D.; Lombardi, R.; Chen, S.N. Established and emerging mechanisms in the pathogenesis of arrhythmogenic cardiomyopathy: A multifaceted disease. *Int. J. Mol. Sci.* **2020**, *21*, 6320. [CrossRef] [PubMed]
64. Ross, S.E.; Hemati, N.; Longo, K.A.; Bennett, C.N.; Lucas, P.C.; Erickson, R.L.; MacDougald, O.A. Inhibition of adipogenesis by Wnt signaling. *Science* **2000**, *289*, 950–953. [CrossRef]
65. Chen, S.; Guttridge, D.C.; You, Z.; Zhang, Z.; Fribley, A.; Mayo, M.W.; Kitajewski, J.; Wang, C.Y. Wnt-1 signaling inhibits apoptosis by activating beta-catenin/T cell factor-mediated transcription. *J. Cell Biol.* **2001**, *152*, 87–96. [CrossRef]
66. Piquer-Gil, M.; Domenech-Dauder, S.; Sepúlveda-Gómez, M.; Machí-Camacho, C.; Braza-Boïls, A.; Zorio, E. Non coding RNAs as regulators of Wnt/β-catenin and Hippo pathways in arrhythmogenic cardiomyopathy. *Biomedicines* **2022**, *10*, 2619. [CrossRef]
67. Yu, F.-X.; Zhao, B.; Guan, K.-L. Hippo pathway in organ size control, tissue homeostasis, and cancer. *Cell* **2015**, *163*, 811–828. [CrossRef]
68. Han, Y. Analysis of the role of the Hippo pathway in cancer. *J. Transl. Med.* **2019**, *17*, 116. [CrossRef]
69. Fu, M.; Hu, Y.; Lan, T.; Guan, K.-L.; Luo, T.; Luo, M. The Hippo signalling pathway and its implications in human health and diseases. *Signal Transduct. Target. Ther.* **2022**, *7*, 376. [CrossRef] [PubMed]
70. Plouffe, S.W.; Meng, Z.; Lin, K.C.; Lin, B.; Hong, A.W.; Chun, J.V.; Guan, K.-L. Characterization of Hippo pathway components by gene inactivation. *Mol. Cell* **2016**, *64*, 993–1008. [CrossRef]
71. Chen, S.N.; Gurha, P.; Lombardi, R.; Ruggiero, A.; Willerson, J.T.; Marian, A.J. The hippo pathway is activated and is a causal mechanism for adipogenesis in arrhythmogenic cardiomyopathy. *Circ. Res.* **2014**, *114*, 454–468. [CrossRef]
72. Heallen, T.; Zhang, M.; Wang, J.; Bonilla-Claudio, M.; Klysik, E.; Johnson, R.L.; Martin, J.F. Hippo pathway inhibits Wnt signaling to restrain cardiomyocyte proliferation and heart size. *Science* **2011**, *332*, 458–461. [CrossRef]
73. Chandra, M.; Miriyala, S.; Panchatcharam, M. PPARγ and its role in cardiovascular diseases. *PPAR Res.* **2017**, *2017*, 6404638. [CrossRef] [PubMed]
74. Ahmadian, M.; Suh, J.M.; Hah, N.; Liddle, C.; Atkins, A.R.; Downes, M.; Evans, R.M. PPARγ signaling and metabolism: The good, the bad and the future. *Nat. Med.* **2013**, *19*, 557–566. [CrossRef]
75. Reisqs, J.-B.; Moreau, A.; Charrabi, A.; Sleiman, Y.; Meli, A.C.; Millat, G.; Briand, V.; Beauverger, P.; Richard, S.; Chevalier, P. The PPARγ pathway determines electrophysiological remodelling and arrhythmia risks in DSC2 arrhythmogenic cardiomyopathy. *Clin. Transl. Med.* **2022**, *12*, e748. [CrossRef] [PubMed]
76. Sommariva, E.; Stadiotti, I.; Casella, M.; Catto, V.; Dello Russo, A.; Carbucicchio, C.; Arnaboldi, L.; De Metrio, S.; Milano, G.; Scopece, A.; et al. Oxidized LDL-dependent pathway as new pathogenic trigger in arrhythmogenic cardiomyopathy. *EMBO Mol. Med.* **2021**, *13*, e14365. [CrossRef]
77. Wei, C.; Saguner, A.; Duru, F.; Marine, J.E.; James, C.A.; Calkins, H.; Judge, D.P.; Shou, W.; Chen, H.-S.V. Po-02-246 pparγ activation mediates cardiac pathologies in two mouse models of arrhythmogenic cardiomyopathy (acm) and its therapeutic implications. *Heart Rhythm.* **2023**, *20*, S286. [CrossRef]
78. Lu, W.; Rao, Y.; Li, Y.; Dai, Y.; Chen, K. The landscape of cell death processes with associated immunogenic and fibrogenic effects in arrhythmogenic cardiomyopathy. *J. Cardiovasc. Dev. Dis.* **2022**, *9*, 301. [CrossRef]
79. Mallat, Z.; Tedgui, A.; Fontaliran, F.; Frank, R.; Durigon, M.; Fontaine, G. Evidence of apoptosis in arrhythmogenic right ventricular dysplasia. *N. Engl. J. Med.* **1996**, *335*, 1190–1196. [CrossRef]

80. Hariharan, V.; Asimaki, A.; Michaelson, J.E.; Plovie, E.; MacRae, C.A.; Saffitz, J.E.; Huang, H. Arrhythmogenic right ventricular cardiomyopathy mutations alter shear response without changes in cell-cell adhesion. *Cardiovasc. Res.* **2014**, *104*, 280–289. [CrossRef] [PubMed]
81. Caspi, O.; Huber, I.; Gepstein, A.; Arbel, G.; Maizels, L.; Boulos, M.; Gepstein, L. Modeling of arrhythmogenic right ventricular cardiomyopathy with human induced pluripotent stem cells. *Circ. Cardiovasc. Genet.* **2013**, *6*, 557–568. [CrossRef]
82. Garcia-Gras, E.; Lombardi, R.; Giocondo, M.J.; Willerson, J.T.; Schneider, M.D.; Khoury, D.S.; Marian, A.J. Suppression of canonical Wnt/beta-catenin signaling by nuclear plakoglobin recapitulates phenotype of arrhythmogenic right ventricular cardiomyopathy. *J. Clin. Investig.* **2006**, *116*, 2012–2021. [CrossRef]
83. Campuzano, O.; Alcalde, M.; Iglesias, A.; Barahona-Dussault, C.; Sarquella-Brugada, G.; Benito, B.; Arzamendi, D.; Flores, J.; Leung, T.K.; Talajic, M.; et al. Arrhythmogenic right ventricular cardiomyopathy: Severe structural alterations are associated with inflammation. *J. Clin. Pathol.* **2012**, *65*, 1077–1083. [CrossRef]
84. Asatryan, B.; Asimaki, A.; Landstrom, A.P.; Khanji, M.Y.; Odening, K.E.; Cooper, L.T.; Marchlinski, F.E.; Gelzer, A.R.; Semsarian, C.; Reichlin, T.; et al. Inflammation and immune response in arrhythmogenic cardiomyopathy: State-of-the-art review. *Circulation* **2021**, *144*, 1646–1655. [CrossRef] [PubMed]
85. Jordan, A.N.; Lyne, J.; De Silva, R.; Wong, T. Myocarditic appearance of arrhythmogenic right ventricular cardiomyopathy. *Circulation* **2010**, *122*, e556–e557. [CrossRef]
86. Patrianakos, A.P.; Protonotarios, N.; Nyktari, E.; Pagonidis, K.; Tsatsopoulou, A.; Parthenakis, F.I.; Vardas, P.E. Arrhythmogenic right ventricular cardiomyopathy/dysplasia and troponin release. Myocarditis or the "hot phase" of the disease? *Int. J. Cardiol.* **2012**, *157*, e26-8. [CrossRef]
87. Lazaros, G.; Anastasakis, A.; Tsiachris, D.; Dilaveris, P.; Protonotarios, N.; Stefanadis, C. Naxos disease presenting with ventricular tachycardia and troponin elevation. *Heart Vessel.* **2009**, *24*, 63–65. [CrossRef]
88. Monda, E.; Palmiero, G.; Rubino, M.; Verrillo, F.; Amodio, F.; Di Fraia, F.; Pacileo, R.; Fimiani, F.; Esposito, A.; Cirillo, A.; et al. Molecular basis of inflammation in the pathogenesis of cardiomyopathies. *Int. J. Mol. Sci.* **2020**, *21*, 6462. [CrossRef] [PubMed]
89. Piek, A.; de Boer, R.A.; Silljé, H.H.W. The fibrosis-cell death axis in heart failure. *Heart Fail. Rev.* **2016**, *21*, 199–211. [CrossRef]
90. Lubos, N.; van der Gaag, S.; Gerçek, M.; Kant, S.; Leube, R.E.; Krusche, C.A. Inflammation shapes pathogenesis of murine arrhythmogenic cardiomyopathy. *Basic. Res. Cardiol.* **2020**, *115*, 42. [CrossRef] [PubMed]
91. Asimaki, A.; Tandri, H.; Duffy, E.R.; Winterfield, J.R.; Mackey-Bojack, S.; Picken, M.M.; Cooper, L.T.; Wilber, D.J.; Marcus, F.I.; Basso, C.; et al. Altered desmosomal proteins in granulomatous myocarditis and potential pathogenic links to arrhythmogenic right ventricular cardiomyopathy. *Circ. Arrhythm. Electrophysiol.* **2011**, *4*, 743–752. [CrossRef]
92. Stadiotti, I.; Catto, V.; Casella, M.; Tondo, C.; Pompilio, G.; Sommariva, E. Arrhythmogenic cardiomyopathy: The guilty party in adipogenesis. *J. Cardiovasc. Transl. Res.* **2017**, *10*, 446–454. [CrossRef]
93. Vencato, S.; Romanato, C.; Rampazzo, A.; Calore, M. Animal models and molecular pathogenesis of arrhythmogenic cardiomyopathy associated with pathogenic variants in intercalated disc genes. *Int. J. Mol. Sci.* **2024**, *25*, 6208. [CrossRef]
94. Kohela, A.; van Rooij, E. Fibro-fatty remodelling in arrhythmogenic cardiomyopathy. *Basic. Res. Cardiol.* **2022**, *117*, 22. [CrossRef]
95. Lefterova, M.I.; Haakonsson, A.K.; Lazar, M.A.; Mandrup, S. PPARγ and the global map of adipogenesis and beyond. *Trends Endocrinol. Metab.* **2014**, *25*, 293–302. [CrossRef]
96. Olofsson, L.E.; Orho-Melander, M.; William-Olsson, L.; Sjöholm, K.; Sjöström, L.; Groop, L.; Carlsson, B.; Carlsson, L.M.S.; Olsson, B. CCAAT/enhancer binding protein alpha (C/EBPalpha) in adipose tissue regulates genes in lipid and glucose metabolism and a genetic variation in C/EBPalpha is associated with serum levels of triglycerides. *J. Clin. Endocrinol. Metab.* **2008**, *93*, 4880–4886. [CrossRef] [PubMed]
97. d'Amati, G.; di Gioia, C.R.T.; Giordano, C.; Gallo, P. Myocyte transdifferentiation. *Arch. Pathol. Lab. Med.* **2000**, *124*, 287–290. [CrossRef]
98. Fujita, S.; Terasaki, F.; Otsuka, K.; Katashima, T.; Kanzaki, Y.; Kawamura, K.; Tanaka, T.; Kitaura, Y. Markedly increased intracellular lipid droplets and disruption of intercellular junctions in biopsied myocardium from a patient with arrhythmogenic right ventricular cardiomyopathy. *Heart Vessel.* **2008**, *23*, 440–444. [CrossRef]
99. Lyon, R.C.; Mezzano, V.; Wright, A.T.; Pfeiffer, E.; Chuang, J.; Banares, K.; Castaneda, A.; Ouyang, K.; Cui, L.; Contu, R.; et al. Connexin defects underlie arrhythmogenic right ventricular cardiomyopathy in a novel mouse model. *Hum. Mol. Genet.* **2014**, *23*, 1134–1150. [CrossRef] [PubMed]
100. Dorn, T.; Kornherr, J.; Parrotta, E.I.; Zawada, D.; Ayetey, H.; Santamaria, G.; Iop, L.; Mastantuono, E.; Sinnecker, D.; Goedel, A.; et al. Interplay of cell-cell contacts and RhoA/MRTF-A signaling regulates cardiomyocyte identity. *EMBO J.* **2018**, *37*, e98133. [CrossRef] [PubMed]
101. Smits, A.M.; Dronkers, E.; Goumans, M.-J. The epicardium as a source of multipotent adult cardiac progenitor cells: Their origin, role and fate. *Pharmacol. Res.* **2018**, *127*, 129–140. [CrossRef] [PubMed]

102. Rizzo, S.; Lodder, E.M.; Verkerk, A.O.; Wolswinkel, R.; Beekman, L.; Pilichou, K.; Basso, C.; Remme, C.A.; Thiene, G.; Bezzina, C.R. Intercalated disc abnormalities, reduced Na(+) current density, and conduction slowing in desmoglein-2 mutant mice prior to cardiomyopathic changes. *Cardiovasc. Res.* **2012**, *95*, 409–418. [CrossRef] [PubMed]
103. Nielsen, M.S.; van Opbergen, C.J.M.; van Veen, T.A.B.; Delmar, M. The intercalated disc: A unique organelle for electromechanical synchrony in cardiomyocytes. *Physiol. Rev.* **2023**, *103*, 2271–2319. [CrossRef]
104. Cerrone, M.; Lin, X.; Zhang, M.; Agullo-Pascual, E.; Pfenniger, A.; Chkourko Gusky, H.; Novelli, V.; Kim, C.; Tirasawadichai, T.; Judge, D.P.; et al. Missense mutations in plakophilin-2 cause sodium current deficit and associate with a Brugada syndrome phenotype. *Circulation* **2014**, *129*, 1092–1103. [CrossRef]
105. Rook, M.B.; Evers, M.M.; Vos, M.A.; Bierhuizen, M.F.A. Biology of cardiac sodium channel Nav1.5 expression. *Cardiovasc. Res.* **2012**, *93*, 12–23. [CrossRef]
106. Sato, P.Y.; Musa, H.; Coombs, W.; Guerrero-Serna, G.; Patiño, G.A.; Taffet, S.M.; Isom, L.L.; Delmar, M. Loss of plakophilin-2 expression leads to decreased sodium current and slower conduction velocity in cultured cardiac myocytes. *Circ. Res.* **2009**, *105*, 523–526. [CrossRef]
107. Severs, N.J.; Coppen, S.R.; Dupont, E.; Yeh, H.-I.; Ko, Y.-S.; Matsushita, T. Gap junction alterations in human cardiac disease. *Cardiovasc. Res.* **2004**, *62*, 368–377. [CrossRef]
108. Söhl, G.; Willecke, K. Gap junctions and the connexin protein family. *Cardiovasc. Res.* **2004**, *62*, 228–232. [CrossRef]
109. Solan, J.L.; Lampe, P.D. Connexin43 phosphorylation: Structural changes and biological effects. *Biochem. J.* **2009**, *419*, 261–272. [CrossRef]
110. Oxford, E.M.; Musa, H.; Maass, K.; Coombs, W.; Taffet, S.M.; Delmar, M. Connexin43 remodeling caused by inhibition of plakophilin-2 expression in cardiac cells. *Circ. Res.* **2007**, *101*, 703–711. [CrossRef] [PubMed]
111. Asimaki, A.; Tandri, H.; Huang, H.; Halushka, M.K.; Gautam, S.; Basso, C.; Thiene, G.; Tsatsopoulou, A.; Protonotarios, N.; McKenna, W.J.; et al. A new diagnostic test for arrhythmogenic right ventricular cardiomyopathy. *N. Engl. J. Med.* **2009**, *360*, 1075–1084. [CrossRef]
112. Calkins, H.; Corrado, D.; Marcus, F. Risk stratification in arrhythmogenic right ventricular cardiomyopathy. *Circulation* **2017**, *136*, 2068–2082. [CrossRef]
113. Arbelo, E.; Protonotarios, A.; Gimeno, J.R.; Arbustini, E.; Barriales-Villa, R.; Basso, C.; Bezzina, C.R.; Biagini, E.; Blom, N.A.; de Boer, R.A.; et al. 2023 ESC Guidelines for the management of cardiomyopathies. *Eur. Heart J.* **2023**, *44*, 3503–3626. [CrossRef]
114. Corrado, D.; Basso, C.; Rizzoli, G.; Schiavon, M.; Thiene, G. Does sports activity enhance the risk of sudden death in adolescents and young adults? *J. Am. Coll. Cardiol.* **2003**, *42*, 1959–1963. [CrossRef]
115. Gasperetti, A.; Targetti, M.; Olivotto, I. Antiarrhythmic drugs in arrhythmogenic right ventricular cardiomyopathy: The importance of optimal beta-blocker dose titration. *Int. J. Cardiol.* **2021**, *338*, 150–151. [CrossRef]
116. Kreimer, F.; Saguner, A.M.; Akin, I.; Milting, H.; Eckardt, L.; El-Battrawy, I. Arrhythmogenic right ventricular cardiomyopathy: Diagnosis, risk stratification, and treatment. *Dtsch. Arztebl. Int.* **2025**, *122*, 229–234.
117. Jordà, P.; Bosman, L.P.; Gasperetti, A.; Mazzanti, A.; Gourraud, J.B.; Davies, B.; Frederiksen, T.C.; Weidmann, Z.M.; Di Marco, A.; Roberts, J.D.; et al. Arrhythmic risk prediction in arrhythmogenic right ventricular cardiomyopathy: External validation of the arrhythmogenic right ventricular cardiomyopathy risk calculator. *Eur. Heart J.* **2022**, *43*, 3041–3052. [CrossRef]
118. Garcia-Montero, M.; Fanous, Y.; Krahn, A.D.; Davies, B.; Cadrin-Tourigny, J.; Roberts, J.D. New insights into genetic right ventricular cardiomyopathies. *Can. J. Cardiol.* **2025**, *41*, 1023–1037. [CrossRef] [PubMed]
119. Lundstrom, K. Viral vectors in gene therapy. *Diseases* **2018**, *6*, 42. [CrossRef] [PubMed]
120. Büning, H.; Srivastava, A. Capsid modifications for targeting and improving the efficacy of AAV vectors. *Mol. Ther. Methods Clin. Dev.* **2019**, *12*, 248–265. [CrossRef] [PubMed]
121. Mundisugih, J.; Ravindran, D.; Kizana, E. Exploring the therapeutic potential of gene therapy in Arrhythmogenic Right Ventricular Cardiomyopathy. *Biomedicines* **2024**, *12*, 1351. [CrossRef] [PubMed]
122. Inoue, H.; Nakamura, S.; Higo, S.; Shiba, M.; Kohama, Y.; Kondo, T.; Kameda, S.; Tabata, T.; Okuno, S.; Ikeda, Y.; et al. Modeling reduced contractility and impaired desmosome assembly due to plakophilin-2 deficiency using isogenic iPS cell-derived cardiomyocytes. *Stem Cell Rep.* **2022**, *17*, 337–351. [CrossRef]
123. Bradford, W.H.; Zhang, J.; Gutierrez-Lara, E.J.; Liang, Y.; Do, A.; Wang, T.-M.; Nguyen, L.; Mataraarachchi, N.; Wang, J.; Gu, Y.; et al. Plakophilin 2 gene therapy prevents and rescues arrhythmogenic right ventricular cardiomyopathy in a mouse model harboring patient genetics. *Nat. Cardiovasc. Res.* **2023**, *2*, 1246–1261. [CrossRef]
124. Kyriakopoulou, E.; Versteeg, D.; de Ruiter, H.; Perini, I.; Seibertz, F.; Döring, Y.; Zentilin, L.; Tsui, H.; van Kampen, S.J.; Tiburcy, M.; et al. Therapeutic efficacy of AAV-mediated restoration of PKP2 in arrhythmogenic cardiomyopathy. *Nat. Cardiovasc. Res.* **2023**, *2*, 1262–1276. [CrossRef]
125. Wu, I.; Zeng, A.; Greer-Short, A.; Aycinena, J.A.; Tefera, A.E.; Shenwai, R.; Farshidfar, F.; Van Pell, M.; Xu, E.; Reid, C.; et al. AAV9:PKP2 improves heart function and survival in a Pkp2-deficient mouse model of arrhythmogenic right ventricular cardiomyopathy. *Commun. Med.* **2024**, *4*, 38.

126. van Opbergen, C.J.M.; Narayanan, B.; Sacramento, C.B.; Stiles, K.M.; Mishra, V.; Frenk, E.; Ricks, D.; Chen, G.; Zhang, M.; Yarabe, P.; et al. AAV-mediated delivery of plakophilin-2a arrests progression of arrhythmogenic right ventricular cardiomyopathy in Murine hearts: Preclinical evidence supporting gene therapy in humans. *Circ. Genom. Precis. Med.* **2024**, *17*, e004305.
127. Lexeo Therapeutics, Inc. *Form 10-K, Annual Report for the Fiscal Year Ended December 31, 2024*; U.S. Securities and Exchange Commission: Washington, DC, USA, 2025.
128. Therapeutics, T. Tenaya Therapeutics Provides Update on RIDGE-1 Gene Therapy Clinical Trial for PKP2-Associated ARVC. Tenaya Therapeutics Investor Relations. Available online: https://investors.tenayatherapeutics.com/node/8511/pdf (accessed on 8 October 2025).
129. Rocket Pharmaceuticals, Inc. Rocket Pharmaceuticals Announces Preliminary Positive Results from Phase 1 Trial of RP-A601 for PKP2-Associated Arrhythmogenic Cardiomyopathy. Rocket Pharmaceuticals Investor Relations. Available online: https://ir.rocketpharma.com/node/12551/pdf (accessed on 8 October 2025).
130. Beurel, E.; Grieco, S.F.; Jope, R.S. Glycogen synthase kinase-3 (GSK3): Regulation, actions, and diseases. *Pharmacol. Ther.* **2015**, *148*, 114–131. [CrossRef]
131. Eldar-Finkelman, H.; Martinez, A. GSK-3 inhibitors: Preclinical and clinical focus on CNS. *Front. Mol. Neurosci.* **2011**, *4*, 32. [CrossRef] [PubMed]
132. Coghlan, M.P.; Culbert, A.A.; Cross, D.A.; Corcoran, S.L.; Yates, J.W.; Pearce, N.J.; Rausch, O.L.; Murphy, G.J.; Carter, P.S.; Roxbee Cox, L.; et al. Selective small molecule inhibitors of glycogen synthase kinase-3 modulate glycogen metabolism and gene transcription. *Chem. Biol.* **2000**, *7*, 793–803. [CrossRef] [PubMed]
133. Snitow, M.E.; Bhansali, R.S.; Klein, P.S. Lithium and therapeutic targeting of GSK-3. *Cells* **2021**, *10*, 255. [CrossRef]
134. Hamstra, S.I.; Kurgan, N.; Baranowski, R.W.; Qiu, L.; Watson, C.J.F.; Messner, H.N.; MacPherson, R.E.K.; MacNeil, A.J.; Roy, B.D.; Fajardo, V.A. Low-dose lithium feeding increases the SERCA2a-to-phospholamban ratio, improving SERCA function in murine left ventricles. *Exp. Physiol.* **2020**, *105*, 666–675. [CrossRef] [PubMed]
135. Šeflová, J.; Cruz-Cortés, C.; Guerrero-Serna, G.; Robia, S.L.; Espinoza-Fonseca, L.M. Mechanisms for cardiac calcium pump activation by its substrate and a synthetic allosteric modulator using fluorescence lifetime imaging. *PNAS Nexus* **2024**, 3, gad453. [CrossRef] [PubMed]
136. Chelko, S.P.; Asimaki, A.; Andersen, P.; Bedja, D.; Amat-Alarcon, N.; DeMazumder, D.; Jasti, R.; MacRae, C.A.; Leber, R.; Kleber, A.G.; et al. Central role for GSK3β in the pathogenesis of arrhythmogenic cardiomyopathy. *JCI Insight* **2016**, *1*, e85923. [CrossRef] [PubMed]

biomolecules

MDPI

Review

Inflammatory Mechanisms in Myocarditis—Recent Therapeutic Strategies

Stergios Soulaidopoulos, Dimitris Tousoulis *, Marios Sagris, Svetlana Aghayan, Konstantinos Platanias, Alexios Giannakodimos, Emilia Lazarou, Konstantinos Tsioufis and George Lazaros

First Cardiology Department, Hippokration Hospital, Athens Medical School, National and Kapodistrian University of Athens, 11527 Athens, Greece; soulaidopoulos@hotmail.com (S.S.); masagris1919@gmail.com (M.S.); dr.svetlana.aghayan@gmail.com (S.A.); konstantinosplatanias99@gmail.com (K.P.); alexisgiannak@hotmail.com (A.G.); emilazarou@gmail.com (E.L.); ktsioufis@gmail.com (K.T.); glaz35@hotmail.com (G.L.)

* Correspondence: drtousoulis@hotmail.com

Abstract

Myocarditis is an inflammatory disease of the heart characterized by a complex interplay between innate and adaptive immune responses. The innate immune system provides first-line defense and includes soluble molecules, including macrophages, neutrophils, dendritic cells, and molecular mediators, but lacks immunological memory. In contrast, the adaptive immune system, via T and B lymphocytes, offers high specificity and long-term memory, which can sometimes target myocardial tissue, causing autoimmune injury. Particularly, acute myocarditis is characterized by dysregulated immune signaling, with cytokines (IL-2, IFN-γ, IL-12, IL-4, IL-10) and chemokines (MCP-1, CXCL4, CXCL10) driving disease progression, while adhesion molecules (ICAM-1, VCAM-1, VAP-1) promote leukocyte trafficking and cardiac inflammation. The balance between pro-inflammatory and regulatory responses determines disease outcomes, ranging from resolution with recovery to fulminant myocarditis or progression to dilated cardiomyopathy. Emerging therapeutic approaches targeting cytokines, chemokines, and adhesion molecules, along with established immunosuppressive treatments, underline the potential for modulating immune responses in myocarditis and, thereby, improving patient outcomes.

Keywords: myocarditis; inflammation; cardiomyopathy; immunosuppression; immunomodulation

1. Introduction

Myocarditis represents an inflammatory disease of the myocardium, which can be triggered by a plethora of infectious and non-infectious causes. The diagnosis of myocarditis may be often quite challenging, owing to its presentation with a wide spectrum of possible clinical scenarios [1]. For many years, endomyocardial biopsy (EMB) was considered the standard method for diagnosing and classifying myocarditis, based on a combination of established histological, immunological, and immunohistochemical criteria [2,3]. The introduction of advanced imaging techniques, namely cardiac magnetic resonance (CMR), in daily clinical practice, has however, changed the way myocarditis is diagnosed in the majority of cases. Particularly, the detection of cardiac inflammation on CMR along with elevated troponin levels indicating myocardial injury in the presence of relevant clinical symptoms and signs, is usually sufficient for the non-invasive confirmation of the diagnosis in the majority of cases [4,5].

Although determining the etiology may not be necessary in cases with mild clinical presentation, precise identification of the underlying cause through an understanding of the activated pathogenetic pathways may be crucial for effective management and improved patient outcomes in cases with more critical clinical presentation. Indeed, different etiologies trigger distinct immune responses involving various cellular and molecular mechanisms [6]. In addition to this, certain inflammatory markers correlate with disease severity and prognosis, enabling improved risk stratification and treatment monitoring [7]. Overall, a deeper understanding of these mechanisms facilitates the development of targeted therapies and helps prevent long-term complications of myocarditis, such as myocardial fibrosis, arrhythmias, and heart failure [8,9].

Within this context, the scope of this article is to review the inflammatory pathogenetic mechanisms involved in myocarditis and to summarize current therapeutic strategies.

2. Epidemiology, Etiopathogenesis, and Diagnosis

Given the infrequent use of EMB for diagnostic confirmation, it is difficult to determine the actual incidence of myocarditis. In addition to this, the diagnosis is frequently missed owing to subtle clinical manifestations. It is estimated that the global incidence of acute or subacute myocarditis is about 500,000 cases per year, being responsible for about 30,000 deaths [10]. Among patients diagnosed with heart failure, the incidence of myocarditis varies between 0.5% and 4%, while it is estimated at about 3% among patients presenting with chest pain at the emergency department [11,12].

A plethora of infectious and non-infectious agents may cause acute myocarditis. Viral infections represent the most common cause, with classic enteroviruses, adenoviruses, parvovirus B19, and human herpesvirus 6 being more frequently detected when EMB is performed [7,13]. However, non-viral pathogens, such as bacteria and parasites, may also cause myocardial injury either directly or indirectly by the activation of catastrophic autoimmune responses [9].

Multiple pharmacological agents, such as antipsychotics immunotherapies, and vaccines, may display toxic effects on the myocardium, which may be characterized by a diversity of the underlying pathogenetic mechanisms [14]. For example, antipsychotic drugs are commonly associated with hypersensitive reactions that manifest as eosinophilic myocarditis [15]. Myocarditis related to immunotherapies, on the other hand, is driven by T cell-mediated immune checkpoint inhibition, presenting as fulminant lymphocytic myocarditis [16]. Finally, vaccine-mediated myocarditis, which has recently gained increasing interest in the context of COVID-19, seems to be associated with strong innate immune responses secondary to vaccine mimicry of myocardium antigens [17,18].

Regardless of the triggering cause, the activation of autoimmunity, characterized by the presence of heart-reactive autoantibodies, is a prominent feature of all forms of acute or chronic myocardial inflammation [19]. In these terms, myocarditis demonstrates a notable association with systemic autoimmune diseases. The incidence of myocarditis, for instance, may reach 18% among patients with systemic lupus erythematosus [20]. Sarcoidosis, which is characterized by the formation of non-necrotizing granulomas in various organs, seems to manifest in 5% of patients as clinically overt inflammatory infiltration of the myocardium, while in approximately 25% of asymptomatic patients, cardiac involvement is verified by autopsy or imaging studies [21]. Giant cell myocarditis is characterized by T-cell-mediated inflammation of the myocardium and is also considered to be of autoimmune origin, given its association with disorders such as Hashimoto thyroiditis, myasthenia gravis, rheumatoid arthritis, and inflammatory bowel disease [22].

The development of viral or autoimmune myocarditis and its progression to dilated cardiomyopathy are likely influenced by genetic factors. Indeed, not all patients exposed to infectious agents or affected by autoimmune diseases develop myocarditis, and only a small proportion ultimately progress to dilated cardiomyopathy. This progression appears to occur predominantly in cases characterized by uncontrolled, persistent chronic inflammation and the presence of pathogenic cardiac autoantibodies targeting myocardial structural, sarcoplasmic, or sarcolemmal proteins.

The recently published ESC guidelines have introduced the term "inflammatory myopericardial syndrome", encompassing a broad spectrum of inflammatory conditions affecting both the myocardium and pericardium [23]. It is recommended to use this term as an umbrella diagnosis during the initial workup, guided by the recognition of diagnostic red flags, including clinical, serological, and imaging biomarkers. The identification of these features should raise clinical suspicion and subsequently be followed by a risk stratification based on the clinical presentation and the assessment of LV function by cardiac imaging. According to the proposed classification, stable and oligosymptomatic, low-risk patients may be safely discharged. In contrast, patients with heart failure symptoms, arrhythmias, or impaired left ventricular function are considered either intermediate or high risk and should be hospitalized for appropriate management, including heart failure therapy or biopsy-guided treatment. Particularly, EMB should be performed in patients with high-risk clinical features, such as advanced heart failure symptoms, arrhythmias, and significant LV dysfunction, to enable precise identification of the histological subtype of cardiac inflammation and allow targeted therapy.

3. Inflammatory Mechanisms

3.1. Innate and Adaptive Immunity Regulation

The innate immune system constitutes the first line of defense against pathogens. It includes physical barriers such as the skin, as well as soluble and cellular components including cytokines, chemokines, complement proteins, macrophages, neutrophils, and dendritic cells. Although rapid and effective, this system does not generate immunological memory. In contrast, the adaptive immune system is characterized by high specificity and the capacity to develop immunological memory, enabling faster and more robust responses upon re-exposure to previously encountered pathogens. The principal regulators of this response are T and B lymphocytes [23].

In acute myocarditis, an imbalance in the coordination between innate and adaptive immune mechanisms has been reported [24]. Following recognition of pathogens by pattern recognition receptors (PRRs)—notably Toll-like receptors (TLRs) and NOD-like receptors (NLRs) [25]—the innate immune system releases cytokines, chemokines, and chemoattractants that recruit mast cells, neutrophils, dendritic cells, monocytes, and macrophages to the injured myocardium [23,24]. Among these, macrophages play a pivotal role through phagocytosis, cytotoxicity, and cytokine release. A key mediator is the NLRP3 inflammasome, a multiprotein complex composed of NLRP3 (NOD-like receptor family, pyrin domain–containing 3), ASC (apoptosis-associated speck-like protein containing a caspase recruitment domain), and pro-caspase-1 [26]. Upon activation by pathogen-associated molecular patterns (PAMPs) or danger-associated molecular patterns (DAMPs)—particularly after myocardial injury—the inflammasome triggers caspase-1 activation and subsequent release of pro-inflammatory cytokines IL-1β and IL-18, amplifying the inflammatory response [26,27].

The adaptive immune response provides a broad repertoire of antigen receptors capable of recognizing diverse pathogens. However, dysregulation may result in misidentification of cardiac antigens as foreign, leading to autoimmunity. While immune activation is generally protective, excessive or unresolved inflammation can promote cardiomyocyte injury, ventricular remodeling, and progressive dysfunction [28,29]. Following myocardial injury, intracellular antigens released from necrotic myocytes may be inadequately cleared and are instead processed by antigen-presenting cells (APCs) [30]. These antigens, when presented via major histocompatibility complex (MHC) molecules, activate antigen-specific T and B lymphocytes, driving their differentiation into cytotoxic T lymphocytes (CTLs), helper T cells (Th), plasma cells, and memory cells. These subsets mediate complementary immune functions: CTLs exert cytotoxic activity, Th cells coordinate cytokine signaling, plasma cells secrete antibodies, and memory cells sustain long-term immunological surveillance [30].

During the acute phase of myocarditis, dysregulated adaptive responses may produce self-reactive T and B lymphocytes that specifically target myocardial tissue. In most cases, regulatory T lymphocytes (Tregs) restore immune homeostasis, leading to gradual recovery of left ventricular (LV) contractility, resolution of arrhythmias, and reduction in inflammatory markers [9]. However, when this regulation is insufficient, persistent inflammation may result in fulminant myocarditis, characterized by cytokine storm and cardiogenic shock, or in chronic inflammation progressing to dilated cardiomyopathy (DCM).

3.2. Molecular Pathways

The molecular mechanisms underlying myocarditis involve adhesion molecules, cytokines, and chemokines, which collectively regulate immune cell activation, recruitment, and tissue injury.

Cytokines play a central role in orchestrating the immune response during myocarditis. They are broadly classified into Th1 cytokines, including IL-2, IFN-γ, IL-12, and TNF-α, and Th2 cytokines, such as IL-4, IL-5, IL-6, and IL-10 [23]. Using myosin-induced myocarditis models in rats, Ayach et al. demonstrated that the acute phase is characterized by a dominant Th1 response, with elevated IL-2 and IFN-γ, whereas during the convalescent phase, Th2 cytokines such as IL-10 predominate [3]. Similarly, Izumi et al. confirmed that IL-2 and IL-12 are pivotal regulators across both acute and convalescent phases of the disease [24].

IFN-γ appears to act protectively, as demonstrated by Eriksson et al., who found that impaired IFN-γ receptor signaling coupled with elevated IL-12 aggravated myocarditis and led to increased mortality [5]. Barin et al. further reported that fatal eosinophilic myocarditis developed in the absence of both IFN-γ and IL-17A [25]. In contrast, IL-12 has been associated with the differentiation of pathogenic $CD8^+$ T-cell effectors, which promote myocardial injury [26]. Consistent with this, mice lacking the IL-12 receptor or the IL-12p40 ligand exhibited markedly reduced IFN-γ expression and were protected from myocarditis [27].

The role of IL-4 and IL-10 remains more complex. Both cytokines promote B-cell antibody production, yet studies report conflicting outcomes in myocarditis. Roffe et al. showed that IL-10 reduced parasite burden and protected against fatal myocarditis in a *Trypanosoma cruzi* mouse model [28], whereas Cen et al. found that IL-10–producing B cells were increased in Coxsackievirus B3-induced acute viral myocarditis [29]. Similarly, Afanasyeva et al. demonstrated that neutralization of IL-4 protected against myocarditis and enhanced IFN-γ expression [30], while other studies indicated that sustained IL-4 and IL-17 elevation predicted progression from viral myocarditis to dilated cardiomyopathy (DCM). Conversely, Li et al. reported that IL-4 suppressed matrix metalloproteinase activity and improved cardiac function in murine myocarditis, highlighting its potential cardioprotective role [31].

Chemokines are another essential component of the immune response, functioning in chemotaxis, leukocyte degranulation, hematopoiesis, and angiogenesis. They are structurally categorized into CXC, CC, XC, and CX3C families based on the positioning of conserved cysteine residues and act through binding to G protein–coupled receptors, which makes them attractive therapeutic targets [32]. In experimental viral myocarditis, Coxsackievirus B3 infection was shown to modulate chemokine expression in cardiomyocytes, with monocyte chemoattractant protein-1 (MCP-1) playing a critical role during the early stages [33]. Furthermore, the natural compound myricetin was shown to suppress autoimmune responses and reduce MCP-1 expression in cardiomyocytes, indicating its potential as a therapeutic agent [34]. Additional pathways have also been implicated. Asaumi, Y. et al. identified the CXCL13/CXCR5 axis as a contributor to myocarditis pathogenesis [35], while Krueger, G.R. et al. revealed that the CXCL4/CXCR3 axis regulates cardiac fibrosis via TGF-β1/Smad2/3 signaling in viral myocarditis [36]. Furthermore, CXCL10 was shown to inhibit viral replication by recruiting natural killer cells in Coxsackievirus B3-induced myocarditis [37]. More recently, Manaresi et al) demonstrated that in immune checkpoint inhibitor (ICI)-related myocarditis, the CXCR3–CXCL9/10 axis plays a critical role in driving myocardial inflammation, and inhibition of this pathway reduced inflammation and improved survival [38].

Adhesion molecules also contribute significantly to disease pathogenesis by regulating leukocyte trafficking, cell–cell interactions, and cell–matrix adhesion. The principal families involved include integrins, selectins, and members of the immunoglobulin superfamily [39]. Intercellular adhesion molecule-1 (ICAM-1), a ligand for lymphocyte function–associated antigen-1 (LFA-1), is markedly upregulated in murine models of acute Coxsackievirus B3-induced myocarditis and remains elevated even during immunosuppressive therapy, suggesting a role in chronic myocardial inflammation. Similarly, vascular cell adhesion molecule-1 (VCAM-1) expression is increased in autoimmune myocarditis, though not directly correlated with disease severity. More recently, vascular adhesion protein-1 (VAP-1), a membrane-bound molecule mediating leukocyte adhesion and transmigration, has been studied as a novel imaging biomarker, with VAP-1–targeted PET successfully identifying myocardial lesions in autoimmune myocarditis [40]. Table 1 summarizes the main pathogenetic pathways involved in acute myocarditis.

Table 1. Main pathogenetic mechanism involved in acute myocarditis.

Feature	Innate Immunity	Adaptive Immunity
Main Cells	Neutrophils, Macrophages, Dendritic cells, NK	B cells, T cells ($CD4^+$, $CD8^+$), Plasma cells
Trigger	PAMPs/DAMPs from myocardial injury	Antigen presentation via MHC
Timing	Hours–Days	Days–Weeks
Key Cytokines	IL-1β, IL-6, TNF-α	IL-2, IFN-γ, IL-4, IL-17
Role	Rapid, non-specific defense; debris clearance	Antigen-specific response; memory formation
Dysregulation	Excess inflammation, tissue damage	Autoimmune injury to myocytes

Abbreviations: NK = natural killer, PAMP = pathogen-associated molecular patterns, DAMP = danger-associated molecular patterns, IL = interleukin, TNF = tumor-necrosis-factor.

4. Therapeutic Strategies for Myocarditis

4.1. General Treatment Measures

Although the therapeutic strategy for myocarditis is directed by the underlying cause of the disease and often requires multispecialty input, general measurements are applied based on the presence of complications and especially the hemodynamic instability of the patient, the reduction in left ventricular ejection fraction, and the appearance of arrhythmias. Acute forms of myocarditis often have uncertain progress, and hospital admission is generally required for clinical observation of the patient, diagnosis investigation, and surveillance of complications [24,25]. Relaxation, monitoring of the vital signs, execution of electrocardiograms, and often pain management with conventional therapies are the basic approach in an uncomplicated form of acute myocarditis with preserved ejection fraction. Conventional pain management includes paracetamol, while use of nonsteroidal anti-inflammatory drugs (NSAIDs) should be administered in cases of pericardium involvement or persistent chest pain, as they are now considered safe for use in this setting [26]. Other anti-inflammatory drugs such as colchicine are recommended in patients with myopericarditis [27]. Physical activity as well as exercise testing are contraindicated during the acute phase of myocarditis, while the duration of exercise restriction ranges between 3 and 6 months [28]. General use of immunosuppressive drugs failed to prove clinical benefits and is restricted to certain indications, such as severe patients with giant cell or non-viral myocarditis. However, when patients present with complications, especially left ventricular systolic dysfunction, cardiac arrhythmias, and hemodynamic instability, admission to intensive care units is considered necessary.

The initial treatment of hemodynamically stable patients with reduced LVEF is based on heart failure (HF) guidelines. Early initiation of HF drugs with proven survival benefits, such as beta-blockers, angiotensin-converting enzyme inhibitors or angiotensin receptor–neprilysin inhibitors, aldosterone antagonists, and sodium–glucose cotransporter 2 inhibitors, is considered crucial and suggests potential benefits [29,30]. Appropriate use of diuretics is also recommended when symptoms and signs of HF congestion are present. No specific recommendations exist regarding transient reduction in LVEF; however, maintenance of HF treatment seems to be beneficial when CMR preserves myocardial Late Gadolinium Enhancement (LGE) [24].

Management of cardiac arrhythmias is considered of greatest importance in the treatment of myocarditis. Myocarditis caused by sarcoidosis, giant cells, or Lyme disease typically leads to both atrioventricular blocks and tachyarrhythmias. Supportive measures are usually applied while conduction disturbances and ventricular arrhythmias tend to be reversible after the acute phase of myocarditis [31]. No specific recommendations are established; close monitoring of the patients, beta-blockers, and antiarrhythmic agents typically consists of the treatment for premature ventricular contractions, supraventricular and ventricular arrhythmias [32]. Implantation of implantable cardioverter-defibrillators (ICDs) is still debatable. ICDs are generally not recommended while the arrhythmogenic risk decreases after the acute phase of the disease. Alternative strategies using wearable defibrillator vests could be applied for the prevention of malignant arrhythmias during the early stage of myocarditis [33]. Secondary prevention of sudden cardiac death using ICDs should be considered in specific cases of the disease, taking into consideration the history of the patients, imaging results, and the type of myocarditis. It is worth mentioning that in the ITAMY registry, the percentage of ICD implantation in patients after acute myocarditis and preserved left ventricular systolic function was only 1.6% [34].

Cardiogenic shock and hemodynamic instability are considered crucial complications of myocarditis, rendering necessary the immediate transport of the patients to intensive care units for administration of inotropic agents and access to mechanical circulatory support systems. Ventricular assist devices (VADs) and extracorporeal membrane oxygenation (ECMO) systems alone or in combination with devices reducing left ventricular afterload and intracavitary pressures could support hemodynamic circulation, offering myocardial recovery, ameliorating biventricular unloading and coronary perfusion, and providing a successful bridge to permanent recovery or long-term left ventricular assist devices and heart transplantation [5]. Inability to wean off the mechanical circulatory support systems signals the possibility for long-term VAD placement and heart transplantation [35].

4.2. Antiviral Treatment for Myocarditis

To date, randomized clinical trials have not confirmed the therapeutic effectiveness of antiviral treatments for all causes of acute viral myocarditis [9]. Regarding patients with latent infections caused by HHV-6 or parvovirus B19 (B19V), no specific antiviral therapy has been established. In cases of acute myocarditis associated with active HHV-6, B19V, adenovirus, or enterovirus infections, antiviral agents such as ganciclovir or valganciclovir have been proposed; however, their clinical efficacy remains uncertain, and evidence is limited to small, heterogeneous studies [23]. Consequently, antiviral treatment in myocarditis remains controversial and should be considered on an individual basis. Interferon beta immunoglobulins may also prove useful [24]. In patients with latent myocarditis caused by Epstein–Barr virus or cytomegalovirus, antiviral therapy targeting HSV may also help lower viral load [24]. Although the effectiveness of aciclovir, ganciclovir, or valaciclovir in myocarditis patients has not been directly assessed, their use may generally be considered in HSV myocardial infections [36]. In cases of enteroviral myocarditis in neonates, treatment with antiviral agents such as pocapavir and pleconaril, along with intravenous immunoglobulin (IVIG) therapy, has shown effectiveness [37]. Concerning B19V, IVIG is commonly administered to patients with severe viremia and associated clinical complications. Emerging antiviral approaches for B19V infection, currently under investigation, include synthetic nucleotide analogs such as cidofovir and brincidofovir, flavonoids, and hydroxyurea [38]. Nevertheless, no established treatments currently exist for B19V-associated inflammatory cardiomyopathy, and it is generally suggested that treatment is not required when low levels of B19V DNA are found in cardiac tissue without evidence of myocardial inflammation [28]. Findings from small observational studies suggest that immunosuppressive therapy may prove useful for patients with low myocardial B19V DNA levels and ongoing cardiac inflammation, as shown in the CaPACITY program. Additionally, patients with B19V RNA positivity have shown improvement following treatment with the antiviral agent telbivudine, likely due to its immunomodulatory effects [39].

Established antiviral therapies are used to manage myocarditis or inflammatory cardiomyopathy linked to HIV, HCV, or influenza infections. These include antiretroviral regimens for HIV-associated myocarditis, as well as a combination of ombitasvir, paritaprevir, ritonavir, and dasabuvir for HCV-related cases; neuraminidase inhibitors such as peramivir and zanamivir may also be administered for influenza-associated myocarditis [40–42]. Moreover, in regard to COVID-19-related myocarditis, numerous antiviral treatment strategies are currently being explored. These include agents that block viral entry into host cells, such as chloroquine, hydroxychloroquine, camostat mesylate, and umifenovir; protease inhibitors, such as lopinavir–ritonavir and darunavir; RNA polymerase inhibitors, like remdesivir; as well as anti-cytokine therapies targeting IL-6 and IL-1β pathways [43].

Experimental studies in animals have demonstrated that both IFN-α and IFN-β reduce viral replication and myocardial injury, with IFN-β proving more effective than IFN-α in fully eradicating the cardiac viral load [44]. Both spontaneous and IFN-β–induced viral clearance have been linked to clinical and hemodynamic improvement in myocarditis patients, while ongoing viral presence is associated with adverse outcomes due to progressive left ventricular dysfunction [45]. Notably, the phase II BICC clinical trial examined the impact of IFN-β therapy on viral clearance in patients with inflammatory cardiomyopathy and persistent myocardial viral infection caused by adenovirus, enterovirus, or B19V; patients with myocarditis positive for enterovirus or adenovirus, confirmed via endomyocardial biopsy, experienced viral clearance following IFN-β treatment. However, this therapy did not result in viral DNA clearance in individuals with B19V-positive myocarditis [46]. Further research is needed to determine whether combining antiviral and immunosuppressive therapies could be beneficial for selected patients with virus-positive myocarditis depending on the stage of the disease.

5. Immunosuppressive and Immunomodulatory Therapy

The role of immunosuppressive and immunomodulatory therapy in myocarditis remains controversial. In the absence of a clearly defined viral etiology, particularly when active viral replication has been excluded, such therapies may be considered in selected cases. Their use is most justified in myocarditis of confirmed autoimmune origin, as established through myocardial biopsy, serological testing, and PCR analysis. Immunosuppressive treatment may also be appropriate in the presence of histological evidence of active myocardial inflammation or in specific subtypes of myocarditis, such as giant cell, sarcoid, eosinophilic, or immune checkpoint inhibitor (ICI)–induced myocarditis. However, the efficacy of these approaches in viral or idiopathic forms remains uncertain, and thus, treatment decisions should be individualized based on the underlying pathophysiology and clinical presentation [23].

5.1. *Glucocorticoids*

Glucocorticoids are the first-line agents in primary autoimmune myocarditis, particularly in giant cell and sarcoid variants. The recommended initial dose is 1 mg/kg/day. Randomized controlled trials have demonstrated improvements in LVEF in patients with virus-negative myocarditis treated with Prednisolone in combination with Azathioprine [47].

5.2. *Azathioprine*

Azathioprine is a purine nucleoside analog that inhibits lymphocyte proliferation. It is predominantly used in combination with glucocorticoids. Clinical studies have shown that this combination results in improved LVEF and a reduction in heart failure symptoms [48,49].

5.3. *Cyclosporine A/Tacrolimus*

These agents act by inhibiting T-cell activity and are particularly effective in giant cell myocarditis. Case reports have documented clinical improvement and symptom reduction when Cyclosporine is administered in combination with glucocorticoids, especially with giant cell myocarditis [50,51].

5.4. *Mycophenolate Mofetil (MMF)*

MMF has seen increasing use in recent years and is generally well tolerated. It is employed in cases of sarcoid myocarditis. In patients with autoimmune, virus-negative myocarditis, substitution of Azathioprine with MMF has been associated with improvements in LVEF, reductions in left ventricular end-diastolic volume, and decreases in cardiac biomarkers such as troponin and NT-pro-BNP [52].

5.5. *Intravenous Immunoglobulin (IVIG)*

IVIG is indicated in viral myocarditis, particularly in pediatric patients and in cases of acute myocarditis, due to its immunomodulatory properties. It exerts its effects by regulating T-cell function and reducing pro-inflammatory cytokines such as TNF-α and IL-6. Meta-analyses have demonstrated reductions in mortality and improvements in LVEF with IVIG therapy [9,53].

5.6. Interferon-α/Thymomodulin

In a randomized trial involving 38 patients with biopsy-confirmed myocarditis, a 2-year follow-up in patients with dilated cardiomyopathy (DCM) demonstrated improved progression-free survival (PFS): 88% in the interferon group and 66% in the thymomodulin group. These agents may serve as adjuncts to standard therapy in cases of persistent myocardial inflammation [54,55].

5.7. Rituximab/Tofacitinib (JAC Inhibitor)

Rituximab has been shown to attenuate the clinical manifestations of myocardial inflammation. There is also a case report indicating successful remission of IVIG-refractory myocarditis following Tofacitinib administration, without adverse effects [52,56].

5.8. Experimental Approaches

Emerging therapies include the use of anti-IL-1β monoclonal antibodies, which have demonstrated potential in reducing myocardial inflammation, fibrosis, and adverse remodeling [57].

5.9. Biomimetic Nanoparticles

These nanoparticles modulate immune cell activity and are coated with autologous cell membranes (e.g., macrophages or T lymphocytes), enabling targeted delivery to inflamed myocardial tissue. By mimicking native cells, they evade immune detection and can deliver therapeutic agents directly to affected areas, thereby reducing the local immune response [58].

6. Conclusions

Myocarditis is a heterogeneous inflammatory disorder with diverse etiologies and pathogenetic mechanisms. Advances in imaging and immunology have changed the traditional diagnostic approach, while a growing understanding of the molecular pathways is reshaping therapeutic strategies. Although supportive management remains the cornerstone of treatment, targeted antiviral, immunosuppressive, and immunomodulatory therapies are expanding the therapeutic landscape. Future research focusing on precision medicine, guided by etiology and immune profiling, is critical for improving patient outcomes and preventing progression to dilated or chronic inflammatory cardiomyopathy.

Author Contributions: S.S. wrote the paper, D.T. conceptualization, supervision and editing, G.L. and K.T. critically revise the paper, M.S., S.A., K.P., A.G. and E.L. data acquisition. All authors have read and agreed to the published version of the manuscript.

Funding: This research received no external funding.

Institutional Review Board Statement: Not applicable.

Informed Consent Statement: Not applicable.

Data Availability Statement: No new data were created or analyzed in this study. Data sharing is not applicable to this article.

Conflicts of Interest: The authors declare no conflicts of interest.

References

1. Kindermann, I.; Barth, C.; Mahfoud, F.; Ukena, C.; Lenski, M.; Yilmaz, A.; Klingel, K.; Kandolf, R.; Sechtem, U.; Cooper, L.T.; et al. Update on myocarditis. *J. Am. Coll. Cardiol.* **2012**, *59*, 779–792. [CrossRef]
2. Thomas Aretz, H. Myocarditis: The Dallas criteria. *Hum. Pathol.* **1987**, *18*, 619–624. [CrossRef] [PubMed]

3. Richardson, P.; McKenna, R.W.; Bristow, M.; Maisch, B.; Mautner, B.; O'Connell, J.; Olsen, E.; Thiene, G.; Goodwin, J.; Gyarfas, I.; et al. Report of the 1995 World Health Organization/International Society and Federation of Cardiology Task Force on the Definition and Classification of cardiomyopathies. *Circulation* **1996**, *93*, 841–842. [CrossRef] [PubMed]
4. Smith, S.C.; Ladenson, J.H.; Mason, J.W.; Jaffe, A.S. Elevations of cardiac troponin I associated with myocarditis: Experimental and clinical correlates. *Circulation* **1997**, *95*, 163–168. [CrossRef] [PubMed]
5. Abdel-Aty, H.; Boyé, P.; Zagrosek, A.; Wassmuth, R.; Kumar, A.; Messroghli, D.; Bock, P.; Dietz, R.; Friedrich, M.G.; Schulz-Menger, J.; et al. Diagnostic performance of cardiovascular magnetic resonance in patients with suspected acute myocarditis: Comparison of different approaches. *J. Am. Coll. Cardiol.* **2005**, *45*, 1815–1822. [CrossRef]
6. Rose, N.R. Myocarditis: Infection versus autoimmunity. *J. Clin. Immunol.* **2009**, *29*, 730–737. [CrossRef]
7. Caforio, A.L.P.; Calabrese, F.; Angelini, A.; Tona, F.; Vinci, A.; Bottaro, S.; Ramondo, A.; Carturan, E.; Iliceto, S.; Thiene, G.; et al. A prospective study of biopsy-proven myocarditis: Prognostic relevance of clinical and aetiopathogenetic features at diagnosis. *Eur. Heart J.* **2007**, *28*, 1326–1333. [CrossRef]
8. Kawai, C. From myocarditis to cardiomyopathy: Mechanisms of inflammation and cell death: Learning from the past for the future. Circulation. *Circulation* **1999**, *99*, 1091–1100. [CrossRef]
9. Tschöpe, C.; Ammirati, E.; Bozkurt, B.; Caforio, A.L.P.; Cooper, L.T.; Felix, S.B.; Hare, J.M.; Heidecker, B.; Heymans, S.; Hübner, N.; et al. Myocarditis and inflammatory cardiomyopathy: Current evidence and future directions. *Nat. Rev. Cardiol.* **2021**, *18*, 169–193. [CrossRef] [PubMed]
10. Roth, G.A.; Mensah, G.A.; Johnson, C.O.; Addolorato, G.; Ammirati, E.; Baddour, L.M.; Barengo, N.C.; Beaton, A.Z.; Benjamin, E.J.; Benziger, J.P.; et al. Global Burden of Cardiovascular Diseases and Risk Factors, 1990–2019: Update From the GBD 2019 Study. *J. Am. Coll. Cardiol.* **2020**, *76*, 2982–3021. [CrossRef]
11. Charpentier, S.; Beaune, S.; Joly, L.M.; Khoury, A.; Duchateau, F.-X.; Briot, R.; Renaud, B.; Ageron, F.-X. Management of chest pain in the French emergency healthcare system: The prospective observational EPIDOULTHO study. *Eur. J. Emerg. Med.* **2018**, *25*, 404–410. [CrossRef]
12. Cooper, L.T., Jr.; Keren, A.; Sliwa, K.; Matsumori, A.; Mensah, G.A. The Global Burden of Myocarditis: Part 1: A Systematic Literature Review for the Global Burden of Diseases, Injuries, and Risk Factors 2010 Study. *Glob. Heart* **2014**, *9*, 121–129. [CrossRef]
13. Veronese, G.; Ammirati, E.; Brambatti, M.; Merlo, M.; Cipriani, M.; Potena, L.; Sormani, P.; Aoki, T.; Sugimura, K.; Sawamura, A.; et al. Viral genome search in myocardium of patients with fulminant myocarditis. *Eur. J. Heart Fail.* **2020**, *22*, 1277–1280. [CrossRef]
14. Nguyen, L.S.; Cooper, L.T.; Kerneis, M.; Funck-Brentano, C.; Silvain, J.; Brechot, N.; Hekimian, G.; Ammirati, E.; Ben M'barek, B.; Redheuil, A.; et al. Systematic analysis of drug-associated myocarditis reported in the World Health Organization pharmacovigilance database. *Nat. Commun.* **2022**, *13*, 25. [CrossRef] [PubMed]
15. Wang, J.-F.; Min, J.-Y.; Hampton, T.G.; Amende, I.; Yan, X.; Malek, S.; Abelmann, W.H.; Green, A.I.; Zeind, J.; Morgan, J.P. Clozapine-induced myocarditis: Role of catecholamines in a murine model. *Eur. J. Pharmacol.* **2008**, *592*, 123–127. [CrossRef]
16. Salem, J.-E.; Allenbach, Y.; Vozy, A.; Brechot, N.; Johnson, D.B.; Moslehi, J.J.; Kerneis, M. Abatacept for Severe Immune Checkpoint Inhibitor–Associated Myocarditis. *N. Engl. J. Med.* **2019**, *380*, 2377–2379. [CrossRef]
17. Heidecker, B.; Dagan, N.; Balicer, R.; Eriksson, U.; Rosano, G.; Coats, A.; Tschöpe, C.; Kelle, S.; Poland, G.A.; Frustaci, A.; et al. Myocarditis following COVID-19 vaccine: Incidence, presentation, diagnosis, pathophysiology, therapy, and outcomes put into perspective. A clinical consensus document supported by the Heart Failure Association of the European Society of Cardiology (ESC) and the ESC Working Group on Myocardial and Pericardial Diseases. *Eur. J. Heart Fail.* **2022**, *24*, 2000–2018. [CrossRef]
18. Cassimatis, D.C.; Atwood, J.; Engler, R.M.; E Linz, P.; Grabenstein, J.D.; Vernalis, M.N. Smallpox vaccination and myopericarditis: A clinical review. *J. Am. Coll. Cardiol.* **2004**, *43*, 1503–1510. [CrossRef]
19. Caforio, A.L.; Mahon, N.J.; McKenna, W.J. Cardiac autoantibodies to myosin and other heart-specific autoantigens in myocarditis and dilated cardiomyopathy. *Autoimmunity* **2001**, *34*, 199–204. [CrossRef]
20. Chang, J.C.; Xiao, R.; Mercer-Rosa, L.; Knight, A.M.; Weiss, P.F. Child-onset systemic lupus erythematosus is associated with a higher incidence of myopericardial manifestations compared to adult-onset disease. *Lupus* **2018**, *27*, 2146–2154. [CrossRef]
21. Vereckei, A.; Besenyi, Z.; Nagy, V.; Radics, B.; Vágó, H.; Jenei, Z.; Katona, G.; Sepp, R. Cardiac Sarcoidosis: A Comprehensive Clinical Review. *Rev. Cardiovasc. Med.* **2024**, *25*, 37. [CrossRef]
22. Narula, N.; McNamara, D.M. Endomyocardial Biopsy and Natural History of Myocarditis. *Heart Fail. Clin.* **2005**, *1*, 391–406. [CrossRef] [PubMed]
23. Schulz-Menger, J.; Collini, V.; Gröschel, J.; Adler, Y.; Brucato, A.; Christian, V.; Ferreira, V.M.; Gandjbakhch, E.; Heidecker, B.; Kerneis, M.; et al. 2025 ESC Guidelines for the management of myocarditis and pericarditis. *Eur. Heart J.* **2025**, ehaf192. [CrossRef] [PubMed]

24. Domínguez, F.; Uribarri, A.; Larrañaga-Moreira, J.M.; Ruiz-Guerrero, L.; Pastor-Pueyo, P.; Gayán-Ordás, J.; Fernández-González, B.; Esteban-Fernández, A.; Barreiro, M.; López-Fernández, S.; et al. Diagnosis and treatment of myocarditis and inflammatory cardiomyopathy. Consensus document of the SEC-Working Group on Myocarditis. *Rev. Esp. Cardiol.* **2024**, *77*, 667–679. [CrossRef]
25. Adedinsewo, D.A.; Morales-Lara, A.C.; Afolabi, B.B.; Kushimo, O.A.; Mbakwem, A.C.; Ibiyemi, K.F.; Ogunmodede, J.A.; Raji, H.O.; Ringim, S.H.; Habib, A.A.; et al. Artificial intelligence guided screening for cardiomyopathies in an obstetric population: A pragmatic randomized clinical trial. *Nat. Med.* **2024**, *30*, 2897–2906. [CrossRef]
26. Mirna, M.; Schmutzler, L.; Topf, A.; Boxhammer, E.; Sipos, B.; Hoppe, U.C.; Lichtenauer, M. Treatment with Non-Steroidal Anti-Inflammatory Drugs (NSAIDs) Does Not Affect Outcome in Patients with Acute Myocarditis or Myopericarditis. *J. Cardiovasc. Dev. Dis.* **2022**, *9*, 32. [CrossRef]
27. Tschöpe, C.; Cooper, L.T.; Torre-Amione, G.; Van Linthout, S. Management of Myocarditis-Related Cardiomyopathy in Adults. *Circ. Res.* **2019**, *124*, 1568–1583. [CrossRef]
28. Caforio, A.L.P.; Pankuweit, S.; Arbustini, E.; Basso, C.; Gimeno-Blanes, J.; Felix, S.B.; Fu, M.; Heliö, T.; Heymans, S.; Jahns, R.; et al. Current state of knowledge on aetiology, diagnosis, management, and therapy of myocarditis: A position statement of the European Society of Cardiology Working Group on Myocardial and Pericardial Diseases. *Eur. Heart J.* **2013**, *34*, 2636–2648. [CrossRef]
29. Ammirati, E.; Moslehi, J.J. Diagnosis and Treatment of Acute Myocarditis: A Review. *JAMA* **2023**, *329*, 1098–1113. [CrossRef] [PubMed]
30. Long, Q.; Li, L.; Yang, H.; Lu, Y.; Yang, H.; Zhu, Y.; Tang, Y.; Liu, C.; Yuan, J. SGLT2 inhibitor, canagliflozin, ameliorates cardiac inflammation in experimental autoimmune myocarditis. *Int. Immunopharmacol.* **2022**, *110*, 109024. [CrossRef] [PubMed]
31. Kragholm, K.H.; Lindgren, F.L.; Zaremba, T.; Freeman, P.; Andersen, N.H.; Riahi, S.; Pareek, M.; Køber, L.; Torp-Pedersen, C.; Søgaard, P.; et al. Mortality and ventricular arrhythmia after acute myocarditis: A nationwide registry-based follow-up study. *Open Heart* **2021**, *8*, e001806. [CrossRef]
32. Ferone, E.B.; Segev, A.; Tempo, E.; Gentile, P.; Elsanhoury, A.; Baggio, C.; Artico, J.; Bhatti, P.; Scott, P.; Bobbio, E.; et al. Current Treatment and Immunomodulation Strategies in Acute Myocarditis. *J. Cardiovasc. Pharmacol.* **2024**, *83*, 364–376. [CrossRef]
33. Chung, M.K. The role of the wearable cardioverter defibrillator in clinical practice. *Cardiol. Clin.* **2014**, *32*, 253–270. [CrossRef]
34. Aquaro, G.D.; Perfetti, M.; Camastra, G.; Monti, L.; Dellegrottaglie, S.; Moro, C.; Pepe, A.; Todiere, G.; Lanzillo, C.; Scatteia, A.; et al. Cardiac MR with Late Gadolinium Enhancement in Acute Myocarditis with Preserved Systolic Function: ITAMY Study. *J. Am. Coll. Cardiol.* **2017**, *70*, 1977–1987. [CrossRef]
35. Asaumi, Y.; Yasuda, S.; Morii, I.; Kakuchi, H.; Otsuka, Y.; Kawamura, A.; Sasako, Y.; Nakatani, T.; Nonogi, H.; Miyazaki, S. Favourable clinical outcome in patients with cardiogenic shock due to fulminant myocarditis supported by percutaneous extracorporeal membrane oxygenation. *Eur. Heart J.* **2005**, *26*, 2185–2192. [CrossRef]
36. Krueger, G.R.; Ablashi, D.V. Human herpesvirus-6: A short review of its biological behavior. *Intervirology* **2003**, *46*, 257–269. [CrossRef]
37. Amdani, S.M.; Kim, H.S.; Orvedahl, A.; John, A.O.; Said, A.; Simpson, K. Successful treatment of fulminant neonatal enteroviral myocarditis in monochorionic diamniotic twins with cardiopulmonary support, intravenous immunoglobulin and pocapavir. *BMJ Case Rep.* **2018**, *2018*, bcr2017224133. [CrossRef]
38. Manaresi, E.; Gallinella, G. Advances in the Development of Antiviral Strategies against Parvovirus B19. *Viruses* **2019**, *11*, 659. [CrossRef]
39. Van Linthout, S.; Elsanhoury, A.; Klein, O.; Sosnowski, M.; Miteva, K.; Lassner, D.; Abou-El-Enein, M.; Pieske, B.; Kühl, U.; Tschöpe, C. Telbivudine in chronic lymphocytic myocarditis and human parvovirus B19 transcriptional activity. *ESC Heart Fail.* **2018**, *5*, 818–829. [CrossRef]
40. Sudano, I.; Spieker, L.E.; Noll, G.; Corti, R.; Weber, R.; Lüscher, T.F. Cardiovascular disease in HIV infection. *Am. Heart J.* **2006**, *151*, 1147–1155. [CrossRef] [PubMed]
41. Baik, S.H.; Jeong, H.S.; Kim, S.J.; Yoon, Y.K.; Sohn, J.W.; Kim, M.J. A Case of Influenza Associated Fulminant Myocarditis Successfully Treated with Intravenous Peramivir. *Infect. Chemother.* **2015**, *47*, 272–277. [CrossRef] [PubMed]
42. Matsumori, A. Role of hepatitis C virus in cardiomyopathies. *Ernst Scher. Res. Found Workshop* **2006**, *55*, 99–120.
43. Sanders, J.M.; Monogue, M.L.; Jodlowski, T.Z.; Cutrell, J.B. Pharmacologic Treatments for Coronavirus Disease 2019 (COVID-19): A Review. *JAMA* **2020**, *323*, 1824–1836. [CrossRef] [PubMed]
44. Wang, Y.X.; da Cunha, V.; Vincelette, J.; White, K.; Velichko, S.; Xu, Y.; Gross, C.; Fitch, R.M.; Halks-Miller, M.; Larsen, B.R.; et al. Antiviral and myocyte protective effects of murine interferon-βand -α2 in coxsackievirus B3-induced myocarditis and epicarditis in Balb/c mice. *Am. J. Physiol. Heart Circ. Physiol.* **2007**, *293*, H69–H76. [CrossRef]
45. Kühl, U.; Pauschinger, M.; Schwimmbeck, P.L.; Seeberg, B.; Lober, C.; Noutsias, M.; Poller, W.; Schultheiss, H.P. Interferon-beta treatment eliminates cardiotropic viruses and improves left ventricular function in patients with myocardial persistence of viral genomes and left ventricular dysfunction. *Circulation* **2003**, *107*, 2793–2798. [CrossRef]

46. Schultheiss, H.-P.; Piper, C.; Sowade, O.; Waagstein, F.; Kapp, J.-F.; Wegscheider, K.; Groetzbach, G.; Pauschinger, M.; Escher, F.; Arbustini, E.; et al. Betaferon in chronic viral cardiomyopathy (BICC) trial: Effects of interferon-β treatment in patients with chronic viral cardiomyopathy. *Clin. Res. Cardiol.* **2016**, *105*, 763–773. [CrossRef]
47. Frustaci, A.; Russo, M.A.; Chimenti, C. Randomized study on the efficacy of immunosuppressive therapy in patients with virus-negative inflammatory cardiomyopathy: The TIMIC study. *Eur. Heart J.* **2009**, *30*, 1995–2002. [CrossRef]
48. Grzechocińska, J.; Tymińska, A.; Giordani, A.S.; Wysińska, J.; Ostrowska, E.; Baritussio, A.; Caforio, A.L.P.; Grabowski, M.; Marcolongo, R.; Ozierański, K. Immunosuppressive Therapy of Biopsy-Proven, Virus-Negative, Autoimmune/Immune-Mediated Myocarditis—Focus on Azathioprine: A Review of Existing Evidence and Future Perspectives. *Biology* **2023**, *12*, 356. [CrossRef]
49. Wojnicz, R.; Nowalany-Kozielska, E.; Wojciechowska, C.; Glanowska, G.; Wilczewski, P.; Niklewski, T.; Zembala, M.; PolonsKi, L.; Rozek, M.M.; Wodniecki, J. Randomized, placebo- controlled study for immunosuppressive treatment of inflammatory dilated cardiomyopathy: Two-year follow-up results. *Circulation* **2001**, *104*, 39–45. [CrossRef] [PubMed]
50. Kandolin, R.; Lehtonen, J.; Salmenkivi, K.; Räisänen-Sokolowski, A.; Lommi, J.; Kupari, M. Diagnosis, treatment, and outcome of giant-cell myocarditis in the era of combined immunosuppression. *Circ. Heart Fail.* **2013**, *6*, 15–22. [CrossRef] [PubMed]
51. Cooper, L.T.J.; Berry, G.J.; Shabetai, R. Idiopathic Giant-Cell Myocarditis—Natural History and Treatment. *N. Engl. J. Med.* **1997**, *336*, 1860–1866. [CrossRef] [PubMed]
52. Elsanhoury, A.; Tschöpe, C.; Van Linthout, S. A Toolbox of Potential Immune-Related Therapies for Inflammatory Cardiomyopathy. *J. Cardiovasc. Transl. Res.* **2020**, *14*, 75–87. [CrossRef]
53. Nagai, T.; Inomata, T.; Kohno, T.; Sato, T.; Tada, A.; Kubo, T.; Nakamura, K.; Oyama-Manabe, N.; Ikeda, Y.; Fujino, T.; et al. JCS 2023 Guideline on the Diagnosis and Treatment of Myocarditis. *Circ. J.* **2023**, *87*, 674–754. [CrossRef]
54. Lu, C.; Qin, F.; Yan, Y.; Liu, T.; Li, J.; Chen, H. Immunosuppressive treatment for myocarditis: A meta-analysis of randomized controlled trials. *J. Cardiovasc. Med.* **2016**, *17*, 631–637. [CrossRef]
55. Winter, M.-P.; Sulzgruber, P.; Koller, L.; Bartko, P.; Goliasch, G.; Niessner, A. Immunomodulatory treatment for lymphocytic myocarditis—A systematic review and meta-analysis. *Heart Fail. Rev.* **2018**, *23*, 573–581. [CrossRef]
56. Xing, Q.; Zhang, Z.; Zhu, B.; Lin, Q.; Shen, L.; Li, F.; Xia, Z.; Zhao, Z. Case Report: Treatment for steroid-refractory immune-related myocarditis with tofacitinib. *Front. Immunol.* **2022**, *13*, 944013. [CrossRef]
57. Wang, Y.; Qi, Z.; Yan, Z.; Ji, N.; Yang, X.; Gao, D.; Hu, L.; Lv, H.; Zhang, J.; Li, M. Mesenchymal Stem Cell Immunomodulation: A Novel Intervention Mechanism in Cardiovascular Disease. *Front. Cell Dev. Biol.* **2022**, *9*, 742088. [CrossRef]
58. Xiong, Y.; Zhang, Z.; Liu, S.; Shen, L.; Zheng, L.; Ding, L.; Liu, L.; Wu, L.; Hu, Z.; Li, L.; et al. T Lymphocyte-Macrophage Hybrid Membrane-Coated Biomimetic Nanoparticles Alleviate Myocarditis via Suppressing Pyroptosis by Targeting Gene Silencing. *Int. J. Nanomed.* **2024**, *19*, 12817–12833. [CrossRef]

biomolecules

MDPI

Review

From Gut to Heart: Targeting Trimethylamine N-Oxide as a Novel Strategy in Heart Failure Management

Zehui Ding [1], Yunfeng Yu [1], Jiaming Wei [2], Ziyan Wang [1], Ruifang Lin [2], Ya Li [3] and Zhihua Guo [1,4,*]

[1] First Clinical College of Chinese Medicine, Hunan University of Chinese Medicine, Changsha 410208, China; dingzehui@stu.hnucm.edu.cn (Z.D.); yuyunfeng@stu.hnucm.edu.cn (Y.Y.); 20212011@stu.hnucm.edu.cn (Z.W.)
[2] College of Chinese Medicine, Hunan University of Chinese Medicine, Changsha 410208, China; 004916@hnucm.edu.cn (J.W.); 20232073@stu.hnucm.edu.cn (R.L.)
[3] College of Pharmacy, Hunan University of Chinese Medicine, Changsha 410208, China; 003872@hnucm.edu.cn
[4] Key Laboratory of Chinese Medicine Intelligent Diagnosis and Treatment of Chronic Diseases in General Universities of Hunan Province, Changsha 410208, China
* Correspondence: 004294@hnucm.edu.cn

Abstract

Heart failure (HF) marks the culmination of numerous cardiac pathologies, presenting a major medical hurdle in prevention and treatment. In recent years, with the advancements in genomics and metabolomics, research has demonstrated that gut microbiota plays a significant role in the pathogenesis of HF. Trimethylamine N-oxide (TMAO) is a gut microbiota-derived metabolite and primarily sourced from foods abundant in choline, L-carnitine, and betaine. Research has shown that patients with HF exhibit higher levels of TMAO. Accumulating evidence has indicated that TMAO directly or indirectly mediates the occurrence and development of HF through multiple mechanisms. Furthermore, TMAO functions as a crucial prognostic marker in HF. Therefore, TMAO emerges as a potential therapeutic target for HF. This article reviews the generation and metabolic pathways of TMAO, emphasizes its pathophysiological mechanisms in HF, and explores promising therapeutic approaches targeting TMAO, offering novel insights and strategies for HF management.

Keywords: trimethylamine N-oxide (TMAO); heart failure; gut microbiota; treatment; diet regulation; probiotics; inhibitors; flavin-containing monooxygenases (FMOs)

1. Introduction

Heart failure (HF) represents a complex clinical syndrome, marking the culmination of diverse cardiac pathologies. It is characterized by symptoms and signs arising from structural or functional impairments in ventricular filling or ejection, accompanied by elevated natriuretic peptide levels or objective evidence of pulmonary or systemic congestion [1,2]. Globally, HF affects more than 56 million individuals, constituting over 1% of the world's population [3,4]. As populations age and cardiovascular disease (CVD) rates rise, the prevalence and mortality of HF continue to increase [2,3]. The primary etiologies include ischemic heart disease (IHD), rheumatic and valvular heart disease, cardiomyopathies, amyloidosis, certain infectious diseases, and endocrine disorders [3]. Based on left ventricular ejection fraction (LVEF), HF is classified into three subtypes: HF with reduced ejection fraction (HFrpEF, LVEF < 40%), HF with mid-range ejection fraction (40% $\leq$ LVEF < 50%), and HF with preserved ejection fraction (HFpEF, LVEF $\geq$ 50%) [2,5]. Multiple mechanisms—such as inflammation, oxidative stress, disrupted energy metabolism, and pathological cardiac remodeling—can initiate or worsen HF. Although HF prediction and diagnosis are relatively straightforward, effective prevention and treatment still pose significant challenges [6]. Despite advances in pharmacotherapy and interventions, HF persists as a leading cause of mortality worldwide, underscoring the need for deeper insights into its pathophysiology and innovative therapeutic targets. Recent research has highlighted the role of the gut–heart axis in HF, proposing that gut microbiota dysbiosis may aggravate myocardial remodeling and cardiac dysfunction. This occurs through metabolic disturbances, immune-mediated systemic inflammation, and intestinal barrier damage, leading to translocation of gut-derived toxins [7–10]. Hence, selective modulation of the gut microbiome emerges as a possible breakthrough in addressing HF.

The gut microbiota can be conceived of as a massive virtual metabolic organ within the human body, encompassing more than 30 trillion microorganisms [11]. Its composition is influenced by host age, sex, diet, antibiotic use, environmental factors, and genetics, among other variables. A balanced gut microbiota is essential for cardiovascular homeostasis, and its disruption could induce atherosclerosis (AS), arrhythmias, and other CVDs [8]. The involvement of gut microbiota in HF has gained increasing attention, with evidence indicating that alterations in microbial communities and their metabolites play a pivotal part in HF occurrence and progression [12,13]. Gut microbiota dysbiosis may worsen HF by impairing metabolic, immune, and barrier functions [14]. Gut microbial metabolites, such as short-chain fatty acids (SCFAs), trimethylamine N-oxide (TMAO), bile acids (BAs), and lipopolysaccharide (LPS; an intrinsic component of the Gram-negative bacterial cell wall that exerts its toxicity through lipid A and is released as endotoxin upon bacterial death and lysis), also exert either beneficial or detrimental effects in HF [15]. For example, SCFAs are generally regarded as advantageous, whereas TMAO and LPS may intensify HF [9].

Further studies reveal the gut microbiota-derived metabolite TMAO as a key mediator connecting HF with intestinal homeostasis, potentially bridging microbial dysbiosis and cardiac pathophysiology. HF patients exhibit elevated TMAO levels [16,17], which may result from intestinal and renal alterations associated with the disease. The gut microbiota composition in HF patients differs significantly from that in healthy individuals, typically showing an increased abundance or proportion of *Firmicutes* (e.g., *Ruminococcus gnavus* [18]) and *Proteobacteria* (e.g., *Escherichia*, *Klebsiella*, *Campylobacter*, and *Shigella* [19,20]) [21]. These shifts likely promote the production of TMAO and its precursor trimethylamine (TMA). Moreover, HF often coincides with elevated intestinal permeability and impaired renal function, which facilitate TMAO entry into circulation and diminish its excretion. TMAO has also been shown to predict the onset and progression of HF in certain instances [22] and serves as a valuable prognostic marker [21,23,24]. Increasing evidence indicates that TMAO plays an important regulatory role in HF [25–28].

Although several studies have offered preliminary insights into how TMAO influences HF and into potential TMAO-targeted therapies, direct evidence of causal involvement remains limited, and comprehensive mechanistic and therapeutic investigations are still needed. Accordingly, this review examines the role of TMAO in HF and focuses on potential therapeutic strategies targeting TMAO, thereby offering new targets and perspectives for HF management.

2. The Generation and Metabolism of TMAO

TMAO, a compound bearing the molecular formula $(CH_3)_3NO$ [29], stands as a gut microbiota-derived metabolite increasingly recognized as a vital contributor to intestinal-origin CVDs [7]. Gut microbiota metabolizes choline, L-carnitine, betaine, and other choline- or TMA-rich compounds into TMA, a transformation catalyzed by specific enzymes notably including choline TMA-lyase CutC and its activator CutD, carnitine monooxygenase CntA/B, betaine reductase, TMAO reductase, and the CntA/B homologs YeaW/X [30–33]. The resulting TMA permeates the intestinal barrier, enters circulation, and subsequently travels via the portal circulation to the liver, where it is rapidly oxidized by hepatic flavin-containing monooxygenases (FMOs) into TMAO [34]. TMAO then circulates systemically [35]. In healthy individuals, over 90% of TMAO is eliminated by the kidneys and rapidly excreted in urine [36]. The generation and metabolic process of TMAO is detailed in Figure 1.

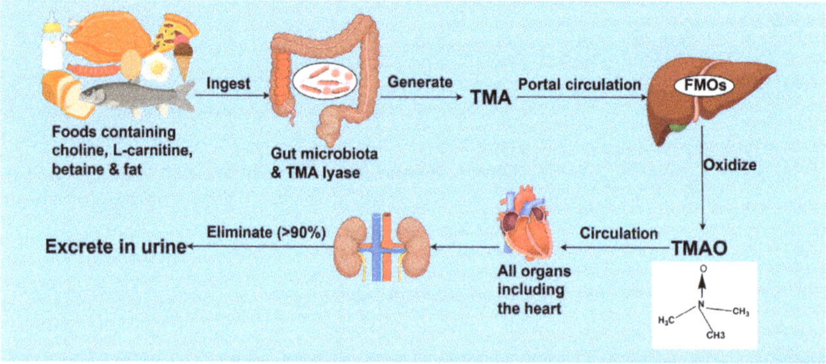

Figure 1. The generation and metabolism of TMAO. TMA: trimethylamine; FMOs: flavin-containing monooxygenases; TMAO: trimethylamine N-oxide. Created by Fidraw (www.figdraw.com, accessed on 6 August 2025).

Systemic TMAO concentrations are influenced by multiple factors, primarily determined by three key processes: intestinal generation, hepatic conversion, and renal elimination. Dysregulation in any of these processes may elevate TMAO levels. Dietary choline is obtained mainly from meat, eggs, dairy products, fish, grains, and foods derived from them [37]. The primary dietary source of L-carnitine is red meat [38]. Betaine occurs naturally in foods like beets, spinach, grains, and shrimp [39]. Gut microbiota metabolize these dietary precursors into TMA via specific key enzymes. However, the gene clusters encoding these enzymes occur only in certain intestinal bacteria [30]. These clusters are more prevalent among the gut bacteria of carnivores and omnivores than herbivores [30], initiating increased TMAO generation [31]. High-fat diets (HFDs) have also been found to elevate TMAO concentrations [32]. Accordingly, TMAO synthesis is influenced by the composition and function of gut microbiota as well as dietary intake. Concurrently, FMO activity affects TMAO generation; for instance, liver diseases such as non-alcoholic fatty liver disease can upregulate FMO expression [40], promoting TMA oxidation. Among various FMOs that convert TMA to TMAO, FMO3 exhibits the highest activity [41]. Knocking out or silencing of hepatic FMO3 in mice decreases circulating TMAO levels [36,42]. Renal insufficiency may also elevate TMAO due to reduced excretion [43]. A limited number of studies suggested that dietary modulation of TMAO levels may not occur through direct regulation of FMO3 activity [44,45]. However, direct evidence remains scarce that elevated TMAO arises solely from increased TMA formation or impaired renal excretion, independent of FMO-mediated oxidation. Consequently, high TMAO levels likely reflect the combined influence of increased TMA precursors, enhanced FMO conversion efficiency, and impaired renal excretion—three dynamically interacting factors that should be considered together.

Moreover, TMAO concentrations correlate positively with age [46,47]. Genetic factors also influence TMAO production and metabolism. For example, mutations in the

FMO3 gene can reduce or abolish enzymatic activity, impairing TMA-to-TMAO conversion and leading to toxic TMA accumulation. This condition, trimethylaminuria (TMAU), is characterized by a fish-like odor in bodily secretions [48] and represents the most direct genetic influence on TMAO metabolism. Host genetics also partially shape gut microbiota composition and function. Individual genetic backgrounds may indirectly modulate TMA production by influencing microbial community structure, thereby contributing to interindividual variation in TMAO levels. A large-scale population study linked variability in TMAO and its precursors to the abundance of *Ruminococcaceae* genera and the *CutC* gene in gut microbiota [33]. Certain TMA-producing bacteria, including some *Firmicutes*, show heritability [49,50]. Additionally, circulating TMAO concentrations vary substantially among healthy adults across geographic regions. In the United States, the median circulating TMAO level in healthy populations was 3.5 μmol/L [51,52], whereas Fang et al. [53] recently documented a lower median of 1.70 μmol/L among healthy Chinese adults. These disparities may arise from genetic, dietary, and ethnic factors.

In conclusion, TMAO levels correlate with various factors, including dietary patterns, gut microbiota structure, activity of TMAO-producing enzymes such as TMA lyase and FMOs, renal function, age, genetics, ethnicity, and geographical region.

3. The Mechanisms by Which TMAO Regulates HF

3.1. Inflammation

Inflammation is closely linked to HF [14]. Higher concentrations of TMAO, such as 600 μmol/L, have been shown to enhance the expression of inflammatory genes and activate proinflammatory cytokines, thus amplifying inflammatory responses [54]. Boini et al. [55] experimentally demonstrated that TMAO stimulated the secretion of inflammatory cytokines, including interleukin (IL)-1β, by activating the nucleotide-binding oligomerization domain-like receptor containing pyrin domain 3 (NLRP3) inflammasome, leading to endothelial inflammation. TMAO also inhibited the deacetylation of superoxide dismutase 2 (SOD2) by suppressing Sirtuin 3 (SIRT3), thereby reducing SOD2 activity. This process promoted mitochondrial reactive oxygen species (mtROS) generation, leading to elevated overall reactive oxygen species (ROS) levels [56,57]. This increased thioredoxin-interacting protein (TXNIP) levels, subsequently activating the NLRP3 inflammasome [57]. Such activation elicited a dose- and time-dependent release of IL-1β and IL-18 [58], culminating in an inflammatory response.

Moreover, TMAO may influence HF by modulating inflammatory pathways. Zhang and colleagues [59] reported that TMAO could possibly exacerbate myocardial inflammation in mice by elevating TNF (tumor necrosis factor)-α and reducing IL-10 levels. Additionally, their study showed that TMAO supplementation (120 mg/kg) in drinking water may counteract the cardioprotective benefits of voluntary exercise [59].

3.2. Abnormal Energy Metabolism

Abnormal energy metabolism plays a pivotal role in the development of HF. In HF, due to the increased dependence of cardiac metabolism on glucose, the primary energy source shifts from fatty acid (FA) oxidation to glucose metabolism [60]. Additionally, hypoxia reduces oxidative phosphorylation (OXPHOS) and enhances glycolysis [61]. These metabolic alterations contribute to an insufficient energy supply, thus worsening the condition of HF.

Elevated TMAO levels are strongly linked to impaired cardiac energy metabolism. HFD-induced TMAO accumulation inhibited mitochondrial activity and consequently compromised energy metabolism [62]. Saaoud et al. [54] revealed that 600 μmol/L TMAO initiated a metabolic switch from OXPHOS to glycolysis. This shift occurred by upregulating mitochondrial regulators such as perilipin 4 (PLIN4), overlapping activity with m-AAA protease 1 (OMA1), and oxoglutarate dehydrogenase L (OGDHL), ultimately disrupting energy homeostasis. Brown adipose tissue (BAT) enhances thermogenesis and supports cardiac function [63,64], but hypoxia induces brown adipocyte apoptosis, resulting in BAT dysfunction [28]. In physiological conditions, BAT absorbs and metabolizes circulating choline, potentially to maintain its cellular membrane integrity. In a murine model of HF, BAT dysfunction led to choline accumulation, which may be partially converted to TMAO via the choline-TMA-TMAO pathway; this increased TMAO synthesis exacerbated HF [28]. This study further revealed that TMAO significantly attenuated creatine phosphate (CP) and adenosine triphosphate (ATP) concentrations in cardiac tissue by suppressing mitochondrial complex IV activity, further impairing cardiac energy metabolism. This led to a vicious cycle of BAT dysfunction and cardiac dysfunction ultimately [28].

In addition, impaired calcium ion (Ca^{2+}) handling in myocardial cells increases the energy demand of the failing heart [65], compromising both systolic and diastolic function and worsening HF. TMAO disrupts Ca^{2+} regulation in cardiac cells. In endothelial cells, 100 μmol/L TMAO suppressed purinergic response-induced increase in Ca^{2+} influx and prolongation of Ca^{2+} signaling, indicating loss of Ca^{2+} homeostasis [66]. High glucose and high fat could also induce TMAO production, possibly predisposing individuals to Ca^{2+} overload [67]. Esposito and coworkers [68] proved that 100 μmol/L TMAO transiently suppressed contractile recovery of myocardial cells by elevating intracellular Ca^{2+} currents.

3.3. *Oxidative Stress*

Oxidative stress contributes to the pathophysiological HF via multiple mechanisms. Studies have demonstrated that TMAO can mediate oxidative stress. TMAO concentrations were remarkably higher in preeclampsia patients. Transplanting fecal microbiota from these patients into antibiotic-treated mice resulted in pronounced oxidative stress in the recipients [69]. In addition, TMAO can provoke oxidative stress by distinct pathways. A study involving mice and healthy human subjects revealed that TMAO promoted oxidative stress by increasing superoxide levels and impairing endothelial nitric oxide synthase (eNOS) activity [46]. Chen et al. [57] confirmed that TMAO, at concentrations of 150, 300, 600, and 900 μmol/L, augmented ROS generation, particularly mtROS, through modulating the SIRT3-SOD2-mtROS signaling cascade, thus fostering oxidative stress. Moreover, TMAO induced oxidative stress through activation of NADPH oxidase 2 (NOX2) [70].

Elevated TMAO levels have been demonstrated to initiate HF through oxidative stress. Recent research revealed that TMAO induced oxidative stress by downregulating piezo type mechanosensitive ion channel component 1 (PIEZO1), thereby exacerbating cardiac diastolic dysfunction in mice suffering from HFpEF [25]. Therefore, decreasing TMAO concentrations and alleviating oxidative stress appear to be a viable therapeutic approach for HF.

3.4. *Myocardial Remodeling*

Myocardial remodeling stands as a significant mechanism underlying the occurrence and progression of HF. Its primary features consist of pathological cardiomyocyte hypertrophy and myocardial fibrosis. TMAO directly influences this process. Research in mice indicated that a Western diet (WD) or TMAO supplementation increased TMAO levels, promoted myocardial fibrosis, and impaired cardiac function [59]. Studies demonstrated that TMAO stimulated myocardial remodeling in Sprague-Dawley rats and triggered hypertrophic responses in rat cardiomyocytes, enlarging cell size while upregulating hypertrophy markers such as atrial natriuretic peptide (ANP) and β-myosin heavy chain (β-MHC). This effect might involve activation of the transforming growth factor (TGF)-β1/Smad3 signaling pathway [71]. Another study indicated that transverse aortic constriction-induced HF in mice correlated with elevated TMAO levels, exacerbated myocardial hypertrophy and fibrosis, and compromised cardiac function. The mechanisms could be associated with activation of the TGF-β1/Smad3 and p65 nuclear factor (NF)-κB signaling pathways [27].

TMAO aggravates HF by affecting adverse cardiac remodeling. Organ et al. [72] discovered that diets containing choline (1.2%) or TMAO (0.12%) markedly increased circulating TMAO and brain natriuretic peptide (BNP) levels in HF mice. Concurrently, left ventricular (LV) fractional shortening and ejection fraction were decreased, accompanied by exacerbated LV remodeling, intensified myocardial fibrosis, and deteriorated cardiac dysfunction, ultimately accelerating HF progression. Subsequent experiments showed that discontinuing TMAO intake or inhibiting TMA lyase effectively lowered the HF marker BNP levels, suppressed numerous profibrotic gene expressions, and ultimately mitigated ventricular remodeling and cardiac dysfunction [73].

3.5. Indirect Mechanisms

TMAO indirectly contributes to the emergence and progression of HF by mediating several mechanisms such as endothelial dysfunction [46], renal insufficiency [74], elevated blood pressure [75], enhanced platelet activity and thrombosis [34], and lipid metabolism disorders [9,31].

3.5.1. Endothelial Dysfunction

Endothelial dysfunction plays a crucial part in AS-related CVDs and ensuing HF. Studies uncovered a distinct association between TMAO and endothelial dysfunction biomarkers in patients suffering from HF [9]. Elevated serum TMAO levels disrupt vascular endothelial integrity and exacerbate endothelial dysfunction, promoting vascular dysfunction, inflammation, fibrosis, and AS [9], ultimately increasing CVD risk [9,54].

Multiple studies have explored the mechanisms by which TMAO causes endothelial dysfunction. In both aged mice and middle-aged or elderly human subjects, TMAO levels and the severity of endothelial dysfunction increased significantly with age [46]. Additionally, TMAO exacerbated endothelial dysfunction by promoting inflammation and oxidative stress, suppressing eNOS activity and impairing NO (nitric oxide)-mediated vasodilation. These findings suggested that elevated TMAO levels during aging likely reduce eNOS-derived NO bioavailability through vascular inflammation and oxidative stress, contributing to age-related endothelial dysfunction [46]. Additional research revealed that TMAO induced endothelial dysfunction via the ROS-TXNIP-NLRP3 inflammasome signaling pathway [27]. Evidence indicated that TMAO fostered endothelial dysfunction and vascular calcification through the activation of the NLRP3 inflammasome and NF-κB signaling [76]. Kim et al. [77] revealed that TMAO (50 and 100 μmol/L) mediated chromatin remodeling in coronary artery endothelial cells by augmenting histone methylation, fostering endothelial-to-myofibroblast transformation and vascular fibrosis, thus inducing ischemic HF [78].

3.5.2. Renal Insufficiency

Studies have shown that patients with deficient renal function have higher blood TMAO levels [9,43], and a notable inverse relationship exists between TMAO levels and estimated glomerular filtration rate [79]. Rising TMAO concentrations may worsen renal insufficiency and raise all-cause mortality in chronic kidney disease patients through inducing progressive renal fibrosis and dysfunction [43].

In HF, neurohormonal activation ultimately results in hypoperfusion and dysfunction in multiple organs, the kidneys among them. Likewise, the significance of defective renal function in the advancement of HF is not negligible. TMAO is linked to the onset and development of both HF and renal insufficiency [8]. On one hand, TMAO promotes water and sodium retention by triggering glomerular interstitial fibrosis and renal dysfunction, indirectly deteriorating HF [43,80]. On the other hand, chronic heart failure (CHF) frequently contributes to progressive renal function impairment [74], which decreases TMAO excretion and increases circulating TMAO concentrations. These elevated TMAO levels further exacerbate HF progression. For instance, in a rat model of cardiorenal syndrome featuring both HF and renal insufficiency, increased blood TMAO levels exacerbated cardiac and renal dysfunction [74].

3.5.3. Elevated Blood Pressure

Elevated blood pressure is a primary inducer of HF. Uncontrolled chronic hypertension (HTN) could induce LV remodeling via biochemical and neurohumoral mechanisms, which may subsequently result in hypertensive heart disease and, in its most severe form, progress to HF.

TMAO is closely associated with enhanced blood pressure. A systematic review and meta-analysis revealed an obvious positive dose-dependent correlation between circulating TMAO levels and HTN prevalence [81]. Jiang et al. [75] reported that 1% TMAO enhanced angiotensin (ANG) II-induced vasoconstriction and acute pressor responses by activating protein kinase R-like endoplasmic reticulum kinase-associated pathways, consequently worsening HTN. Another research indicated that higher TMAO concentrations (30 μmol/L) increased systolic blood pressure in both mice and human subjects by inducing advanced glycation end-product accumulation and oxidative stress [82]. These observations suggest that TMAO could function as a potential biomarker for HTN.

3.5.4. Increased Platelet Activity and Thrombosis

Excessive platelet activation may facilitate thrombosis. Research has shown that elevated platelet activity and thrombosis play vital roles in the onset and advancement of HF [83]. Plasma TMAO might trigger excessive platelet activation and thrombosis in humans [42,84]. Clinical studies further indicated that TMAO predicted the likelihood of thrombotic events, like myocardial infarction (MI) and stroke [34,85]. In mice, plasma TMAO concentrations correlated positively with thrombosis rates, and dietary supplementation with choline or TMAO increased platelet activity and accelerated thrombosis [34,84,85]. Similarly, in humans, choline intake resulted in elevated TMAO levels, which enhanced platelet reactivity, ultimately facilitating thrombosis [86]. Zhu and colleagues [34] confirmed that 100 μmol/L TMAO potentiated platelet hyperreactivity by stimulating intracellular Ca^{2+} release, thereby raising thrombosis risk.

Additionally, TMAO can contribute to CVDs via increased platelet activity and thrombosis [87]. Hence, attenuating TMAO production may serve as a therapeutic target to curb excessive platelet activation and thrombosis, improving clinical outcomes in CVDs, including HF.

3.5.5. Abnormal Lipid Metabolism

Abnormal lipid metabolism emerges as a key pathogenic factor in HF, and correcting lipid metabolism is crucial for its treatment [88]. TMAO accumulation has been confirmed to cause abnormal lipid metabolism [9,89], consequently increasing the risk of diverse CVDs, HF among them. Elevated TMAO disrupts lipid metabolism through mechanisms such as inhibiting reverse cholesterol transport (RCT), promoting macrophage foam-cell formation, and modifying BA synthesis.

First, TMAO suppresses RCT and subsequently disrupts cholesterol metabolism. TMAO and its precursors, choline and carnitine, impede RCT in the body through a gut microbiota-dependent mechanism [31]. Second, TMAO facilitates cholesterol accumulation in macrophages and the formation of foam cells [9,90]. Wang et al. [91] discovered that

TMAO enhanced phagocytic activity and cholesterol content in peritoneal macrophages by upregulating scavenger receptor A and CD36, thereby promoting foam-cell formation. Third, TMAO also restricts the production of BA. In apolipoprotein E knockout ($ApoE^{-/-}$) mice fed a diet containing 0.3% TMAO, TMAO altered the BA profile and suppressed hepatic BA synthesis by specifically inhibiting the classical pathway, leading to abnormal lipid metabolism and remarkably increased blood lipid concentrations [89]. Also, according to He and coworkers [92], TMAO increased hepatic cholesterol accumulation by reducing fecal excretion of acidic sterols, contributing to hyperlipidemia.

In summary, TMAO directly or indirectly modulates HF through multiple pathophysiological mechanisms, including inflammation, abnormal energy metabolism, oxidative stress, and myocardial remodeling [25,28,59,72] (Figure 2). Moreover, these pathways often interact, creating a pernicious cycle that accelerates the advancement of HF even more [93].

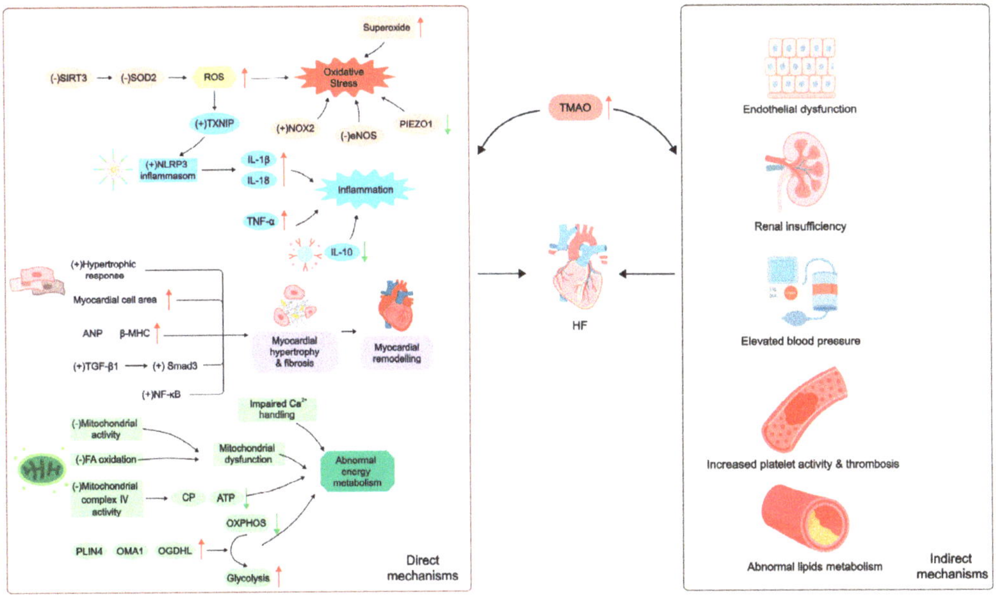

Figure 2. The mechanisms by which TMAO regulates HF. An upward arrow (↑) indicates an increase. A downward arrow (↓) indicates a decrease. A plus sign (+) indicates promotion. A minus sign (−) indicates inhibition. SIRT3: Sirtuin 3; SOD2: superoxide dismutase 2; ROS: reactive oxygen species; TXNIP: thioredoxin-interacting protein; NOX2: NADPH oxidase 2; eNOS: endothelial nitric oxide synthase; PIEZO1: piezo type mechanosensitive ion channel component 1; NLRP3: nucleotide-binding oligomerization domain-like receptor containing pyrin domain 3; IL: interleukin; TNF: tumor necrosis factor; ANP: atrial natriuretic peptide; β-MHC: β-myosin heavy chain; TGF: transforming growth factor; NF: nuclear factor; Ca^{2+}: calcium ion; FA: fatty acid; CP: creatine phosphate; ATP: adenosine triphosphate; PLIN4: perilipin 4; OMA1: overlapping activity with m-AAA protease 1; OGDHL: oxoglutarate dehydrogenase L; OXPHOS: oxidative phosphorylation; TMAO: trimethylamine N-oxide; HF: heart failure.

4. Potential Therapeutic Approaches

Multiple studies have demonstrated that elevated TMAO levels are independently associated with an increased risk of adverse outcomes in patients with HFpEF [22,94,95] and HFrEF [94,96,97], suggesting its potential as a prognostic biomarker. Targeting TMAO has therefore emerged as a viable therapeutic approach. In HFpEF, strategies to lower TMAO—such as dietary modification, physical activities, inhibitors, or medications like sodium-glucose cotransporter-2 (SGLT2) inhibitors—may reduce systemic inflammation, improve metabolism, and mitigate myocardial remodeling. For HFrEF, interventions including aspirin, angiotensin-converting enzyme inhibitors (ACEIs), or statins can lower TMAO, which may help alleviate vascular inflammation and atherosclerosis, improve myocardial ischemia, and lead to better clinical outcomes.

4.1. Diet Regulation

Current evidence indicates that dietary intervention could effectively lessen circulating TMAO levels, probably aiding in the prevention and treatment of HF [21]. As noted, consuming foods abundant in TMA precursors—including choline, L-carnitine, and betaine—or high-fat foods increases TMAO concentrations in humans. In animal studies, a choline-rich diet significantly increased TMAO and BNP levels, worsened ventricular remodeling, impaired cardiac function, and further aggravated HF [72]. Mo et al. [32] demonstrated that an HFD elevated plasma TMAO concentrations in naturally aging rats. In healthy rats, prolonged HFD intake triggered low-grade systemic inflammation and diminished beneficial SCFA levels, while long-term fructose consumption heightened oxidative stress, elevated blood pressure, and expanded *Enterobacter* and *Escherichia coli* populations [98]. The WD, distinguished by high amounts of TMA precursors, fats, and sugars [9,47,62], is strongly associated with increased TMAO concentrations [52]. Consuming WD alters gut microbiota composition, diminishes its diversity, and consequently promotes TMAO production, rising blood TMAO concentrations, even in healthy people [99]. The WD has been proven to elevate circulating TMAO levels, thus inducing cardiac fibrosis and systemic inflammation. These pathological processes can subsequently cause cardiac dysfunction, heightening the risk of developing AS, coronary artery diseases (CADs), and HF [9,21,47,62]. Furthermore, animal experiments demonstrated that a high-salt diet may reduce the ratio of SCFAs to TMAO, lower *Lactobacillus* abundance, and exacerbate immune and inflammatory responses [100]. Additionally, insufficient dietary fiber could worsen gut microbiota dysbiosis [101]. Hence, it is advisable to steer clear of overindulging in foods rich in TMA precursors, fats, sugars, or salts [99], while augmenting dietary fiber intake.

Studies have shown that consuming red meat raises TMAO concentrations. Decreasing consumption of choline- or L-carnitine-rich foods such as red meat can hinder TMAO synthesis, ultimately reducing HF incidence [7,52]. A plant-based diet primarily consists of fruits, vegetables, seeds, nuts, legumes, whole grains, and herbal plants. It may adhere strictly to a vegetarian diet or incorporate moderate quantities of animal-derived foods like fish, seafood, dairy products, and eggs [102]. This dietary pattern has been demonstrated to modulate gut microbiota, augment the generation of advantageous metabolites such as SCFAs, and diminish TMAO levels [103–105]. The Mediterranean diet (MD), a well-established healthy pattern, emphasizes ample fruits, vegetables, nuts, and whole grains, along with moderate meat, eggs, sugar, olive oil, and wine [93]. It is low in saturated FAs, salt, and phosphates, but rich in unsaturated FAs, antioxidants (such as polyphenols, vitamins, and flavonoids), nitrates, and dietary fiber [9,37,67]. Healthy diets like the MD could modulate gut microbiota composition and its metabolites, influencing the occurrence and progression of HF. Studies have indicated that MD mitigates oxidative stress and inflammation while enhancing antioxidant capacity and endothelial function by augmenting the gut microbial diversity and abundance, raising SCFA production, and reducing TMAO concentrations. These effects collectively improve cardiac performance and lower CVD risk [9,106,107]. Adherence to the MD has been linked to decreased HF incidence and mortality [108], and it has exhibited additional benefits such as restoring gut microbial balance and lessening TMAO concentrations [109]. Estruch and colleagues [110] demonstrated that MD consumption, especially when enriched with extra-virgin olive oil, effectively attenuated TMAO levels, thus diminishing major cardiovascular events such as MI and stroke. Kaluza et al. [111] uncovered that certain foods, including vegetables, fruits, and nuts, lessened HF risk in individuals with smoking addictions by modifying gut microbiota, suppressing oxidative stress, and inhibiting cell death. Merques and coworkers [112] revealed that a high-fiber diet regulated gut microbiota, lowered blood pressure, and ameliorated cardiac fibrosis and LV hypertrophy, thereby contributing to the prevention of HF. Furthermore, obese patients following a low-calorie diet may experience decreased TMAO concentrations [113,114]. In summary, healthy eating serves as a practical and economical way to manage HF, mainly by modulating gut microbiota and their metabolites [93].

4.2. *Physical Activities*

Tailored exercise training can enhance exercise capacity and quality of life in patients with HF, improve survival rates, and lower the likelihood of HF-related hospitalization [115,116]. Additionally, physical activities have been shown to modulate gut microbiota [117], leading to lower TMAO levels. For example, sprint training decreased serum and urinary TMAO concentrations in male participants [118,119]. Physical activity also notably reduced plasma TMAO levels in a study of 16 obese adults [113]. Zhang et al. [120] discovered that voluntary wheel running in mice alleviated gut microbiota dysbiosis and decreased serum levels of TMAO and its precursors, TMA and betaine.

Physical activities confer beneficial effects on the heart by downregulating TMAO concentrations. After being fed a WD, mice that engaged in voluntary exercise demonstrated reduced plasma TMAO levels, diminished myocardial inflammation and fibrosis, and enhanced cardiac function compared with sedentary controls [59]. Brandao and colleagues [121] found that a combined exercise regimen—incorporating both strength and aerobic training—decreased TMAO levels, thereby mitigating cardiovascular risk and strengthening physical function in obese women.

4.3. *Probiotics, Prebiotics, and Synbiotics*

Probiotics, prebiotics, and synbiotics serve as prevalent therapeutic avenues for preserving gut microbiota homeostasis [122]. Probiotics, live microorganisms capable of adjusting the quantity and composition of gut microbiota, confer health benefits [35]. Prebiotics consist of non-digestible substrates selectively utilized by beneficial microorganisms in the host to promote their growth and/or activity, ultimately improving host health [122,123]. Synbiotics combine live microorganisms with substrates selectively metabolized by host microbiota, effectively integrating the benefits of both probiotics and prebiotics to foster host well-being. The synergistic and complementary effects of live microorganisms and substrates allow synbiotics to effectually influence microbiota composition and immune function [124].

Research has shown that probiotics, prebiotics, or synbiotics can remodel the gut microbiota in animals, thereby reducing gut-derived TMA synthesis, inhibiting or blocking the TMA/TMAO pathway, and ultimately decreasing TMAO levels [7]. Probiotics, including *Lactobacillus plantarum* ZDY04, *Bifidobacterium breve* Bb4, *Bifidobacterium longum* BL1 and BL7, and *Lacticaseibacillus rhamnosus* L34, may decrease TMAO concentrations by adjusting gut microbiota and restricting TMA generation [44,125,126]. Wang et al. [125] discovered that *Bifidobacterium* lessened the abundance of TMA-producing *Ruminococcaceae* UCG-009 and UCG-010, thereby lowering plasma TMAO concentrations. Among *lactobacilli* strains with robust adherence capability, *Lactobacillus amylovorus* LAM1345 and *Lactiplantibacillus plantarum* LP1145 individually reduced serum TMAO in choline-fed mice. Whereas, a multistrain formula (MF) containing these two strains plus *Limosilactobacillus fermentum* LF33 exhibited the most pronounced reduction, likely due to the additive and synergistic actions in MF [127]. Furthermore, following a 12-week synbiotic intervention decreased serum TMAO and improved metabolic profiles in patients with dyslipidemia [128].

Probiotics, prebiotics, and synbiotics hold potential in treating CVDs, like HF, via various pathophysiological mechanisms [129,130]. The gut microbiota of HF patients showed reduced probiotic abundance and elevated harmful bacteria [131]. Several probiotics, such as *Lactobacillus plantarum* ZDY04 [44], *Lactobacillus rhamnosus* GG strain [132], and *Bifidobacterium lactis* Probio-M8 [133], reduced TMAO concentrations, aid in CVD management, and diminish adverse cardiovascular events, indicating TMAO inhibition contributed to their cardioprotective effects. Sánchez-Quintero et al. [134] transplanted gut microbiota from patients with IHD into mice, elevating TMAO concentrations in recipients. Prebiotic essential oils from parsley (*Petroselinum crispum*) and rosemary (*Rosmarinus officinalis*) lessened plasma TMAO levels by reshaping gut microbiota in these mice. Similarly, thyme (*Thymus vulgaris*) and oregano (*Origanum vulgare*) essential oils functioned as effective prebiotics and produced comparable effects in gnotobiotic mice colonized with gut microbiota from CAD patients [135]. Synbiotics also reduced NT-proBNP levels and ameliorated the inflammatory status in patients with CHF [130]. As a result, probiotics, prebiotics, synbiotics, and their derivatives may offer safe and effective strategies for HF prevention and treatment through mechanisms including modulating gut microbiota, reducing TMAO levels, and maintaining host intestinal homeostasis [93].

Archaea, ancient single-celled prokaryotes inhabiting extreme environments and morphologically similar to bacteria [136], are also present in the human gastrointestinal tract. Certain archaea partially convert gut-derived TMA into methane, reducing TMA content [137]. Ramezani and coworkers [138] found that intestinal colonization with *Methanobrevibacter smithii*, a methanogenic archae, reduced plasma TMAO in ApoE$^{-/-}$ mice, consequently attenuating AS progression. Although current research on archaea-based therapies for CVDs is still limited, this approach appears highly promising.

4.4. Inhibitors

As previously mentioned, choline TMA lyase—which metabolizes choline into TMA—and FMOs—which oxidize TMA to produce TMAO—play critical parts in the TMAO synthesis and thus represent crucial targets for impeding TMAO production. Current methods for reducing TMAO generation typically focus on inhibiting TMA lyase or FMOs employing pharmacological inhibitors or genetic modification.

Choline TMA lyase (CutC), encoded by adjacent genes within a gene cluster, is a glycosyl radical enzyme. Both CutC and its activating protein CutD are critical for TMA biosynthesis in the gut microbiota [139]. 3,3-dimethyl-1-butanol (DMB), a choline analog, is the most widely studied inhibitor of this enzyme. It is naturally present in vinegar, extra virgin olive oil, wine, and grape seed oil; notably, extra virgin olive oil and wine are components of the MD [93,140]. Hence, DMB may partly account for the TMAO-lowering effects associated with the MD. Wang and colleagues [140] first showed in vitro and in vivo that DMB can partially inhibit TMA lyase activity in different microorganisms, blocking TMA formation from various substrates, including choline and carnitine, ultimately decreasing plasma TMAO concentrations. Additionally, DMB is non-toxic and non-lethal. Nevertheless, it does not inhibit TMA generation from certain substrates such as γ-butyrobetaine, nor does it affect TMA oxidation by FMO3 [140]. Based on DMB, the Hazen team developed a new generation of TMA lyase inhibitors: fluoromethylcholine (FMC) and iodomethylcholine (IMC). FMC and IMC irreversibly impede CutC/D, thereby effectively and continuously attenuating host TMA and TMAO levels. Furthermore, they exhibit poor absorption by the host, leave commensal bacteria unaffected, demonstrate limited systemic exposure, and are non-toxic and associated with a reduced risk of side effects [85]. Meanwhile, FMO inhibitors lower plasma TMAO by blocking hepatic TMA oxidation [141]. Methimazole, a high-affinity substrate of FMO3, acts as a competitive inhibitor, whereas indoles are non-substrate competitive FMO3 inhibitors. Both have been shown to suppress FMO3 activity and decrease TMAO concentrations [28,142]. However, the impact of methimazole on the thyroid cannot be overlooked. Therefore, inhibiting important TMAO-producing enzymes, specifically TMA lyase and FMO3, offers a plausible approach to lessening TMAO levels and ultimately managing CVDs like HF.

TMA lyase inhibitors have been extensively employed in animal models to hinder TMAO generation and ameliorate cardiovascular impairments. Studies have shown that DMB may attenuate inflammation, cardiomyocyte hypertrophy, and fibrosis, and improve cardiac function in mice with compromised hearts or HF, mainly through TMAO reduction. Furthermore, chronic DMB treatment exhibits no overt toxicity [25,27]. In mice fed WDs, DMB ameliorated excessive TMAO levels, vascular dysfunction, reduced exercise tolerance, and frailty. It also alleviated cardiac inflammation and fibrosis, consequently improving cardiac function and ultimately diminishing CVDs likelihood [143]. IMC, another TMA lyase inhibitor, reduced TMAO and BNP levels in HF mice, downregulated profibrotic gene expression, and mitigated TMAO-induced cardiac remodeling and dysfunction. Importantly, IMC also proved pharmacological safety in relevant tests [73]. FMC and IMC have demonstrated notable curative effects in lowering plasma TMAO levels and suppressing choline-triggered platelet hyperreactivity and thrombosis, without increasing bleeding risk [85]. Although DMB, IMC, and FMC have shown preliminary efficacy and safety in HF models, clinical data remain scarce. Moreover, Yoshida et al. [28] reported that either administering the FMO inhibitor methimazole or depleting the *FMO2* gene lessened TMAO levels in HF mice, alleviating cardiac fibrosis and dysfunction. Antisense oligonucleotide-targeted inhibition [84] or knockout [144] of the FMO3 gene effectively lowered TMAO levels and subsequently mitigated TMAO-induced platelet hyperreactivity and thrombosis in mice. These results suggest that pharmacological or genetic inhibition of FMOs to decrease TMAO synthesis may offer a therapeutic avenue for HF. Nevertheless, clinical translation constitutes challenges, considering that FMOs participate in the oxidative metabolism of diverse drugs and chemicals in the body besides TMAO. Inhibition or deficiency of FMOs can cause hepatotoxicity and lead to liver diseases [145]. Additionally, loss of FMO3 function in humans may result in harmful TMA buildup, triggering side effects including TMAU [48]. Therefore, potential adverse effects must be thoroughly evaluated before FMO inhibitors are applied clinically.

In conclusion, targeted inhibition of TMAO-producing enzymes represents a promising tactic for addressing HF, whereas significant gaps remain between experimental studies and clinical application, warranting further comprehensive investigation [35].

4.5. Modern Medicines

4.5.1. Antibiotics

Antibiotics constitute a conventional strategy for correcting gut microbiota imbalance. They could affect diseases driven by gut microbiota, as they modify its abundance, composition, and metabolic products. Furthermore, HF onset and progression are intimately linked to gut microbial dysbiosis. Research has revealed that antibiotics can ameliorate myocardial ischemia and HF by modulating gut microbiota and its metabolites, suppressing inflammation and oxidative stress, mitigating mitochondrial damage, and bolstering cardiac function [146,147]. In addition, antibiotics decreased TMAO levels by reconstructing gut microbiota, inhibiting the proliferation of harmful gut microbiota, and restricting the conversion of dietary precursors such as choline, L-carnitine, and betaine [35]. In both healthy humans [87] and mice [91], broad-spectrum antibiotics significantly suppressed the dietary phosphatidylcholine-induced increase in plasma TMAO by modulating gut microbiota. Li and coworkers [71] found that antibiotics diminished myocardial hypertrophy and fibrosis in rats by impeding TMAO production. Moreover, depleting gut microbiota with antibiotics attenuated TMAO generation, thereby alleviating ANG II-induced HTN [75]. Broad-spectrum antibiotic therapy also lowered TMAO levels, which decreased macrophage cholesterol content and obstructed the progression of dietary choline-induced AS in mice [91].

However, antibiotic treatment is contentious. Human trials demonstrated that plasma TMAO levels declined markedly during antibiotic administration but rebounded after discontinuation [87]. In a mouse model of MI, antibiotic combination treatment depleted gut microbiota, considerably decreased SCFAs, compromised post-infarction repair, impaired cardiac function, and increased mortality [148]. Additionally, long-term antibiotic use may deplete beneficial gut microbiota and promote antibiotic resistance [149]. Given the numerous detrimental impacts, such as gut microbiota dysbiosis, toxicity, and resistance, a careful risk-benefit assessment is essential before prescribing antibiotics [35]. Further studies are needed to evaluate the potential benefits and safety of rational antibiotic use in patients with HF [93].

4.5.2. Aspirin

Aspirin presents a first-line medication for CVDs and exhibits notable therapeutic impacts on HF with AS [150]. Hazen et al. [151] demonstrated in clinical studies that low-dose aspirin significantly blunted the plasma TMAO elevation caused by a high-choline diet, thereby inhibiting thrombosis and attenuating AS progression. These results were corroborated in a subsequent investigation by Zhu and colleagues [86]. The underlying mechanism may involve aspirin's capacity to adjust gut microbiota and suppress TMA lyase activity [151–153].

4.5.3. Antidiabetic Drugs

Type 2 diabetes mellitus (T2DM) constitutes an independent risk factor for HF, and proactive management of T2DM can effectively reduce the likelihood of developing HF. Metformin, a cornerstone medication for T2DM, confers cardioprotective benefits [154]. Studies confirmed that metformin significantly lowered plasma TMAO concentrations and inhibited TMA generation by gut microbiota in a mouse model of T2DM, independently of choline TMA-lyase activity [155]. Su et al. [156] further discovered that metformin reduced excessive serum TMAO levels in mice fed choline by modifying gut microbial communities involved in choline-TMA conversion. Clinical studies have established the cardiovascular benefits of SGLT2 inhibitors [157], which are now guideline-recommended therapies for HF [158]. Mindrescu et al. [159] observed that empagliflozin altered gut microbiota composition in patients with T2DM, promoting beneficial bacteria such as *Bifidobacterium* and *Lactobacillus* while suppressing proinflammatory genera like *Escherichia* and *Streptococcus*. Experimental studies showed that dapagliflozin attenuated ferroptosis in cardiomyocytes following myocardial ischemia–reperfusion injury in diabetic rats, potentially through gut microbiota modulation and decreased TMAO production [160]. A recent clinical trial found that six-month treatment with either α-glucosidase inhibitor acarbose or the dipeptidyl peptidase-4 inhibitor vildagliptin significantly reduced circulating TMAO levels in overweight or obese patients with T2DM [161]. Nonetheless, no current evidence demonstrates that other antidiabetic agents directly affect TMAO levels.

Therefore, barring any contraindications, antidiabetic drugs could be a viable option to lessen TMAO levels.

4.5.4. Statins

Statins, widely prescribed lipid-lowering agents, exhibit pleiotropic effects such as anti-atherosclerotic, endothelial-improving, anti-inflammatory, antioxidant, antifibrotic, and myocardial energy-enhancing properties [162–164]. These multifaceted actions may benefit HF prognosis [94,162,165]. Additionally, Li et al. [166] demonstrated that atorvastatin administration significantly lowered circulating TMAO levels in dyslipidemia patients. Similarly, rosuvastatin therapy in 112 individuals with suspected atherosclerotic cardiovascular disease (ASCVD) improved lipid profiles and notably lessened TMAO concentrations [167].

However, the use of statins in patients with HF remains complex and multifaceted. Although beneficial in many settings, their overall utility in HF patients is not fully established. For instance, different clinical trials in HFpEF populations have not consistently shown significant survival benefits with statin therapy [168,169]. Current guidelines recommend statins for HF patients with concurrent ASCVD, hyperlipidemia, or high cardiovascular risk [2,150]. Patients developing new-onset HF during ongoing statin therapy should maintain treatment unless contraindicated. Clinicians must vigilantly monitor for statin-associated myopathies or hepatotoxicity, necessitating immediate withdrawal upon such adverse events. In hepatic impairment, agents like pravastatin with non-CYP450 metabolism are preferable, whereas high-dose statins should be avoided. Notably, severe liver disease absolutely contraindicates simvastatin, atorvastatin, and lovastatin.

In conclusion, statin use in HF requires careful evaluation of clinical status, HF subtype, and comorbidities. The widespread application of statins in HF still necessitates further research to validate their efficacy and safety. Treatment decisions should therefore be individualized according to patient characteristics and guided by prevailing evidence and guidelines.

4.5.5. ACEIs

ACEIs and angiotensin receptor blockers (ARBs) are highly cost-effective therapies for HF. Studies showed that ACEIs ameliorated host intestinal dysbiosis [170,171]. Konop et al. [172] reported that the ACEI enalapril reduced plasma TMAO levels in rats, suggesting TMAO may serve as a therapeutic target for ACEIs in HF.

Although no direct evidence currently links ARBs to TMAO modulation, these agents are known to influence gut microbiota. Studies showed that ARB treatment reduced blood pressure and alleviated vascular and intestinal injury by altering the diversity, composition, and abundance of gut microbial communities [170,173]. This drug–host–microbiome interaction may extend to angiotensin receptor–neprilysin inhibitors (ARNIs), which incorporate ARB and neprilysin-inhibiting activities. As a foundational HF therapy [174], sacubitril/valsartan belongs to the ARNI class. Research indicated that it ameliorated gut microbiota dysbiosis in murine models of diabetic nephropathy [175].

The above research findings indicate that reducing TMAO levels via modern medicines, including antibiotics, aspirin, antidiabetic drugs, statins, ACEIs, etc., represents a feasible method for managing HF.

4.6. *Phytomedicines*

Phytomedicines can modulate gut microbiota and their metabolites, reduce TMAO concentrations, and exert certain preventive and therapeutic effects on CVDs such as HF [176]. Phytomedicines, distinguished by multi-component, multi-target, and multi-pathway attributes, could alleviate drug resistance via compensatory mechanisms and reduce adverse drug reactions to some extent [177]. They thus provide underlying new strategies for inhibiting TMAO production and treating CVDs.

Research indicated that berberine (BBR), a bioactive compound derived from *Rhizoma coptidis*, showed efficacy in managing HF [178]. BBR reduced TMAO generation in animal models and AS patients by remodeling gut microbiota, downregulating TMA-producing enzymes and FMO3 [179], and suppressing the choline-TMA-TMAO pathway [180], thereby curbing TMAO-induced AS. Importantly, BBR exhibited no hepatorenal toxicity [179]. The findings suggest that BBR possesses significant research and development potential. Resveratrol (RES), a natural extract from plants like *Polygonum cuspidatum*, lowered TMAO levels by decreasing TMA generation via remodeling gut microbiota [181], and was shown to alleviate HF in mice [182]. Garlic (*Allium sativum*), renowned for its antibacterial properties, contains allicin, which has been proven effective in addressing CVDs, including HF [183]. In vitro and in vivo studies confirmed that allicin and fresh garlic juice rich in allicin suppressed the TMAO synthesis through gut microbiota modulation [184]. Recent investigation has revealed that aged garlic oligosaccharides mitigate AS in $ApoE^{-/-}$ mice fed a high-fat and high-cholesterol diet by regulating gut microbiota and decreasing TMAO levels [185]. Hawthorn (*Crataegus*) and its extracts also exhibited therapeutic potential in HF [186,187]. He et al. [92] reported that hawthorn fruit extract lowered TMAO concentrations in mice dose-dependently, mediated by anti-inflammatory and antioxidant properties. Gypenosides (GYP), the primary constituent of *Gynostemma pentaphyllum*, lessened plasma TMAO levels by reshaping gut microbiota and inhibiting TMA lyase activity [188]. Moreover, GYP exerted cardioprotective effects and enhanced cardiac function by promoting mitochondrial autophagy, ultimately ameliorating HF [189]. Ji and coworkers [190] discovered that enemas with rhubarb (*Rhei Radix et Rhizoma*) modulated gut microbiota, decreasing serum TMAO and TMA concentrations and downregulating inflammatory markers in rats. Additional evidence indicated that the main constituents of rhubarb, emodin [191,192] and rhein [193,194], ameliorated pathological cardiac hypertrophy. These findings indicate that rhubarb and its extracts have promise as an innovative therapeutic agent for lowering TMAO concentrations and aiding in managing HF. *Lycium barbarum* polysaccharides (LBPs), important ingredients of *Fructus lycii*, could modulate gut microbiota, reduce intestinal permeability, downregulate inflammatory cytokines, significantly lessen serum TMAO levels, and subsequently improve LV function in mice [195]. Puerarin (PU), derived from *Puerariae lobatae Radix*, enhanced cardiac function in rats suffering from adriamycin-induced HF [196]. A systematic review and meta-analysis showed that combining PU injection with conventional drug therapy was safer and more effective than conventional drug therapy alone in acute heart failure [197]. Recent investigation has demonstrated that PU attenuates TMAO concentrations by altering gut microbiota composition and hindering TMA synthesis, thus exerting anti-AS effects [198]. Polymethoxyflavones (PMFs) from citrus (*Citrus sinensis* L.) peel decreased TMAO levels by inhibiting CutC/D and FMO3 activity, ultimately reducing CVD risk [199,200]. In addition, 5-Demethylnobiletin, a natural PMF, ameliorated isoproterenol-induced cardiac injury in mice [201].

In summary, TMAO stands out as a potential novel therapeutic target for HF, as illustrated in Figure 3.

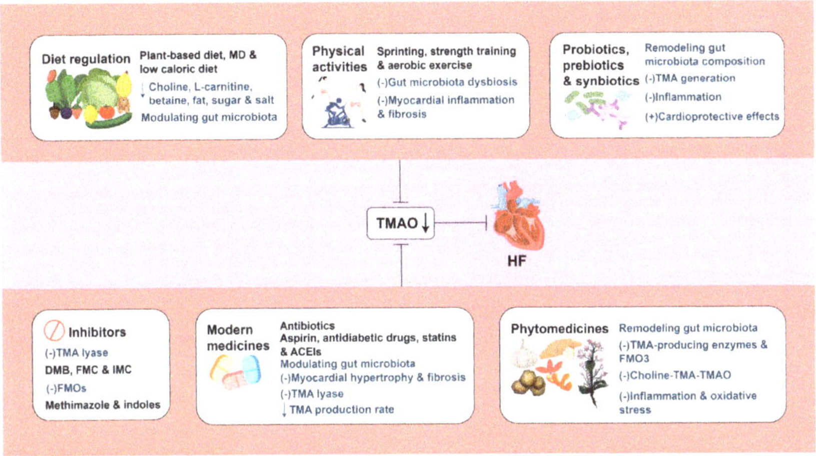

Figure 3. TMAO represents a promising therapeutic target for HF. A downward arrow (↓) indicates a decrease. A plus sign (+) indicates promotion. A minus sign (−) indicates inhibition. MD: Mediterranean diet; TMA: trimethylamine; TMAO: trimethylamine N-oxide; HF: heart failure; DMB: 3,3-dimethyl-1-butanol; FMC: fluoromethylcholine; IMC: iodomethylcholine; FMOs: flavin-containing monooxygenases; ACEIs: Angiotensin-converting enzyme inhibitors. Created by Fidraw (www.figdraw.com, accessed on 10 October 2025).

5. Conclusions and Prospects

HF, a potentially fatal syndrome, poses major obstacles in terms of management, thereby imposing a considerable burden on healthcare systems and affected families. Research has revealed complex relationships between HF pathogenesis and dysregulation of the gut–heart axis. Specifically, patients with HF exhibit gut microbiota dysbiosis with concomitant elevation of the microbial metabolite TMAO. Furthermore, elevated TMAO levels are strongly linked to the occurrence, progression, and prognosis of HF. Therefore, the disparity between TMAO generation and metabolism could constitute a significant mechanism underlying HF. TMAO regulates the advancement of HF via multiple pathways, including inflammation, abnormal energy metabolism, oxidative stress, myocardial remodeling, and so on. Studies have confirmed that decreasing TMAO levels is advantageous in managing HF. Hence, TMAO emerges as a viable target for preventing and treating HF. Several interventions—including healthy diets, regular physical activities, probiotics, prebiotics, synbiotics, inhibitors, antibiotics, aspirin, antidiabetic drugs, statins, ACEIs, and phytomedicines—have shown promising effects in lowering TMAO concentrations and mitigating HF.

However, current research exploring the precise role of TMAO in HF pathogenesis remains limited, particularly in human studies. Many intervention trials often face constraints such as limited sample sizes and insufficient generalizability, necessitating further validation. Furthermore, current research into the mechanisms and treatment of heart failure lacks investigations across distinct phenotypes such as HFpEF and HFrEF, which obscures the phenotype-specific effects of TMAO. Dietary studies have largely been conducted in developed countries, frequently neglecting regional, ethnic, cultural, and customary disparities. Exercise protocols often include sprinting or strength training, which may be unsuitable for HF patients; more research is needed to identify safe and effective activities for those with early-stage HF or at a high risk of developing HF. Clinical trials on other preventive or therapeutic tactics are few, leaving mechanisms and safety in humans incompletely characterized. Interventions using probiotics, prebiotics, and synbiotics encounter limitations, including restricted therapeutic targets and uncertain efficacy, while microbial diversity and environmental complexity complicate the identification of beneficial strains. Although TMA lyase inhibitors have shown preliminary promise, their clinical safety and efficacy require further evaluation. Considering that mutations in the human *FMO* gene may result in conditions including hepatitis and TMAU, utmost prudence is required during the clinical research and development of FMO inhibitors. Antibiotic use is controversial because of potential toxic side effects and the risk of promoting antimicrobial resistance. Therefore, further clinical studies are urgently needed to fully assess the benefits, risks, and prognostic impact of antibiotic therapy. Initial studies have demonstrated that aspirin, antidiabetic drugs, statins, and ACEIs can reduce TMAO levels and HF risk. Phytomedicines possess advantages, including multi-target effects and a lower propensity for drug resistance compared to conventional drugs. Nevertheless, robust clinical data are lacking to fully substantiate the efficacy and underlying mechanisms of these agents. Moreover, the effects of commonly used HF medications—including ARBs, sacubitril/valsartan, and β-adrenergic antagonists—on TMAO levels remain poorly understood.

Furthermore, although cardiac transplantation effectively treats end-stage HF, patients receiving this procedure and its accompanying immunosuppressive therapy often exhibit elevated TMAO levels [202,203]. Several mechanisms may explain this phenomenon. Elevated TMAO precursors were associated with acute rejection and greater atherosclerotic burden [203]. And higher TMAO levels correlated with increased risk of adverse outcomes, including those post-transplantation [97,204], implying that TMAO may already be raised prior to surgery. Immunosuppressed patients are also more vulnerable to infections, and perioperative management often requires broad-spectrum antibiotics. Additionally, heart transplant recipients commonly have multiple comorbidities necessitating polypharmacy. These medications can induce intestinal dysbiosis, which may elevate circulating levels of gut microbiota-derived TMAO [205,206]. Consequently, improved hemodynamics after heart transplantation do not lower TMAO concentrations. In summary, heart transplantation does not mitigate TMAO as a pathological factor in HF. Future studies should further examine the clinical implications of elevated TMAO levels in patients receiving cardiac transplantation and associated immunosuppressive therapy.

Despite these shortcomings and limitations, targeting TMAO remains a promising innovative strategy for preventing and treating HF. More systematic clinical studies are warranted, particularly large-scale, multicenter, randomized controlled trials with sufficient observation durations, clear phenotypes, unambiguous interventions, and precise endpoints. Such efforts would help clarify the mechanisms, efficacy, and safety of TMAO-targeted intervention, providing fresh insights into HF pathogenesis and a scientific foundation for lowering TMAO as a novel strategy in its management.

Author Contributions: Conceptualization, Z.D., Y.Y. and Z.W.; methodology, J.W.; software, Z.D.; investigation, Z.D.; resources, Y.L.; writing—original draft preparation, Z.D.; writing—review and editing, Y.Y., J.W. and R.L.; visualization, Z.D.; supervision, Z.G. and Y.L.; project administration, Z.G.; funding acquisition, Z.G. All authors have read and agreed to the published version of the manuscript.

Funding: This research was funded by the National Natural Science Foundation of China (82174343), the Natural Science Foundation of Hunan Province (2025JJ50658), Key Research and Development Plan of Hunan Province (2022SK2012), Key Subject of Scientific Research Projects of Traditional Chinese Medicine in Hunan Province (A2023017), Hunan Provincial Health Commission Research Project (D202303017861), Innovation Project for Graduate Students at Hunan University of Traditional Chinese Medicine (2024CX019), the Postdoctoral Fellowship Program of CPSF (GZC20252636), the China Postdoctoral Science Foundation (2025M773968 and 2025T181082), and National Natural Science Foundation of China for Young Scientists (82405372).

Institutional Review Board Statement: Not applicable.

Informed Consent Statement: Not applicable.

Data Availability Statement: Not applicable.

Acknowledgments: The graphical abstract, Figures 1 and 3 were created by Figdraw (www.figdraw.com, accessed on 6 August 2025 and 10 October 2025).

Conflicts of Interest: The authors declare no conflicts of interest.

Abbreviations

The following abbreviations are used in this manuscript:

HF	Heart failure
CVD	Cardiovascular disease
IHD	Ischemic heart disease
LVEF	Left ventricular ejection fraction
HFrpEF	HF with reduced ejection fraction
HFpEF	HF with preserved ejection fraction
AS	Atherosclerosis
SCFAs	Short-chain fatty acids
TMAO	Trimethylamine N-oxide
BAs	Bile acids
LPS	Lipopolysaccharide
TMA	Trimethylamine
FMOs	Flavin-containing monooxygenases
TMAU	Trimethylaminuria
HFD	High-fat diet
IL	Interleukin
NLRP3	Nucleotide-binding oligomerization domain-like receptor containing pyrin domain 3
SIRT3	Sirtuin 3
SOD2	Superoxide dismutase 2
mtROS	Mitochondrial reactive oxygen species
ROS	Reactive oxygen species
TXNIP	Thioredoxin-interacting protein
TNF	Tumor necrosis factor
FA	Fatty acid
OXPHOS	Oxidative phosphorylation
PLIN4	Perilipin 4
OMA1	Overlapping activity with m-AAA protease 1
OGDHL	Oxoglutarate dehydrogenase L
BAT	Brown adipose tissue
CP	Creatine phosphate
ATP	Adenosine triphosphate
Ca^{2+}	Calcium ion
eNOS	Endothelial nitric oxide synthase
NOX2	NADPH oxidase 2
PIEZO1	Piezo type mechanosensitive ion channel component 1
WD	Western diet
ANP	atrial natriuretic peptide
β-MHC	β-myosin heavy chain
TGF	Transforming growth factor
NF	Nuclear factor
BNP	Brain natriuretic peptide
LV	Left ventricular

NO	Nitric oxide
CHF	Chronic heart failure
HTN	Hypertension
ANG	Angiotensin
MI	Myocardial infarction
RCT	Reverse cholesterol transport
$ApoE^{-/-}$	Apolipoprotein E knockout
CADs	Coronary artery diseases
SGLT2	Sodium-glucose cotransporter-2
ACEIs	Angiotensin-converting enzyme inhibitors
MD	Mediterranean diet
MF	Multistrain formula
DMB	3,3-dimethyl-1-butanol
FMC	Fluoromethylcholine
IMC	Iodomethylcholine
T2DM	Type 2 diabetes mellitus
ASCVD	Atherosclerotic cardiovascular disease
ARBs	Angiotensin receptor blockers
ARNIs	Angiotensin receptor–neprilysin inhibitors
BBR	Berberine
RES	Resveratrol
GYP	Gypenosides
LBPs	Lycium barbarum polysaccharides
PU	Puerarin
PMFs	Polymethoxyflavones

References

1. Bozkurt, B.; Coats, A.J.S.; Tsutsui, H.; Abdelhamid, C.M.; Adamopoulos, S.; Albert, N.; Anker, S.D.; Atherton, J.; Böhm, M.; Butler, J.; et al. Universal Definition and Classification of Heart Failure: A Report of the Heart Failure Society of America, Heart Failure Association of the European Society of Cardiology, Japanese Heart Failure Society and Writing Committee of the Universal Definition of Heart Failure: Endorsed by the Canadian Heart Failure Society, Heart Failure Association of India, Cardiac Society of Australia and New Zealand, and Chinese Heart Failure Association. *Eur. J. Heart Fail.* **2021**, *23*, 352–380. [CrossRef] [PubMed]
2. Heidenreich, P.A.; Bozkurt, B.; Aguilar, D.; Allen, L.A.; Byun, J.J.; Colvin, M.M.; Deswal, A.; Drazner, M.H.; Dunlay, S.M.; Evers, L.R.; et al. 2022 AHA/ACC/HFSA Guideline for the Management of Heart Failure: A Report of the American College of Cardiology/American Heart Association Joint Committee on Clinical Practice Guidelines. *Circulation* **2022**, *145*, e895–e1032. [CrossRef]
3. Khan, M.S.; Shahid, I.; Bennis, A.; Rakisheva, A.; Metra, M.; Butler, J. Global Epidemiology of Heart Failure. *Nat. Rev. Cardiol.* **2024**, *21*, 717–734. [CrossRef]
4. Yan, T.; Zhu, S.; Yin, X.; Xie, C.; Xue, J.; Zhu, M.; Weng, F.; Zhu, S.; Xiang, B.; Zhou, X.; et al. Burden, Trends, and Inequalities of Heart Failure Globally, 1990 to 2019: A Secondary Analysis Based on the Global Burden of Disease 2019 Study. *J. Am. Heart Assoc.* **2023**, *12*, e027852. [CrossRef]
5. McDonagh, T.A.; Metra, M.; Adamo, M.; Gardner, R.S.; Baumbach, A.; Böhm, M.; Burri, H.; Butler, J.; Čelutkienė, J.; Chioncel, O.; et al. 2021 ESC Guidelines for the Diagnosis and Treatment of Acute and Chronic Heart Failure. *Eur. Heart. J.* **2021**, *42*, 3599–3726. [CrossRef]
6. Whellan, D.J.; Lindenfeld, J. Easy to Predict, Difficult to Prevent. *JACC Heart Fail.* **2017**, *5*, 226–228. [CrossRef]
7. Karmazyn, M.; Gan, X.T. Probiotics as Potential Treatments to Reduce Myocardial Remodelling and Heart Failure via the Gut-Heart Axis: State-of-the-Art Review. *Mol. Cell Biochem.* **2023**, *478*, 2539–2551. [CrossRef]
8. Tang, W.H.W.; Bäckhed, F.; Landmesser, U.; Hazen, S.L. Intestinal Microbiota in Cardiovascular Health and Disease: JACC State-of-the-Art Review. *J. Am. Coll. Cardiol.* **2019**, *73*, 2089–2105. [CrossRef] [PubMed]
9. Hemmati, M.; Kashanipoor, S.; Mazaheri, P.; Alibabaei, F.; Babaeizad, A.; Asli, S.; Mohammadi, S.; Gorgin, A.H.; Ghods, K.; Yousefi, B.; et al. Importance of Gut Microbiota Metabolites in the Development of Cardiovascular Diseases (CVD). *Life Sci.* **2023**, *329*, 121947. [CrossRef] [PubMed]
10. Salzano, A.; Cassambai, S.; Yazaki, Y.; Israr, M.Z.; Bernieh, D.; Wong, M.; Suzuki, T. The Gut Axis Involvement in Heart Failure: Focus on Trimethylamine N-Oxide. *Cardiol. Clin.* **2022**, *40*, 161–169. [CrossRef]

11. Sender, R.; Fuchs, S.; Milo, R. Revised Estimates for the Number of Human and Bacteria Cells in the Body. *PLoS Biol.* **2016**, *14*, e1002533. [CrossRef]
12. Kaye, D.M.; Shihata, W.A.; Jama, H.A.; Tsyganov, K.; Ziemann, M.; Kiriazis, H.; Horlock, D.; Vijay, A.; Giam, B.; Vinh, A.; et al. Deficiency of Prebiotic Fiber and Insufficient Signaling Through Gut Metabolite-Sensing Receptors Leads to Cardiovascular Disease. *Circulation* **2020**, *141*, 1393–1403. [CrossRef] [PubMed]
13. Trøseid, M.; Andersen, G.Ø.; Broch, K.; Hov, J.R. The Gut Microbiome in Coronary Artery Disease and Heart Failure: Current Knowledge and Future Directions. *eBioMedicine* **2020**, *52*, 102649. [CrossRef]
14. Mann, D.L. Innate Immunity and the Failing Heart: The Cytokine Hypothesis Revisited. *Circ. Res.* **2015**, *116*, 1254–1268. [CrossRef] [PubMed]
15. Kim, M.; Huda, M.N.; Bennett, B.J. Sequence Meets Function-Microbiota and Cardiovascular Disease. *Cardiovasc. Res.* **2022**, *118*, 399–412. [CrossRef]
16. Guivala, S.J.; Bode, K.A.; Okun, J.G.; Kartal, E.; Schwedhelm, E.; Pohl, L.V.; Werner, S.; Erbs, S.; Thiele, H.; Büttner, P. Interactions between the Gut Microbiome, Associated Metabolites and the Manifestation and Progression of Heart Failure with Preserved Ejection Fraction in ZSF1 Rats. *Cardiovasc. Diabetol.* **2024**, *23*, 299. [CrossRef]
17. Kozhevnikova, M.V.; Kakotkina, A.V.; Korobkova, E.O.; Kuznetsov, I.V.; Shestakova, K.M.; Moskaleva, N.E.; Appolonova, S.A.; Belenkov, Y.N. Metabolomic Panel for the Diagnosis of Heart Failure with Preserved Ejection Fraction. *Int. J. Mol. Sci.* **2025**, *26*, 2102. [CrossRef]
18. Qin, J.; Li, R.; Raes, J.; Arumugam, M.; Burgdorf, K.S.; Manichanh, C.; Nielsen, T.; Pons, N.; Levenez, F.; Yamada, T.; et al. A Human Gut Microbial Gene Catalogue Established by Metagenomic Sequencing. *Nature* **2010**, *464*, 59–65. [CrossRef]
19. Pasini, E.; Aquilani, R.; Testa, C.; Baiardi, P.; Angioletti, S.; Boschi, F.; Verri, M.; Dioguardi, F. Pathogenic Gut Flora in Patients with Chronic Heart Failure. *JACC Heart Fail.* **2016**, *4*, 220–227. [CrossRef] [PubMed]
20. Emoto, T.; Hayashi, T.; Tabata, T.; Yamashita, T.; Watanabe, H.; Takahashi, T.; Gotoh, Y.; Kami, K.; Yoshida, N.; Saito, Y.; et al. Metagenomic Analysis of Gut Microbiota Reveals Its Role in Trimethylamine Metabolism in Heart Failure. *Int. J. Cardiol.* **2021**, *338*, 138–142. [CrossRef]
21. Suzuki, T.; Yazaki, Y.; Voors, A.A.; Jones, D.J.L.; Chan, D.C.S.; Anker, S.D.; Cleland, J.G.; Dickstein, K.; Filippatos, G.; Hillege, H.L.; et al. Association with Outcomes and Response to Treatment of Trimethylamine N-Oxide in Heart Failure: Results from BIOSTAT-CHF. *Eur. J. Heart Fail.* **2019**, *21*, 877–886. [CrossRef]
22. Tang, W.H.W.; Lemaitre, R.N.; Jensen, P.N.; Wang, M.; Wang, Z.; Li, X.S.; Nemet, I.; Lee, Y.; Lai, H.T.M.; de Oliveira Otto, M.C.; et al. Trimethylamine N-Oxide and Related Gut Microbe-Derived Metabolites and Incident Heart Failure Development in Community-Based Populations. *Circ. Heart Fail.* **2024**, *17*, e011569. [CrossRef] [PubMed]
23. Schuett, K.; Kleber, M.E.; Scharnagl, H.; Lorkowski, S.; März, W.; Niessner, A.; Marx, N.; Meinitzer, A. Trimethylamine-N-Oxide and Heart Failure with Reduced Versus Preserved Ejection Fraction. *J. Am. Coll. Cardiol.* **2017**, *70*, 3202–3204. [CrossRef]
24. Salzano, A.; Israr, M.Z.; Yazaki, Y.; Heaney, L.M.; Kanagala, P.; Singh, A.; Arnold, J.R.; Gulsin, G.S.; Squire, I.B.; McCann, G.P.; et al. Combined Use of Trimethylamine N-Oxide with BNP for Risk Stratification in Heart Failure with Preserved Ejection Fraction: Findings from the DIAMONDHFpEF Study. *Eur. J. Prev. Cardiol.* **2020**, *27*, 2159–2162. [CrossRef]
25. Chen, Q.; Zhang, H.; Chen, Y.; Peng, Y.; Yao, Y.; Xue, H.; Guo, Q.; Tian, D.; Xiao, L.; Teng, X.; et al. Trimethylamine N-Oxide Induces Cardiac Diastolic Dysfunction by down-Regulating Piezo1 in Mice with Heart Failure with Preserved Ejection Fraction. *Life Sci.* **2025**, *369*, 123554. [CrossRef]
26. Chen, K.; Zheng, X.; Feng, M.; Li, D.; Zhang, H. Gut Microbiota-Dependent Metabolite Trimethylamine N-Oxide Contributes to Cardiac Dysfunction in Western Diet-Induced Obese Mice. *Front. Physiol.* **2017**, *8*, 139. [CrossRef]
27. Wang, G.; Kong, B.; Shuai, W.; Fu, H.; Jiang, X.; Huang, H. 3,3-Dimethyl-1-Butanol Attenuates Cardiac Remodeling in Pressure-Overload-Induced Heart Failure Mice. *J. Nutr. Biochem.* **2020**, *78*, 108341. [CrossRef]
28. Yoshida, Y.; Shimizu, I.; Shimada, A.; Nakahara, K.; Yanagisawa, S.; Kubo, M.; Fukuda, S.; Ishii, C.; Yamamoto, H.; Ishikawa, T.; et al. Brown Adipose Tissue Dysfunction Promotes Heart Failure via a Trimethylamine N-Oxide-Dependent Mechanism. *Sci. Rep.* **2022**, *12*, 14883. [CrossRef] [PubMed]
29. Xie, H.; Jiang, J.; Cao, S.; Xu, X.; Zhou, J.; Zhang, R.; Huang, B.; Lu, P.; Peng, L.; Liu, M. The Role of Gut Microbiota-Derived Trimethylamine N-Oxide in the Pathogenesis and Treatment of Mild Cognitive Impairment. *Int. J. Mol. Sci.* **2025**, *26*, 1373. [CrossRef]
30. Rath, S.; Rud, T.; Pieper, D.H.; Vital, M. Potential TMA-Producing Bacteria Are Ubiquitously Found in Mammalia. *Front. Microbiol.* **2019**, *10*, 2966. [CrossRef] [PubMed]
31. Koeth, R.A.; Wang, Z.; Levison, B.S.; Buffa, J.A.; Org, E.; Sheehy, B.T.; Britt, E.B.; Fu, X.; Wu, Y.; Li, L.; et al. Intestinal Microbiota Metabolism of L-Carnitine, a Nutrient in Red Meat, Promotes Atherosclerosis. *Nat. Med.* **2013**, *19*, 576–585. [CrossRef] [PubMed]
32. Mo, X.; Cheng, R.; Shen, L.; Sun, Y.; Wang, P.; Jiang, G.; Wen, L.; Li, X.; Peng, X.; Liao, Y.; et al. High-Fat Diet Induces Sarcopenic Obesity in Natural Aging Rats through the Gut-Trimethylamine N-Oxide-Muscle Axis. *J. Adv. Res.* **2025**, *70*, 405–422. [CrossRef] [PubMed]

33. Andreu-Sánchez, S.; Ahmad, S.; Kurilshikov, A.; Beekman, M.; Ghanbari, M.; van Faassen, M.; van den Munckhof, I.C.L.; Steur, M.; Harms, A.; Hankemeier, T.; et al. Unraveling Interindividual Variation of Trimethylamine N-Oxide and Its Precursors at the Population Level. *iMeta* **2024**, *3*, e183. [CrossRef]
34. Zhu, W.; Gregory, J.C.; Org, E.; Buffa, J.A.; Gupta, N.; Wang, Z.; Li, L.; Fu, X.; Wu, Y.; Mehrabian, M.; et al. Gut Microbial Metabolite TMAO Enhances Platelet Hyperreactivity and Thrombosis Risk. *Cell* **2016**, *165*, 111–124. [CrossRef] [PubMed]
35. Lv, S.; Wang, Y.; Zhang, W.; Shang, H. Trimethylamine Oxide: A Potential Target for Heart Failure Therapy. *Heart* **2022**, *108*, 917–922. [CrossRef]
36. Bennett, B.J.; de Aguiar Vallim, T.Q.; Wang, Z.; Shih, D.M.; Meng, Y.; Gregory, J.; Allayee, H.; Lee, R.; Graham, M.; Crooke, R.; et al. Trimethylamine-N-Oxide, a Metabolite Associated with Atherosclerosis, Exhibits Complex Genetic and Dietary Regulation. *Cell Metab.* **2013**, *17*, 49–60. [CrossRef]
37. Wiedeman, A.M.; Barr, S.I.; Green, T.J.; Xu, Z.; Innis, S.M.; Kitts, D.D. Dietary Choline Intake: Current State of Knowledge Across the Life Cycle. *Nutrients* **2018**, *10*, 1513. [CrossRef]
38. Koeth, R.A.; Lam-Galvez, B.R.; Kirsop, J.; Wang, Z.; Levison, B.S.; Gu, X.; Copeland, M.F.; Bartlett, D.; Cody, D.B.; Dai, H.J.; et al. L-Carnitine in Omnivorous Diets Induces an Atherogenic Gut Microbial Pathway in Humans. *J. Clin. Investig.* **2019**, *129*, 373–387. [CrossRef]
39. Nikrandt, G.; Chmurzynska, A. Decoding Betaine: A Critical Analysis of Therapeutic Potential Compared with Marketing Hype-A Narrative Review. *J. Nutr.* **2024**, *154*, 3167–3176. [CrossRef]
40. Jiang, H.; Hu, Y.; Zhang, Z.; Chen, X.; Gao, J. Identification of Metabolic Biomarkers Associated with Nonalcoholic Fatty Liver Disease. *Lipids Health Dis.* **2023**, *22*, 150. [CrossRef]
41. Zhen, J.; Zhou, Z.; He, M.; Han, H.-X.; Lv, E.-H.; Wen, P.-B.; Liu, X.; Wang, Y.-T.; Cai, X.-C.; Tian, J.-Q.; et al. The Gut Microbial Metabolite Trimethylamine N-Oxide and Cardiovascular Diseases. *Front. Endocrinol.* **2023**, *14*, 1085041. [CrossRef] [PubMed]
42. Duttaroy, A.K. Role of Gut Microbiota and Their Metabolites on Atherosclerosis, Hypertension and Human Blood Platelet Function: A Review. *Nutrients* **2021**, *13*, 144. [CrossRef] [PubMed]
43. Tang, W.H.W.; Wang, Z.; Kennedy, D.J.; Wu, Y.; Buffa, J.A.; Agatisa-Boyle, B.; Li, X.S.; Levison, B.S.; Hazen, S.L. Gut Microbiota-Dependent Trimethylamine N-Oxide (TMAO) Pathway Contributes to Both Development of Renal Insufficiency and Mortality Risk in Chronic Kidney Disease. *Circ. Res.* **2015**, *116*, 448–455. [CrossRef] [PubMed]
44. Qiu, L.; Tao, X.; Xiong, H.; Yu, J.; Wei, H. Lactobacillus Plantarum ZDY04 Exhibits a Strain-Specific Property of Lowering TMAO via the Modulation of Gut Microbiota in Mice. *Food Funct.* **2018**, *9*, 4299–4309. [CrossRef]
45. Iglesias-Carres, L.; Chadwick-Corbin, S.A.; Sweet, M.G.; Neilson, A.P. Dietary Phenolics and Their Microbial Metabolites Are Poor Inhibitors of Trimethylamine Oxidation to Trimethylamine N-Oxide by Hepatic Flavin Monooxygenase 3. *J. Nutr. Biochem.* **2023**, *120*, 109428. [CrossRef]
46. Brunt, V.E.; Gioscia-Ryan, R.A.; Casso, A.G.; VanDongen, N.S.; Ziemba, B.P.; Sapinsley, Z.J.; Richey, J.J.; Zigler, M.C.; Neilson, A.P.; Davy, K.P.; et al. Trimethylamine-N-Oxide Promotes Age-Related Vascular Oxidative Stress and Endothelial Dysfunction in Mice and Healthy Humans. *Hypertension* **2020**, *76*, 101–112. [CrossRef]
47. Rath, S.; Rox, K.; Kleine Bardenhorst, S.; Schminke, U.; Dörr, M.; Mayerle, J.; Frost, F.; Lerch, M.M.; Karch, A.; Brönstrup, M.; et al. Higher Trimethylamine-N-Oxide Plasma Levels with Increasing Age Are Mediated by Diet and Trimethylamine-Forming Bacteria. *mSystems* **2021**, *6*, e0094521. [CrossRef]
48. Flaherty, C.C.; Phillips, I.R.; Janmohamed, A.; Shephard, E.A. Living with Trimethylaminuria and Body and Breath Malodour: Personal Perspectives. *BMC Public. Health* **2024**, *24*, 222. [CrossRef]
49. Hall, A.B.; Tolonen, A.C.; Xavier, R.J. Human Genetic Variation and the Gut Microbiome in Disease. *Nat. Rev. Genet.* **2017**, *18*, 690–699. [CrossRef]
50. Kurilshikov, A.; Medina-Gomez, C.; Bacigalupe, R.; Radjabzadeh, D.; Wang, J.; Demirkan, A.; Le Roy, C.I.; Raygoza Garay, J.A.; Finnicum, C.T.; Liu, X.; et al. Large-Scale Association Analyses Identify Host Factors Influencing Human Gut Microbiome Composition. *Nat. Genet.* **2021**, *53*, 156–165. [CrossRef]
51. Tang, W.H.W.; Wang, Z.; Fan, Y.; Levison, B.; Hazen, J.E.; Donahue, L.M.; Wu, Y.; Hazen, S.L. Prognostic Value of Elevated Levels of Intestinal Microbe-Generated Metabolite Trimethylamine-N-Oxide in Patients with Heart Failure: Refining the Gut Hypothesis. *J. Am. Coll. Cardiol.* **2014**, *64*, 1908–1914. [CrossRef]
52. Wang, Z.; Bergeron, N.; Levison, B.S.; Li, X.S.; Chiu, S.; Jia, X.; Koeth, R.A.; Li, L.; Wu, Y.; Tang, W.H.W.; et al. Impact of Chronic Dietary Red Meat, White Meat, or Non-Meat Protein on Trimethylamine N-Oxide Metabolism and Renal Excretion in Healthy Men and Women. *Eur. Heart J.* **2019**, *40*, 583–594. [CrossRef]
53. Fang, Q.; Lei, Y.; Wu, H.; Li, C.; Jiang, J.; Wang, S.; Wu, Y.; Chen, L.; Ouyang, D.; Li, X.; et al. Plasma Reference Interval of Trimethylamine-N-Oxide in Healthy Adults: A Multicenter Study Using Trimethylamine-N-Oxide Assay Kit for Analysis and Validation. *Clin. Chim. Acta* **2025**, *571*, 120223. [CrossRef] [PubMed]

54. Saaoud, F.; Liu, L.; Xu, K.; Cueto, R.; Shao, Y.; Lu, Y.; Sun, Y.; Snyder, N.W.; Wu, S.; Yang, L.; et al. Aorta- and Liver-Generated TMAO Enhances Trained Immunity for Increased Inflammation via ER Stress/Mitochondrial ROS/Glycolysis Pathways. *JCI Insight* **2023**, *8*, e158183. [CrossRef]
55. Boini, K.M.; Puchchakayala, G.; Zhang, Y.; Koka, S. TMAO Activates Carotid Endothelial Inflammasomes Leading to Enhanced Neointimal Formation in Nlrp3 Mice. *FASEB J.* **2020**, *34*, 1. [CrossRef]
56. Chen, Y.; Zhang, J.; Lin, Y.; Lei, Q.; Guan, K.-L.; Zhao, S.; Xiong, Y. Tumour Suppressor SIRT3 Deacetylates and Activates Manganese Superoxide Dismutase to Scavenge ROS. *EMBO Rep.* **2011**, *12*, 534–541. [CrossRef]
57. Chen, M.-L.; Zhu, X.-H.; Ran, L.; Lang, H.-D.; Yi, L.; Mi, M.-T. Trimethylamine-N-Oxide Induces Vascular Inflammation by Activating the NLRP3 Inflammasome Through the SIRT3-SOD2-mtROS Signaling Pathway. *J. Am. Heart Assoc.* **2017**, *6*, e006347. [CrossRef]
58. Abderrazak, A.; Syrovets, T.; Couchie, D.; El Hadri, K.; Friguet, B.; Simmet, T.; Rouis, M. NLRP3 Inflammasome: From a Danger Signal Sensor to a Regulatory Node of Oxidative Stress and Inflammatory Diseases. *Redox Biol.* **2015**, *4*, 296–307. [CrossRef] [PubMed]
59. Zhang, H.; Meng, J.; Yu, H. Trimethylamine N-Oxide Supplementation Abolishes the Cardioprotective Effects of Voluntary Exercise in Mice Fed a Western Diet. *Front. Physiol.* **2017**, *8*, 944. [CrossRef]
60. Kolwicz, S.C.; Purohit, S.; Tian, R. Cardiac Metabolism and Its Interactions with Contraction, Growth, and Survival of Cardiomyocytes. *Circ. Res.* **2013**, *113*, 603–616. [CrossRef]
61. Bedi, K.C.; Snyder, N.W.; Brandimarto, J.; Aziz, M.; Mesaros, C.; Worth, A.J.; Wang, L.L.; Javaheri, A.; Blair, I.A.; Margulies, K.B.; et al. Evidence for Intramyocardial Disruption of Lipid Metabolism and Increased Myocardial Ketone Utilization in Advanced Human Heart Failure. *Circulation* **2016**, *133*, 706–716. [CrossRef]
62. Yoo, W.; Zieba, J.K.; Foegeding, N.J.; Torres, T.P.; Shelton, C.D.; Shealy, N.G.; Byndloss, A.J.; Cevallos, S.A.; Gertz, E.; Tiffany, C.R.; et al. High-Fat Diet-Induced Colonocyte Dysfunction Escalates Microbiota-Derived Trimethylamine N-Oxide. *Science* **2021**, *373*, 813–818. [CrossRef]
63. Ikeda, K.; Maretich, P.; Kajimura, S. The Common and Distinct Features of Brown and Beige Adipocytes. *Trends Endocrinol. Metab.* **2018**, *29*, 191–200. [CrossRef]
64. Tahara, A.; Tahara, N.; Maeda-Ogata, S.; Bekki, M.; Sugiyama, Y.; Honda, A.; Abe, T.; Yamagishi, S.-I.; Fukumoto, Y. Brown Adipose Tissue Activation in Severe Heart Failure. *Eur. Heart J.* **2020**, *41*, 2415. [CrossRef] [PubMed]
65. Gorski, P.A.; Ceholski, D.K.; Hajjar, R.J. Altered Myocardial Calcium Cycling and Energetics in Heart Failure--a Rational Approach for Disease Treatment. *Cell Metab.* **2015**, *21*, 183–194. [CrossRef] [PubMed]
66. Querio, G.; Antoniotti, S.; Geddo, F.; Levi, R.; Gallo, M.P. Trimethylamine N-Oxide (TMAO) Impairs Purinergic Induced Intracellular Calcium Increase and Nitric Oxide Release in Endothelial Cells. *Int. J. Mol. Sci.* **2022**, *23*, 3982. [CrossRef]
67. Singh, R.B.; Fedacko, J.; Pella, D.; Fatima, G.; Elkilany, G.; Moshiri, M.; Hristova, K.; Jakabcin, P.; Vaňova, N. High Exogenous Antioxidant, Restorative Treatment (Heart) for Prevention of the Six Stages of Heart Failure: The Heart Diet. *Antioxidants* **2022**, *11*, 1464. [CrossRef] [PubMed]
68. Esposito, S.; McGuinness, L.R.; Sharma, P.; Chadwick, A.E.; Rainbow, R.D. Trimethylamine N-Oxide (TMAO) Acutely Alters Ionic Currents but Does Not Increase Cardiac Cell Death. *Front. Physiol.* **2025**, *16*, 1505813. [CrossRef]
69. Wang, J.; Gao, Y.; Ren, S.; Li, J.; Chen, S.; Feng, J.; He, B.; Zhou, Y.; Xuan, R. Gut Microbiota-Derived Trimethylamine N-Oxide: A Novel Target for the Treatment of Preeclampsia. *Gut Microbes* **2024**, *16*, 2311888. [CrossRef]
70. Lee, J.; Lee, J.; Kim, K.; Lee, J.; Jung, Y.; Hyeon, J.S.; Seo, A.; Jin, W.; Weon, B.; Shin, N.; et al. Antibiotic-Induced Intestinal Microbiota Depletion Can Attenuate the Acute Kidney Injury to Chronic Kidney Disease Transition via NADPH Oxidase 2 and Trimethylamine-N-Oxide Inhibition. *Kidney Int.* **2024**, *105*, 1239–1253. [CrossRef]
71. Li, Z.; Wu, Z.; Yan, J.; Liu, H.; Liu, Q.; Deng, Y.; Ou, C.; Chen, M. Gut Microbe-Derived Metabolite Trimethylamine N-Oxide Induces Cardiac Hypertrophy and Fibrosis. *Lab. Investig.* **2019**, *99*, 346–357. [CrossRef]
72. Organ, C.L.; Otsuka, H.; Bhushan, S.; Wang, Z.; Bradley, J.; Trivedi, R.; Polhemus, D.J.; Tang, W.H.W.; Wu, Y.; Hazen, S.L.; et al. Choline Diet and Its Gut Microbe-Derived Metabolite, Trimethylamine N-Oxide, Exacerbate Pressure Overload-Induced Heart Failure. *Circ. Heart Fail.* **2016**, *9*, e002314. [CrossRef]
73. Organ, C.L.; Li, Z.; Sharp, T.E.; Polhemus, D.J.; Gupta, N.; Goodchild, T.T.; Tang, W.H.W.; Hazen, S.L.; Lefer, D.J. Nonlethal Inhibition of Gut Microbial Trimethylamine N-Oxide Production Improves Cardiac Function and Remodeling in a Murine Model of Heart Failure. *J. Am. Heart Assoc.* **2020**, *9*, e016223. [CrossRef]
74. Zou, D.; Li, Y.; Sun, G. Attenuation of Circulating Trimethylamine N-Oxide Prevents the Progression of Cardiac and Renal Dysfunction in a Rat Model of Chronic Cardiorenal Syndrome. *Front. Pharmacol.* **2021**, *12*, 751380. [CrossRef] [PubMed]
75. Jiang, S.; Shui, Y.; Cui, Y.; Tang, C.; Wang, X.; Qiu, X.; Hu, W.; Fei, L.; Li, Y.; Zhang, S.; et al. Gut Microbiota Dependent Trimethylamine N-Oxide Aggravates Angiotensin II-Induced Hypertension. *Redox Biol.* **2021**, *46*, 102115. [CrossRef]
76. Zhang, X.; Li, Y.; Yang, P.; Liu, X.; Lu, L.; Chen, Y.; Zhong, X.; Li, Z.; Liu, H.; Ou, C.; et al. Trimethylamine-N-Oxide Promotes Vascular Calcification Through Activation of NLRP3 (Nucleotide-Binding Domain, Leucine-Rich-Containing Family, Pyrin

Domain-Containing-3) Inflammasome and NF-κB (Nuclear Factor κB) Signals. *Arterioscler. Thromb. Vasc. Biol.* **2020**, *40*, 751–765. [CrossRef] [PubMed]
77. Kim, S.-J.; Bale, S.; Verma, P.; Wan, Q.; Ma, F.; Gudjonsson, J.E.; Hazen, S.L.; Harms, P.W.; Tsou, P.-S.; Khanna, D.; et al. Gut Microbe-Derived Metabolite Trimethylamine N-Oxide Activates PERK to Drive Fibrogenic Mesenchymal Differentiation. *iScience* **2022**, *25*, 104669. [CrossRef] [PubMed]
78. Hatamnejad, M.R.; Medzikovic, L.; Dehghanitafti, A.; Rahman, B.; Vadgama, A.; Eghbali, M. Role of Gut Microbial Metabolites in Ischemic and Non-Ischemic Heart Failure. *Int. J. Mol. Sci.* **2025**, *26*, 2242. [CrossRef] [PubMed]
79. Cheng, E.; Hung, S.-C.; Lin, T.-Y. Association of Trimethylamine N-Oxide and Metabolites with Kidney Function Decline in Patients with Chronic Kidney Disease. *Clin. Nutr.* **2025**, *44*, 239–247. [CrossRef]
80. Dannenberg, L.; Zikeli, D.; Benkhoff, M.; Ahlbrecht, S.; Kelm, M.; Levkau, B.; Polzin, A. Targeting the Human Microbiome and Its Metabolite TMAO in Cardiovascular Prevention and Therapy. *Pharmacol. Ther.* **2020**, *213*, 107584. [CrossRef]
81. Ge, X.; Zheng, L.; Zhuang, R.; Yu, P.; Xu, Z.; Liu, G.; Xi, X.; Zhou, X.; Fan, H. The Gut Microbial Metabolite Trimethylamine N-Oxide and Hypertension Risk: A Systematic Review and Dose-Response Meta-Analysis. *Adv. Nutr.* **2020**, *11*, 66–76. [CrossRef]
82. Brunt, V.E.; Casso, A.G.; Gioscia-Ryan, R.A.; Sapinsley, Z.J.; Ziemba, B.P.; Clayton, Z.S.; Bazzoni, A.E.; VanDongen, N.S.; Richey, J.J.; Hutton, D.A.; et al. Gut Microbiome-Derived Metabolite Trimethylamine N-Oxide Induces Aortic Stiffening and Increases Systolic Blood Pressure with Aging in Mice and Humans. *Hypertension* **2021**, *78*, 499–511. [CrossRef]
83. Tang, X.; Wang, P.; Zhang, R.; Watanabe, I.; Chang, E.; Vinayachandran, V.; Nayak, L.; Lapping, S.; Liao, S.; Madera, A.; et al. KLF2 Regulates Neutrophil Activation and Thrombosis in Cardiac Hypertrophy and Heart Failure Progression. *J. Clin. Investig.* **2022**, *132*, e147191. [CrossRef]
84. Zhu, W.; Buffa, J.A.; Wang, Z.; Warrier, M.; Schugar, R.; Shih, D.M.; Gupta, N.; Gregory, J.C.; Org, E.; Fu, X.; et al. Flavin Monooxygenase 3, the Host Hepatic Enzyme in the Metaorganismal Trimethylamine N-Oxide-Generating Pathway, Modulates Platelet Responsiveness and Thrombosis Risk. *J. Thromb. Haemost.* **2018**, *16*, 1857–1872. [CrossRef]
85. Roberts, A.B.; Gu, X.; Buffa, J.A.; Hurd, A.G.; Wang, Z.; Zhu, W.; Gupta, N.; Skye, S.M.; Cody, D.B.; Levison, B.S.; et al. Development of a Gut Microbe-Targeted Nonlethal Therapeutic to Inhibit Thrombosis Potential. *Nat. Med.* **2018**, *24*, 1407–1417. [CrossRef] [PubMed]
86. Zhu, W.; Wang, Z.; Tang, W.H.W.; Hazen, S.L. Gut Microbe-Generated Trimethylamine N-Oxide From Dietary Choline Is Prothrombotic in Subjects. *Circulation* **2017**, *135*, 1671–1673. [CrossRef] [PubMed]
87. Tang, W.H.W.; Wang, Z.; Levison, B.S.; Koeth, R.A.; Britt, E.B.; Fu, X.; Wu, Y.; Hazen, S.L. Intestinal Microbial Metabolism of Phosphatidylcholine and Cardiovascular Risk. *N. Engl. J. Med.* **2013**, *368*, 1575–1584. [CrossRef] [PubMed]
88. Mao, H.; Angelini, A.; Li, S.; Wang, G.; Li, L.; Patterson, C.; Pi, X.; Xie, L. CRAT Links Cholesterol Metabolism to Innate Immune Responses in the Heart. *Nat. Metab.* **2023**, *5*, 1382–1394. [CrossRef]
89. Ding, L.; Chang, M.; Guo, Y.; Zhang, L.; Xue, C.; Yanagita, T.; Zhang, T.; Wang, Y. Trimethylamine-N-Oxide (TMAO)-Induced Atherosclerosis Is Associated with Bile Acid Metabolism. *Lipids Health Dis.* **2018**, *17*, 286. [CrossRef]
90. Komaroff, A.L. The Microbiome and Risk for Atherosclerosis. *JAMA* **2018**, *319*, 2381–2382. [CrossRef] [PubMed]
91. Wang, Z.; Klipfell, E.; Bennett, B.J.; Koeth, R.; Levison, B.S.; Dugar, B.; Feldstein, A.E.; Britt, E.B.; Fu, X.; Chung, Y.-M.; et al. Gut Flora Metabolism of Phosphatidylcholine Promotes Cardiovascular Disease. *Nature* **2011**, *472*, 57–63. [CrossRef]
92. He, Z.; Kwek, E.; Hao, W.; Zhu, H.; Liu, J.; Ma, K.Y.; Chen, Z.-Y. Hawthorn Fruit Extract Reduced Trimethylamine-N-Oxide (TMAO)-Exacerbated Atherogenesis in Mice via Anti-Inflammation and Anti-Oxidation. *Nutr. Metab.* **2021**, *18*, 6. [CrossRef]
93. Zhang, Y.; Wang, Y.; Ke, B.; Du, J. TMAO: How Gut Microbiota Contributes to Heart Failure. *Transl. Res.* **2021**, *228*, 109–125. [CrossRef] [PubMed]
94. Ortega-Hernández, S.; González-Sosa, S.; Conde-Martel, A.; Trullàs, J.C.; Llàcer, P.; Pérez-Silvestre, J.; Arévalo-Lorido, J.C.; Casado, J.; Formiga, F.; Manzano, L.; et al. Prognostic Impact of Statins in Heart Failure with Preserved Ejection Fraction. *J. Clin. Med.* **2024**, *13*, 5844. [CrossRef] [PubMed]
95. Kinugasa, Y.; Nakamura, K.; Kamitani, H.; Hirai, M.; Yanagihara, K.; Kato, M.; Yamamoto, K. Trimethylamine N-Oxide and Outcomes in Patients Hospitalized with Acute Heart Failure and Preserved Ejection Fraction. *ESC Heart Fail.* **2021**, *8*, 2103–2110. [CrossRef]
96. Gui, X.Y.; Rabkin, S.W. C-Reactive Protein, Interleukin-6, Trimethylamine-N-Oxide, Syndecan-1, Nitric Oxide, and Tumor Necrosis Factor Receptor-1 in Heart Failure with Preserved Versus Reduced Ejection Fraction: A Meta-Analysis. *Curr. Heart Fail. Rep.* **2023**, *20*, 1–11. [CrossRef] [PubMed]
97. Wei, H.; Zhao, M.; Huang, M.; Li, C.; Gao, J.; Yu, T.; Zhang, Q.; Shen, X.; Ji, L.; Ni, L.; et al. FMO3-TMAO Axis Modulates the Clinical Outcome in Chronic Heart-Failure Patients with Reduced Ejection Fraction: Evidence from an Asian Population. *Front. Med.* **2022**, *16*, 295–305. [CrossRef]
98. Ramos-Romero, S.; Hereu, M.; Atienza, L.; Casas, J.; Jáuregui, O.; Amézqueta, S.; Dasilva, G.; Medina, I.; Nogués, M.R.; Romeu, M.; et al. Mechanistically Different Effects of Fat and Sugar on Insulin Resistance, Hypertension, and Gut Microbiota in Rats. *Am. J. Physiol. Endocrinol. Metab.* **2018**, *314*, E552–E563. [CrossRef]

99. Matsiras, D.; Bezati, S.; Ventoulis, I.; Verras, C.; Parissis, J.; Polyzogopoulou, E. Gut Failure: A Review of the Pathophysiology and Therapeutic Potentials in the Gut-Heart Axis. *J. Clin. Med.* **2023**, *12*, 2567. [CrossRef]
100. Miranda, P.M.; De Palma, G.; Serkis, V.; Lu, J.; Louis-Auguste, M.P.; McCarville, J.L.; Verdu, E.F.; Collins, S.M.; Bercik, P. High Salt Diet Exacerbates Colitis in Mice by Decreasing Lactobacillus Levels and Butyrate Production. *Microbiome* **2018**, *6*, 57. [CrossRef]
101. Desai, M.S.; Seekatz, A.M.; Koropatkin, N.M.; Kamada, N.; Hickey, C.A.; Wolter, M.; Pudlo, N.A.; Kitamoto, S.; Terrapon, N.; Muller, A.; et al. A Dietary Fiber-Deprived Gut Microbiota Degrades the Colonic Mucus Barrier and Enhances Pathogen Susceptibility. *Cell* **2016**, *167*, 1339–1353.e21. [CrossRef]
102. Hargreaves, S.M.; Raposo, A.; Saraiva, A.; Zandonadi, R.P. Vegetarian Diet: An Overview through the Perspective of Quality of Life Domains. *Int. J. Environ. Res. Public. Health* **2021**, *18*, 4067. [CrossRef]
103. Shen, X.; Tilves, C.; Kim, H.; Tanaka, T.; Spira, A.P.; Chia, C.W.; Talegawkar, S.A.; Ferrucci, L.; Mueller, N.T. Plant-Based Diets and the Gut Microbiome: Findings from the Baltimore Longitudinal Study of Aging. *Am. J. Clin. Nutr.* **2024**, *119*, 628–638. [CrossRef]
104. Soldán, M.; Argalášová, Ľ.; Hadvinová, L.; Galileo, B.; Babjaková, J. The Effect of Dietary Types on Gut Microbiota Composition and Development of Non-Communicable Diseases: A Narrative Review. *Nutrients* **2024**, *16*, 3134. [CrossRef] [PubMed]
105. Alzughayyar, D.-K.N.; Weber, R.-M.; Husain, S.; Schoch, N.; Englert, H. Impact of the Healthy Lifestyle Community Program (HLCP-3) on Trimethylamine N-Oxide (TMAO) and Risk Profile Parameters Related to Lifestyle Diseases During the Six Months Following an Intervention Study. *Nutrients* **2025**, *17*, 298. [CrossRef]
106. Merra, G.; Noce, A.; Marrone, G.; Cintoni, M.; Tarsitano, M.G.; Capacci, A.; De Lorenzo, A. Influence of Mediterranean Diet on Human Gut Microbiota. *Nutrients* **2020**, *13*, 7. [CrossRef]
107. Ghosh, T.S.; Rampelli, S.; Jeffery, I.B.; Santoro, A.; Neto, M.; Capri, M.; Giampieri, E.; Jennings, A.; Candela, M.; Turroni, S.; et al. Mediterranean Diet Intervention Alters the Gut Microbiome in Older People Reducing Frailty and Improving Health Status: The NU-AGE 1-Year Dietary Intervention across Five European Countries. *Gut* **2020**, *69*, 1218–1228. [CrossRef]
108. Tektonidis, T.G.; Åkesson, A.; Gigante, B.; Wolk, A.; Larsson, S.C. Adherence to a Mediterranean Diet Is Associated with Reduced Risk of Heart Failure in Men. *Eur. J. Heart Fail.* **2016**, *18*, 253–259. [CrossRef] [PubMed]
109. De Filippis, F.; Pellegrini, N.; Vannini, L.; Jeffery, I.B.; La Storia, A.; Laghi, L.; Serrazanetti, D.I.; Di Cagno, R.; Ferrocino, I.; Lazzi, C.; et al. High-Level Adherence to a Mediterranean Diet Beneficially Impacts the Gut Microbiota and Associated Metabolome. *Gut* **2016**, *65*, 1812–1821. [CrossRef] [PubMed]
110. Estruch, R.; Ros, E.; Salas-Salvadó, J.; Covas, M.-I.; Corella, D.; Arós, F.; Gómez-Gracia, E.; Ruiz-Gutiérrez, V.; Fiol, M.; Lapetra, J.; et al. Primary Prevention of Cardiovascular Disease with a Mediterranean Diet Supplemented with Extra-Virgin Olive Oil or Nuts. *N. Engl. J. Med.* **2018**, *378*, e34. [CrossRef]
111. Kaluza, J.; Levitan, E.B.; Michaëlsson, K.; Wolk, A. Anti-Inflammatory Diet and Risk of Heart Failure: Two Prospective Cohort Studies. *Eur. J. Heart Fail.* **2020**, *22*, 676–682. [CrossRef]
112. Marques, F.Z.; Nelson, E.; Chu, P.-Y.; Horlock, D.; Fiedler, A.; Ziemann, M.; Tan, J.K.; Kuruppu, S.; Rajapakse, N.W.; El-Osta, A.; et al. High-Fiber Diet and Acetate Supplementation Change the Gut Microbiota and Prevent the Development of Hypertension and Heart Failure in Hypertensive Mice. *Circulation* **2017**, *135*, 964–977. [CrossRef]
113. Erickson, M.L.; Malin, S.K.; Wang, Z.; Brown, J.M.; Hazen, S.L.; Kirwan, J.P. Effects of Lifestyle Intervention on Plasma Trimethylamine N-Oxide in Obese Adults. *Nutrients* **2019**, *11*, 179. [CrossRef]
114. Battillo, D.J.; Malin, S.K. Impact of Caloric Restriction and Exercise on Trimethylamine N-Oxide Metabolism in Women with Obesity. *Nutrients* **2023**, *15*, 1455. [CrossRef]
115. Bozkurt, B.; Fonarow, G.C.; Goldberg, L.R.; Guglin, M.; Josephson, R.A.; Forman, D.E.; Lin, G.; Lindenfeld, J.; O'Connor, C.; Panjrath, G.; et al. Cardiac Rehabilitation for Patients with Heart Failure: JACC Expert Panel. *J. Am. Coll. Cardiol.* **2021**, *77*, 1454–1469. [CrossRef] [PubMed]
116. Taylor, R.S.; Dalal, H.M.; Zwisler, A.-D. Cardiac Rehabilitation for Heart Failure: "Cinderella" or Evidence-Based Pillar of Care? *Eur. Heart J.* **2023**, *44*, 1511–1518. [CrossRef] [PubMed]
117. Allen, J.M.; Mailing, L.J.; Niemiro, G.M.; Moore, R.; Cook, M.D.; White, B.A.; Holscher, H.D.; Woods, J.A. Exercise Alters Gut Microbiota Composition and Function in Lean and Obese Humans. *Med. Sci. Sports Exerc.* **2018**, *50*, 747–757. [CrossRef] [PubMed]
118. Pechlivanis, A.; Papaioannou, K.G.; Tsalis, G.; Saraslanidis, P.; Mougios, V.; Theodoridis, G.A. Monitoring the Response of the Human Urinary Metabolome to Brief Maximal Exercise by a Combination of RP-UPLC-MS and (1)H NMR Spectroscopy. *J. Proteome Res.* **2015**, *14*, 4610–4622. [CrossRef] [PubMed]
119. Pechlivanis, A.; Kostidis, S.; Saraslanidis, P.; Petridou, A.; Tsalis, G.; Veselkov, K.; Mikros, E.; Mougios, V.; Theodoridis, G.A. 1H NMR Study on the Short- and Long-Term Impact of Two Training Programs of Sprint Running on the Metabolic Fingerprint of Human Serum. *J. Proteome Res.* **2013**, *12*, 470–480. [CrossRef]
120. Zhang, Y.; Wang, G.; Li, R.; Liu, R.; Yu, Z.; Zhang, Z.; Wan, Z. Trimethylamine N-Oxide Aggravated Cognitive Impairment from APP/PS1 Mice and Protective Roles of Voluntary Exercise. *Neurochem. Int.* **2023**, *162*, 105459. [CrossRef]

121. Brandao, C.F.C.; Krempf, M.; Giolo de Carvalho, F.; Aguesse, A.; Junqueira-Franco, M.V.M.; Batitucci, G.; de Freitas, E.C.; Noronha, N.Y.; Rodrigues, G.d.S.; Junqueira, G.P.; et al. Sphingolipid and Trimethylamine-N-Oxide (TMAO) Levels in Women with Obesity after Combined Physical Training. *Metabolites* **2024**, *14*, 398. [CrossRef]
122. Sanders, M.E.; Merenstein, D.J.; Reid, G.; Gibson, G.R.; Rastall, R.A. Probiotics and Prebiotics in Intestinal Health and Disease: From Biology to the Clinic. *Nat. Rev. Gastroenterol. Hepatol.* **2019**, *16*, 605–616. [CrossRef]
123. Hiel, S.; Gianfrancesco, M.A.; Rodriguez, J.; Portheault, D.; Leyrolle, Q.; Bindels, L.B.; Gomes da Silveira Cauduro, C.; Mulders, M.D.G.H.; Zamariola, G.; Azzi, A.-S.; et al. Link between Gut Microbiota and Health Outcomes in Inulin -Treated Obese Patients: Lessons from the Food4Gut Multicenter Randomized Placebo-Controlled Trial. *Clin. Nutr.* **2020**, *39*, 3618–3628. [CrossRef] [PubMed]
124. Swanson, K.S.; Gibson, G.R.; Hutkins, R.; Reimer, R.A.; Reid, G.; Verbeke, K.; Scott, K.P.; Holscher, H.D.; Azad, M.B.; Delzenne, N.M.; et al. The International Scientific Association for Probiotics and Prebiotics (ISAPP) Consensus Statement on the Definition and Scope of Synbiotics. *Nat. Rev. Gastroenterol. Hepatol.* **2020**, *17*, 687–701. [CrossRef] [PubMed]
125. Wang, Q.; Guo, M.; Liu, Y.; Xu, M.; Shi, L.; Li, X.; Zhao, J.; Zhang, H.; Wang, G.; Chen, W. Bifidobacterium Breve and Bifidobacterium Longum Attenuate Choline-Induced Plasma Trimethylamine N-Oxide Production by Modulating Gut Microbiota in Mice. *Nutrients* **2022**, *14*, 1222. [CrossRef]
126. Tungsanga, S.; Panpetch, W.; Bhunyakarnjanarat, T.; Udompornpitak, K.; Katavetin, P.; Chancharoenthana, W.; Chatthanathon, P.; Somboonna, N.; Tungsanga, K.; Tumwasorn, S.; et al. Uremia-Induced Gut Barrier Defect in 5/6 Nephrectomized Mice Is Worsened by Candida Administration through a Synergy of Uremic Toxin, Lipopolysaccharide, and (1→3)-β-D-Glucan, but Is Attenuated by Lacticaseibacillus Rhamnosus L34. *Int. J. Mol. Sci.* **2022**, *23*, 2511. [CrossRef]
127. Ramireddy, L.; Tsen, H.-Y.; Chiang, Y.-C.; Hung, C.-Y.; Wu, S.-R.; Young, S.-L.; Lin, J.-S.; Huang, C.-H.; Chiu, S.-H.; Chen, C.-C.; et al. Molecular Identification and Selection of Probiotic Strains Able to Reduce the Serum TMAO Level in Mice Challenged with Choline. *Foods* **2021**, *10*, 2931. [CrossRef] [PubMed]
128. Salamat, S.; Jahan-Mihan, A.; Tabandeh, M.R.; Mansoori, A. Randomized Clinical Trial Evaluating the Efficacy of Synbiotic Supplementation on Serum Endotoxin and Trimethylamine N-Oxide Levels in Patients with Dyslipidaemia. *Arch. Med. Sci. Atheroscler. Dis.* **2024**, *9*, e18–e25. [CrossRef]
129. Pourrajab, B.; Naderi, N.; Janani, L.; Mofid, V.; Hajahmadi, M.; Dehnad, A.; Shidfar, F. Comparison of Probiotic Yogurt and Ordinary Yogurt Consumption on Serum Pentraxin3, NT-proBNP, oxLDL, and ApoB100 in Patients with Chronic Heart Failure: A Randomized, Triple-Blind, Controlled Trial. *Food Funct.* **2020**, *11*, 10000–10010. [CrossRef]
130. Shoaei Matin, S.; Shidfar, F.; Naderi, N.; Amin, A.; Hosseini-Baharanchi, F.S.; Dehnad, A. The Effect of Synbiotic Consumption on Serum NTproBNP, hsCRP and Blood Pressure in Patients with Chronic Heart Failure: A Randomized, Triple-Blind, Controlled Trial. *Front. Nutr.* **2021**, *8*, 822498. [CrossRef]
131. Cui, H.; Han, S.; Dai, Y.; Xie, W.; Zheng, R.; Sun, Y.; Xia, X.; Deng, X.; Cao, Y.; Zhang, M.; et al. Gut Microbiota and Integrative Traditional Chinese and Western Medicine in Prevention and Treatment of Heart Failure. *Phytomedicine* **2023**, *117*, 154885. [CrossRef]
132. Liu, J.; Li, T.; Wu, H.; Shi, H.; Bai, J.; Zhao, W.; Jiang, D.; Jiang, X. Lactobacillus Rhamnosus GG Strain Mitigated the Development of Obstructive Sleep Apnea-Induced Hypertension in a High Salt Diet via Regulating TMAO Level and CD4+ T Cell Induced-Type I Inflammation. *Biomed. Pharmacother.* **2019**, *112*, 108580. [CrossRef] [PubMed]
133. Sun, B.; Ma, T.; Li, Y.; Yang, N.; Li, B.; Zhou, X.; Guo, S.; Zhang, S.; Kwok, L.-Y.; Sun, Z.; et al. Bifidobacterium Lactis Probio-M8 Adjuvant Treatment Confers Added Benefits to Patients with Coronary Artery Disease via Target Modulation of the Gut-Heart/-Brain Axes. *mSystems* **2022**, *7*, e0010022. [CrossRef]
134. Sánchez-Quintero, M.J.; Delgado, J.; Medina-Vera, D.; Becerra-Muñoz, V.M.; Queipo-Ortuño, M.I.; Estévez, M.; Plaza-Andrades, I.; Rodríguez-Capitán, J.; Sánchez, P.L.; Crespo-Leiro, M.G.; et al. Beneficial Effects of Essential Oils from the Mediterranean Diet on Gut Microbiota and Their Metabolites in Ischemic Heart Disease and Type-2 Diabetes Mellitus. *Nutrients* **2022**, *14*, 4650. [CrossRef] [PubMed]
135. Sánchez-Quintero, M.J.; Delgado, J.; Martín Chaves, L.; Medina-Vera, D.; Murri, M.; Becerra-Muñoz, V.M.; Estévez, M.; Crespo-Leiro, M.G.; Paz López, G.; González-Jiménez, A.; et al. Multi-Omics Approach Reveals Prebiotic and Potential Antioxidant Effects of Essential Oils from the Mediterranean Diet on Cardiometabolic Disorder Using Humanized Gnotobiotic Mice. *Antioxidants* **2023**, *12*, 1643. [CrossRef]
136. Eme, L.; Doolittle, W.F. Archaea. *Curr. Biol.* **2015**, *25*, R851–R855. [CrossRef]
137. Hoegenauer, C.; Hammer, H.F.; Mahnert, A.; Moissl-Eichinger, C. Methanogenic Archaea in the Human Gastrointestinal Tract. *Nat. Rev. Gastroenterol. Hepatol.* **2022**, *19*, 805–813. [CrossRef]
138. Ramezani, A.; Nolin, T.D.; Barrows, I.R.; Serrano, M.G.; Buck, G.A.; Regunathan-Shenk, R.; West, R.E.; Latham, P.S.; Amdur, R.; Raj, D.S. Gut Colonization with Methanogenic Archaea Lowers Plasma Trimethylamine N-Oxide Concentrations in Apolipoprotein e$^{-/-}$ Mice. *Sci. Rep.* **2018**, *8*, 14752. [CrossRef] [PubMed]

139. Craciun, S.; Balskus, E.P. Microbial Conversion of Choline to Trimethylamine Requires a Glycyl Radical Enzyme. *Proc. Natl. Acad. Sci. USA* **2012**, *109*, 21307–21312. [CrossRef]
140. Wang, Z.; Roberts, A.B.; Buffa, J.A.; Levison, B.S.; Zhu, W.; Org, E.; Gu, X.; Huang, Y.; Zamanian-Daryoush, M.; Culley, M.K.; et al. Non-Lethal Inhibition of Gut Microbial Trimethylamine Production for the Treatment of Atherosclerosis. *Cell* **2015**, *163*, 1585–1595. [CrossRef]
141. Zixin, Y.; Lulu, C.; Xiangchang, Z.; Qing, F.; Binjie, Z.; Chunyang, L.; Tai, R.; Dongsheng, O. TMAO as a Potential Biomarker and Therapeutic Target for Chronic Kidney Disease: A Review. *Front. Pharmacol.* **2022**, *13*, 929262. [CrossRef]
142. Chen, S.; Henderson, A.; Petriello, M.C.; Romano, K.A.; Gearing, M.; Miao, J.; Schell, M.; Sandoval-Espinola, W.J.; Tao, J.; Sha, B.; et al. Trimethylamine N-Oxide Binds and Activates PERK to Promote Metabolic Dysfunction. *Cell Metab.* **2019**, *30*, 1141–1151.e5. [CrossRef]
143. Brunt, V.E.; Greenberg, N.T.; Sapinsley, Z.J.; Casso, A.G.; Richey, J.J.; VanDongen, N.S.; Gioscia-Ryan, R.A.; Ziemba, B.P.; Neilson, A.P.; Davy, K.P.; et al. Suppression of Trimethylamine N-Oxide with DMB Mitigates Vascular Dysfunction, Exercise Intolerance, and Frailty Associated with a Western-Style Diet in Mice. *J. Appl. Physiol.* **2022**, *133*, 798–813. [CrossRef]
144. Shih, D.M.; Zhu, W.; Schugar, R.C.; Meng, Y.; Jia, X.; Miikeda, A.; Wang, Z.; Zieger, M.; Lee, R.; Graham, M.; et al. Genetic Deficiency of Flavin-Containing Monooxygenase 3 (FMO3) Protects Against Thrombosis but Has Only a Minor Effect on Plasma Lipid Levels-Brief Report. *Arterioscler. Thromb. Vasc. Biol.* **2019**, *39*, 1045–1054. [CrossRef]
145. Agarwal, M.; Roth, K.; Yang, Z.; Sharma, R.; Maddipati, K.; Westrick, J.; Petriello, M.C. Loss of Flavin-Containing Monooxygenase 3 Modulates Dioxin-like Polychlorinated Biphenyl 126-Induced Oxidative Stress and Hepatotoxicity. *Environ. Res.* **2024**, *250*, 118492. [CrossRef] [PubMed]
146. Riba, A.; Deres, L.; Eros, K.; Szabo, A.; Magyar, K.; Sumegi, B.; Toth, K.; Halmosi, R.; Szabados, E. Doxycycline Protects against ROS-Induced Mitochondrial Fragmentation and ISO-Induced Heart Failure. *PLoS ONE* **2017**, *12*, e0175195. [CrossRef] [PubMed]
147. Lam, V.; Su, J.; Hsu, A.; Gross, G.J.; Salzman, N.H.; Baker, J.E. Intestinal Microbial Metabolites Are Linked to Severity of Myocardial Infarction in Rats. *PLoS ONE* **2016**, *11*, e0160840. [CrossRef] [PubMed]
148. Tang, T.W.H.; Chen, H.-C.; Chen, C.-Y.; Yen, C.Y.T.; Lin, C.-J.; Prajnamitra, R.P.; Chen, L.-L.; Ruan, S.-C.; Lin, J.-H.; Lin, P.-J.; et al. Loss of Gut Microbiota Alters Immune System Composition and Cripples Postinfarction Cardiac Repair. *Circulation* **2019**, *139*, 647–659. [CrossRef]
149. Peng, J.; Xiao, X.; Hu, M.; Zhang, X. Interaction between Gut Microbiome and Cardiovascular Disease. *Life Sci.* **2018**, *214*, 153–157. [CrossRef]
150. Virani, S.S.; Newby, L.K.; Arnold, S.V.; Bittner, V.; Brewer, L.C.; Demeter, S.H.; Dixon, D.L.; Fearon, W.F.; Hess, B.; Johnson, H.M.; et al. 2023 AHA/ACC/ACCP/ASPC/NLA/PCNA Guideline for the Management of Patients with Chronic Coronary Disease: A Report of the American Heart Association/American College of Cardiology Joint Committee on Clinical Practice Guidelines. *Circulation* **2023**, *148*, e9–e119. [CrossRef]
151. Hazen, S.L. Treating and Preventing Disease with TMA and TMAO Lowering Agents 2016. US9694020B2, 4 July 2017.
152. Zhao, W.; Li, Z.; Yu, M.L.; Liu, Y.; Liu, C.C.; Jia, X.J.; Liu, M.Q.; Li, Y.G. Aspirin Inhibits Rotavirus Replication and Alters Rat Gut Microbial Composition. *Virol. J.* **2023**, *20*, 237. [CrossRef]
153. Li, T.; Ding, N.; Guo, H.; Hua, R.; Lin, Z.; Tian, H.; Yu, Y.; Fan, D.; Yuan, Z.; Gonzalez, F.J.; et al. A Gut Microbiota-Bile Acid Axis Promotes Intestinal Homeostasis upon Aspirin-Mediated Damage. *Cell Host Microbe* **2024**, *32*, 191–208.e9. [CrossRef]
154. Xu, Z.; Pan, Z.; Jin, Y.; Gao, Z.; Jiang, F.; Fu, H.; Chen, X.; Zhang, X.; Yan, H.; Yang, X.; et al. Inhibition of PRKAA/AMPK (Ser485/491) Phosphorylation by Crizotinib Induces Cardiotoxicity via Perturbing Autophagosome-Lysosome Fusion. *Autophagy* **2024**, *20*, 416–436. [CrossRef]
155. Kuka, J.; Videja, M.; Makrecka-Kuka, M.; Liepins, J.; Grinberga, S.; Sevostjanovs, E.; Vilks, K.; Liepinsh, E.; Dambrova, M. Metformin Decreases Bacterial Trimethylamine Production and Trimethylamine N-Oxide Levels in Db/Db Mice. *Sci. Rep.* **2020**, *10*, 14555. [CrossRef]
156. Su, C.; Li, X.; Yang, Y.; Du, Y.; Zhang, X.; Wang, L.; Hong, B. Metformin Alleviates Choline Diet-Induced TMAO Elevation in C57BL/6J Mice by Influencing Gut-Microbiota Composition and Functionality. *Nutr. Diabetes* **2021**, *11*, 27. [CrossRef]
157. Piperis, C.; Marathonitis, A.; Anastasiou, A.; Theofilis, P.; Mourouzis, K.; Giannakodimos, A.; Tryfou, E.; Oikonomou, E.; Siasos, G.; Tousoulis, D. Multifaceted Impact of SGLT2 Inhibitors in Heart Failure Patients: Exploring Diverse Mechanisms of Action. *Biomedicines* **2024**, *12*, 2314. [CrossRef]
158. Beghini, A.; Sammartino, A.M.; Papp, Z.; von Haehling, S.; Biegus, J.; Ponikowski, P.; Adamo, M.; Falco, L.; Lombardi, C.M.; Pagnesi, M.; et al. 2024 Update in Heart Failure. *ESC Heart Fail* **2025**, *12*, 8–42. [CrossRef]
159. Mindrescu, N.M.; Guja, C.; Jinga, V.; Ispas, S.; Curici, A.; Danciulescu Miulescu, R.E.; Nelson Twakor, A.; Pantea Stoian, A.M. SGLT-2 Inhibitors and Metabolic Outcomes: A Primary Data Study Exploring the Microbiota-Diabetes Connection. *Metabolites* **2025**, *15*, 411. [CrossRef] [PubMed]
160. Wang, L.; Wang, Y.; Xu, H.; Li, W. Effect of Dapagliflozin on Ferroptosis through the Gut Microbiota Metabolite TMAO during Myocardial Ischemia-Reperfusion Injury in Diabetes Mellitus Rats. *Sci. Rep.* **2024**, *14*, 13851. [CrossRef] [PubMed]

205. Yuzefpolskaya, M.; Bohn, B.; Nasiri, M.; Zuver, A.M.; Onat, D.D.; Royzman, E.A.; Nwokocha, J.; Mabasa, M.; Pinsino, A.; Brunjes, D.; et al. Gut Microbiota, Endotoxemia, Inflammation, and Oxidative Stress in Patients with Heart Failure, Left Ventricular Assist Device, and Transplant. *J. Heart Lung Transplant.* **2020**, *39*, 880–890. [CrossRef] [PubMed]
206. Maier, L.; Pruteanu, M.; Kuhn, M.; Zeller, G.; Telzerow, A.; Anderson, E.E.; Brochado, A.R.; Fernandez, K.C.; Dose, H.; Mori, H.; et al. Extensive Impact of Non-Antibiotic Drugs on Human Gut Bacteria. *Nature* **2018**, *555*, 623–628. [CrossRef] [PubMed]

biomolecules

MDPI

Review

Aortic Stiffness and Alzheimer's Disease: The Medin Connection

Filippos Triposkiadis [1,2,*], Andrew Xanthopoulos [1,2], Harisios Boudoulas [3] and Dirk L. Brutsaert [4]

[1] European University Cyprus, 1516 Nicosia, Cyprus
[2] Department of Cardiology, University Hospital of Larissa, Faculty of Medicine, School of Health Sciences, University of Thessaly, Biopolis, 41500 Larissa, Greece
[3] Division of Cardiovascular Medicine, The Ohio State University, Columbus, OH 43210, USA
[4] Department of Medicine & Cardiology, University of Antwerp, 2000 Antwerpen, Belgium
* Correspondence: ftriposkiadis@gmail.com

Abstract

Aging is associated with aortic stiffening (AoSt), a condition characterized by diminished aortic elasticity that predisposes individuals to cognitive decline, including Alzheimer's disease (AD). Emerging evidence implicates medin, which is derived from milk fat globule-EGF factor 8 protein (MFG-E8), as a key link between AoSt and AD. Medin aggregates into aortic medial amyloid (AMA), which is found in approximately 97% of Caucasian individuals aged 50 and above, contributing to vascular inflammation, calcification, and loss of arterial elasticity. These changes may promote hyperpulsatile cerebral blood flow and impair glymphatic clearance, resulting in increased deposition of neurotoxic proteins, such as amyloid-β (Aβ) and possibly medin, which colocalizes with vascular Aβ in the brain. Medin enhances Aβ aggregation, generating heterologous fibrils, and thereby contributes to cerebrovascular dysfunction and neuroinflammation. This interaction (cross-seeding) may deteriorate amyloid pathology in both the vasculature and the parenchyma in AD. Furthermore, medin per se causes endothelial dysfunction, increases oxidative stress, and activates glial cells, promoting the development of a pro-inflammatory environment that enhances cognitive decline. In this manuscript, we contend that medin might act as a bridge connecting the age-related increase in aortic stiffness to AD, and therefore, medin might present a novel therapeutic target within this context. This hypothesis deserves experimental and clinical validation.

Keywords: aging; aorta; stiffening; Alzheimer's; medin; hypothesis

1. Introduction

Aging is characterized by aortic stiffening (AoSt), the limited capacity of the aorta to expand and recoil as result of oscillations in pressure [1]. AoSt has been linked to several morbidities that predispose individuals to focal or global cerebral dysfunction, such as systolic hypertension [2,3]. Increased left ventricular (LV) afterload associated with reduced coronary perfusion pressure promotes LV remodeling [4,5] and kidney damage [6,7]. Importantly, several of these morbidities often coexist in the same individual [8].

The accumulation of amyloid fibrils seems to be universal in the elderly [9]. Aortic medial amyloid (AMA) is the most common human amyloid, as it has been found in approximately 97% of Caucasian individuals above 50 years of age [10]. AMA is derived from a proteolytic fragment of milk fat globule-EGF factor 8 protein (MFG-E8), also known as lactadherin [11]. Medin, a peptide consisting of 50 amino acids that is cleaved from the second discoidin-like domain of MFG-E8, constitutes the primary component of AMA and

https://doi.org/10.3390/biom15081148

shares 16% of its sequence with amyloid β (Aβ) [12,13], which accumulates in Alzheimer's disease (AD).

AD is a debilitating neurodegenerative disease that affects millions of individuals globally, and is characterized by memory loss, cognitive decline, and behavioral disturbances [14]. Two types of brain lesions define the neuropathological diagnosis of AD [15]. The first type, *Aβ plaques*, has become a unifying term for heterogeneous formations, many of which contain massed, fibrillar polymers of Aβ, but some of which lack the defining amyloid features [16]. The second type, intracellular *neurofibrillary tangles*, contains tau protein as the primary protein constituent, which is critically essential for the manifestation of dementia [17].

Accumulating evidence supports the association of AoSt with dementia-related pathology [18–23] and contributes to the pathogenesis of AD [24]. In this manuscript, we summarize the structure and function of the young and aging aorta, and contend that medin may be the link between increased AoSt and the development of AD in the elderly.

2. The Effect of Aging on the Aorta

2.1. The Young Aorta

The wall of the aorta comprises three main layers: the tunica intima, tunica media, and tunica adventitia [25]. The internal elastic lamina separates the tunica intima from the tunica media, whereas the external elastic lamina separates the tunica media from the tunica adventitia.

The endothelium of the intima, which lies between the circulating blood and the underlying cellular milieu, is a key mediator of cardiovascular function and displays transcriptional plasticity in response to normal and abnormal stimuli [26,27]. The endothelium not only serves as a barrier, but additionally actively regulates organ development, homeostasis, and tissue regeneration [28–30]. Studies employing single-cell RNA sequencing have revealed the presence of specialized subpopulations of endothelial cells (ECs) with distinct marker genes, suggesting functional specialization [31]. ECs sense various stimuli, including fluid shear stress and cytokines, and actively control the contraction and relaxation of vascular smooth muscle cells (VSMCs) located in the media through the secretion of various mediators [32].

The aortic tunica media, which virtually determines aortic biomechanics [33], consists of VSMCs and an extracellular matrix (ECM) containing elastic lamellae and discontinuous basement membrane components, along with interstitial collagens [34] with a specialized structure. Elastin dominates the mechanical behavior of aortic tissue under normal pressure fluctuations, working together with VSMCs of the contractile phenotype to allow the large aortic deformations necessary for cardiovascular functioning. The collagen does not contribute to biomechanics until a high deformation level is reached, which is assumed to exceed the level of physiological functioning. A biomechanically normal functioning ascending aorta distends during LV systole and recoils during LV diastole, allowing the storage of energy in the vessel from LV systole and energy redistribution during LV diastole by elastic recoil, which provides antegrade flow [35]. This phenomenon, the *Windkessel effect*, is crucial for the maintenance of pulse pressure within the physiological range (Figure 1).

In healthy young adults, the LV–aortic interactions are optimized to deliver cardiac output with a suitable pulsatile hydraulic load on the LV [36,37]. The low-impedance, compliant aorta interconnects with conduit arteries, such as the carotid, which have higher stiffness. The resulting impedance mismatches, accompanied by wave reflections at the aorta–brain interfaces, restrict the transmission of immoderate pulsatile energy into the cerebral microcirculation, protecting the brain tissue [38,39]. Besides impedance mismatch, other mechanisms contributing to protection against excessive pulsatile energy transmission include vessel size and area [40,41]. Optimal pulsatile energy transmission to the brain occurs when the human heart rate is close to normal, and it is stable across a wide range of aortic arch stiffnesses and LV contractility [42].

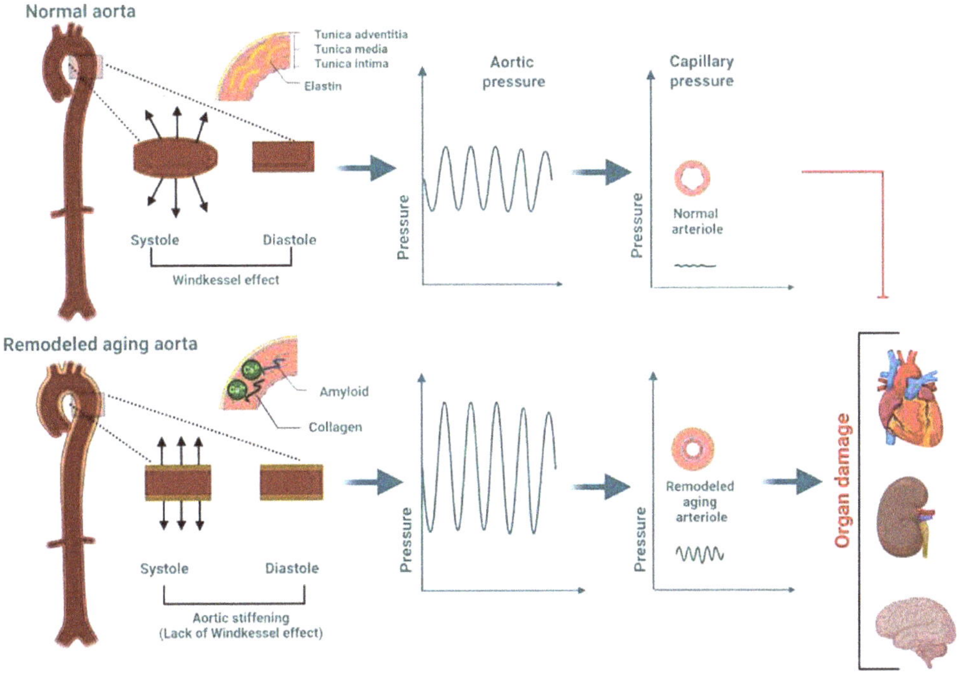

Figure 1. In the young, a biomechanically normal functioning ascending aorta expands during left ventricular (LV) systole and recoils during LV diastole, allowing vessel energy storage from LV systole and energy redistribution during LV diastole through elastic recoil, thereby providing antegrade flow. This phenomenon, *the Windkessel effect*, is crucial for the maintenance of pulse pressure within the physiological range. The hallmark of the aging aorta is remodeling of the extracellular matrix (ECM), including the dissolution of elastic fibers and accumulation and cross-linking of collagens. Other characteristic features include vascular calcification and medin amyloid deposition, predominantly in the tunica media. These structural abnormalities develop in the setting of an inflammatory environment (inflammaging) and eventually lead to enhanced aortic stiffness (AoSt) in the elderly. Elevated AoSt augments the LV pulsatile load (elevation of systolic blood pressure (BP), decrease in diastolic BP, and widening of pulse pressure), which is transmitted in the microvasculature, contributing to organ damage, particularly to the heart, the kidney, and the brain. Created in BioRender (https://www.biorender.com/).

2.2. The Aging Aorta

The hallmark of the aging aorta is ECM remodeling, including fragmentation of elastic fibers and excessive deposition of cross-linked collagens [43]. Vascular calcification and the presence of AMA, predominantly in the tunica media of the thoracic aorta and other upper-body arteries, are also characteristic features [44,45]. These structural abnormalities develop in the setting of an inflammatory environment (*inflammaging*) and eventually lead to enhanced AoSt in the elderly [41].

2.2.1. Inflammaging

During aging, cells in the body undergo senescence, characterized by a dysfunctional phenotype, the so-called senescence-associated secretory phenotype (SASP) [46].

Senescent ECs are flattened and enlarged and exhibit increased polyploidy, decreased nitric oxide (NO) production, and increased secretion of several pro-inflammatory cytokines [47]. As a result, the senescence of ECs accelerates vascular aging by increasing intima media layer thickness, the deposition of collagen, and often the thickness of the vascular lumen, and by reducing elastin deposition [48].

Senescent VSMCs in the aging aorta secrete inflammatory cytokines, chemokines, growth factors, and factors of matrix remodeling that change the tissue environment and participate in the development of the chronic, sterile, low-grade inflammation that characterizes inflammaging [49,50]. The latter results in damage to or loss of elastic fibers through the recruitment of several processes (e.g., enzymatic degradation, oxidative damage, glycation, calcification, etc.), which act on top of mechanical fatigue [51]. Further, significant collagen changes occur in the aging aortic wall, including an increase in its amount and a change in its structure, as well as an increase in cross-linking between collagen fibers [52]. Inflammaging is accelerated by several factors, including sex, race, coexisting morbidities, such as hypertension, obesity, and diabetes mellitus, and lifestyle (e.g., excessive alcohol consumption, smoking, and physical inactivity) [53], and it is tightly linked with aortic calcification and AMA [54–56].

Several lines of evidence indicate that inflammaging promotes aortic media calcification by mediating the development of a VSMC synthetic phenotype [57,58]. Calcification occurs at the intima and the media of the arterial wall. Calcification of the intima is linked with atherosclerosis, whereas that of the media has been associated with arteriosclerosis, an age-associated disease [59]. Senescent VSMCs undergo calcification following the uptake of exosomes derived from the endothelium [60,61] or of microparticles [62] released upon endothelial stimulation by chemical messengers of inflammation. In this regard, TNF (tumor necrosis factor)-α stimulates endothelial cells to release microparticles that contain BMP2 (bone morphogenetic protein 2), which subsequently undergo phagocytosis by VSMCs and enhance osteogenesis [63]. CRP (C-reactive protein), another messenger of inflammation, has also been implicated in promoting age-associated VSMC osteogenic trans-differentiation [64]. Finally, DNA damage of diverse etiology facilitates the generation of reactive oxygen and nitrogen species [65,66], which, in turn, promotes the loss of VSMC contractile function and the enhancement of VSMC osteogenic differentiation and calcification [67].

2.2.2. Medin

Experimental studies in mice have demonstrated that medin aggregates develop in the aorta and brain vasculature age-dependently, and that genetically induced deficiency in MFG-E8 abolishes vascular accumulation [68]. Medin has a high affinity to elastin and is tightly linked to arterial inflammation, dissolution of elastic fibers, calcification, and AMA generation [69].

Medin Amyloid Formation

Medin is a soluble monomer consisting of a core created by three β-strands, with smaller and unsteady strands located at the C-termini [13], and medin aggregates are drenched with β-sheet structures, at a proportion close to 65% [12,70]. Protein synthesis in VSMCs proceeds age-dependently, and clarifying the molecular aspects of proteome variation would significantly enhance our comprehension of human biology and disease [70,71]. Senescent VSMCs enhance medin accumulation in the media via increased release of small extracellular vesicles (sEVs) resulting from upregulation of sphingomyelin phosphodiesterase 3 (SMPD3). Moreover, there is increased ECM medin on the EV surface and EV binding to heparan sulfate proteoglycan 2 (HSPG2) in the ECM [72] (Figure 2). The increased expression of HSPG2 in the senescent ECM fosters the generation of medin structures resembling medin, which differ from the small, round aggregates found in the ECM from young SMCs. Thus, senescent SMCs may generate an environment that facilitates medin aggregation and AMA formation. AMA, in turn, may affect the tissue distribution of serum amyloid A (SAA), as evidenced by experimental studies displaying that serum amyloid A fibrillation may be promoted by medin amyloid-like fibrils [73]. SAA proteins are typical elements of the human blood that are conserved during evolution, and are predominantly synthesized in the liver [74]. SAA fragments form insoluble fibrils that characterize "secondary" amyloidosis and may impair organ morphology and function, leading to organ failure.

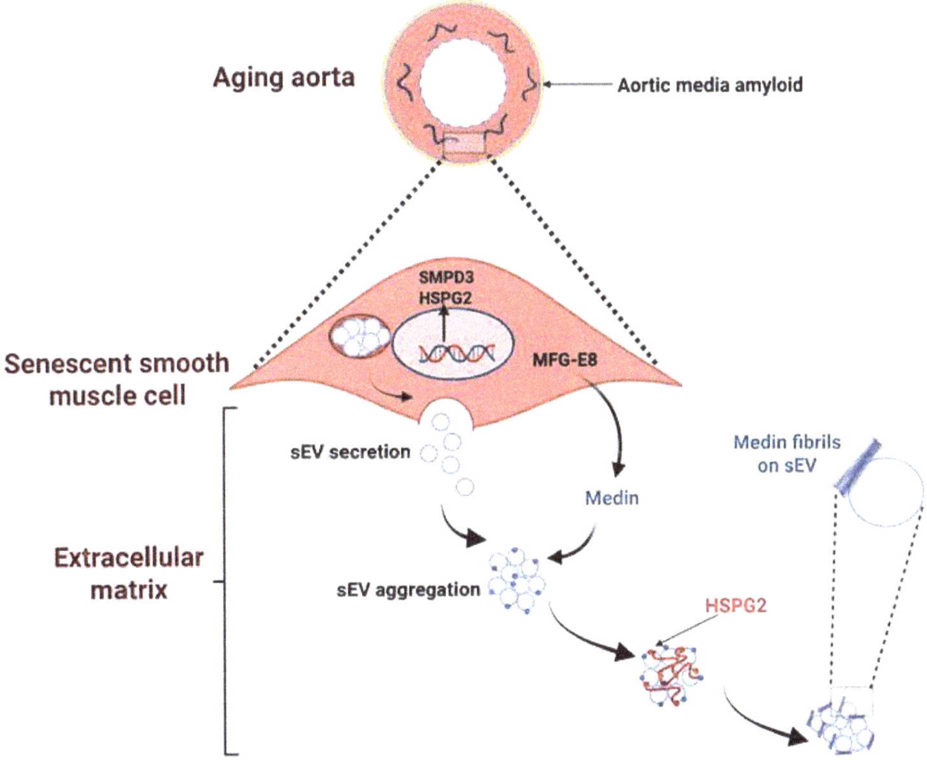

Figure 2. Aortic medial amyloid (AMA) is composed of medin secreted by senescent vascular smooth muscle cells (VSMCs). Small extracellular vesicles (sEVs) seem to mediate medin amyloid accumulation in the extracellular matrix (ECM). SMPD3 (sphingomyelin phosphodiesterase 3) triggers membrane curvature through sphingomyelin hydrolysis to ceramide, thereby influencing sEV release. The abundance of the proteoglycan HSPG2 (heparan sulfate proteoglycan 2) is elevated in the senescent ECM and colocalizes with sEVs and medin. Thus, VSMC-derived EVs and HSPG2 in the ECM mediate medin accumulation and AMA development associated with aging. Figure based on data from references [31,62,72]. Created in BioRender.

The aggregation kinetics of medin [75,76] are represented by a sigmoid curve (Figure 3) consisting of (a) a *lag phase* corresponding to monomer nucleation into oligomers and protofibrils, (b) a growth phase of rapid fibril elongation, and c) an equilibrium plateau generated by mature fibrils. This aggregation pattern can be found in several amyloids, such as Aβ, human islet amyloid polypeptide (hIAPP), and α-synuclein [77,78]. Although the medin monomer predominantly adopts random coil and β-sheet structures, partial helix structures have also been reported [13,79,80].

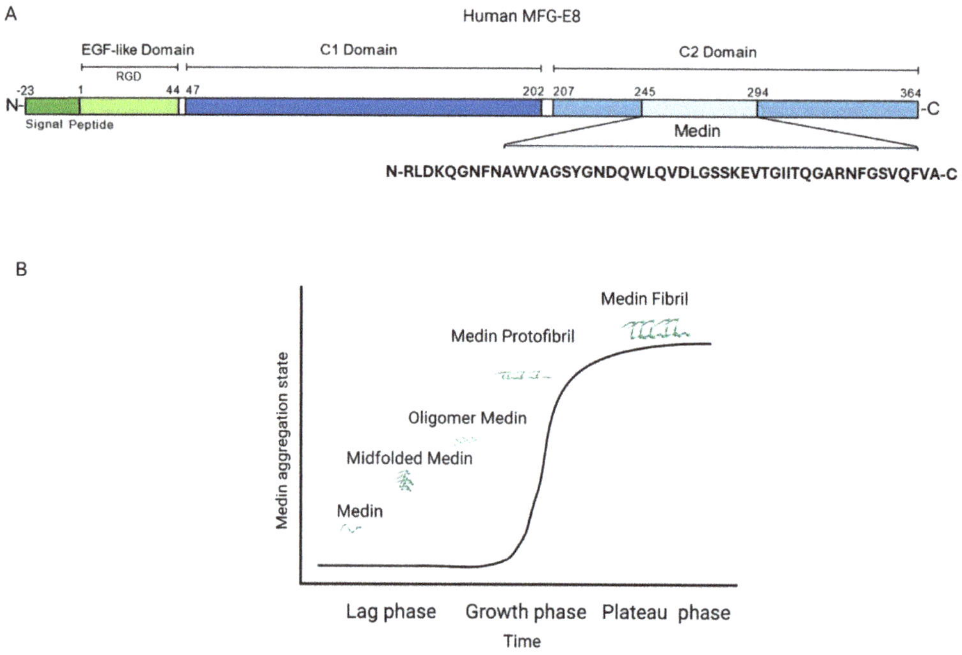

Figure 3. MFG-E8, medin, and medin amyloid formation. (**A**) MFG-E8, also termed lactadherin, contains an EGF-like domain on the N-terminal and two domains exhibiting homology to blood coagulation factor V/VIII (C1 and C2 domains, respectively) on the C-terminal. Medin is a peptide consisting of 50 amino acids cleaved from the C2 domain of MFG-E8. (**B**) The three-phase medin aggregation state: the lag phase, growth phase, and plateau phase. MFG-E8, milk fat globule EGF VIII; EGF, epidermal growth factor; C1, discoidin-like domain 1; C2, discoidin-like domain 2; RGD, an Arg-Gly-Asp motif. Adapted with permission from Ref. [69].

Medin and MFG-E8 interact with elastic fibers, making it likely that elastin contributes to AMA formation [81]. Computational studies suggest that the start of medin folding requires β-sheet formation around medin30–41 and medin42–50, followed by the wrapping up of other segments by their β-sheet boundaries [82]. Two medin peptides may readily aggregate into a β-sheet-rich dimer, which has a strong aggregation tendency. The dimerization of medin enhances β-sheet conformations and simultaneously leads to the generation of β-barrel oligomers. Asp (25) most likely drives the medin assembly by stabilizing the fibrillar conformation of the peptide in a way that is reminiscent of that of Asp (23) on Aβ aggregation [12].

For unknown reasons, amyloid deposits remain unrecognized by the immune system, leading to organ dysfunction. Phagocytosis assays have demonstrated that collagen associated with amyloid hinders phagocytosis and amyloid clearance [83]. In systemic light chain (LC) amyloidosis (AL), helical superstructures involving amyloid and collagen in cardiac fibrils, which likely stabilize the fibrils and decrease their clearance, have been reported [84]. Finally, in studies on mice, treatment of amyloid extracts with collagenase enhanced clearance compared with controls, coinciding with an increase in the accumulation of immune cells in the amyloid lesion [83]. Thus, collagen decreases the clearance of AMA, and dissociation of AMA from collagen might facilitate immunorecognition and clearance of pathologic amyloid deposits.

The Effects of Medin on the Aorta

Although amyloid fibers are crucial for amyloid pathology [85], it seems that intermediates synthesized during aggregation may be more critical for disease development. Indeed, oligomers are toxic [86] and increase membrane permeability, leading to cellular dysfunction [87]. Nevertheless, not all amyloid oligomers are cytotoxic, and those that are toxic seem to have structures that facilitate their toxicity [88].

Based on the typical clinical features of amyloid diseases, it has been suggested that toxic oligomers of diverse amyloid proteins may have several morphological characteristics in common [89]. Toxic oligomers are rich in β-sheets, display an external diameter of 7–12 nm and an internal diameter of 1.5–2.5 nm, and consist of 20–60 monomers each [90,91]. Since they resemble the cytolytic toxins produced by Clostridium perfringens, which form β-barrel pores and have the capacity for membrane incorporation following the *"amyloid-pore"* hypothesis, they have been called β-barrel intermediates and considered to be the toxic oligomers in amyloid aggregation [92]. β-barrel proteins function as monomers or dimers, and additionally, they may interact with each other, forming higher-order oligomers or even aggregates. On the pathway of fibrillogenesis, the β-barrel state can be intermediate ("on-pathway state"), or it can be formed as a result of the assembly of partially unfolded monomers ("off-pathway state") (Figure 4) [93].

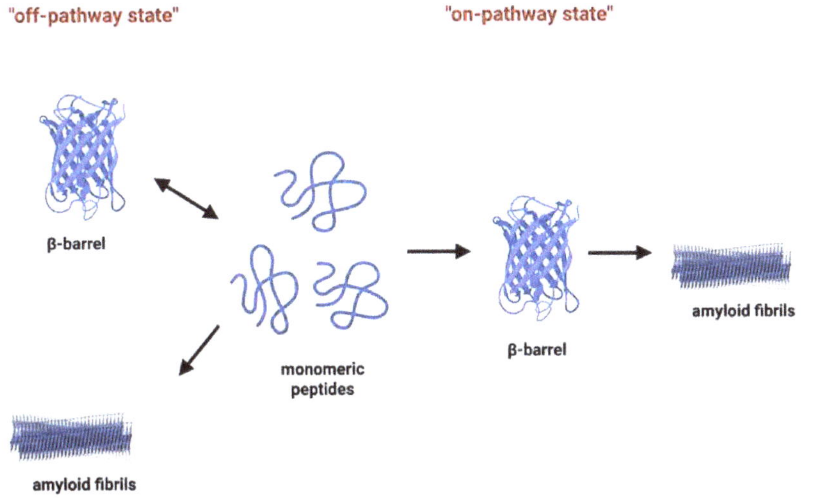

Figure 4. β-barrel and amyloid fibril formation. Short, differing peptides may generate stable oligomers resembling β-barrels (so-called "cylindrins"). On the pathway of fibrillogenesis, the β-barrel state can be intermediate ("on-pathway state"), or it can be formed as a result of the assembly of partially unfolded monomers ("off-pathway state"). The β-barrel oligomers are toxic and involved in the pathogenesis of various diseases. Adapted with permission from Ref. [93].

Medin contributes to inflammaging by increasing oxidative stress, as evidenced by the enhancement of superoxide and reduction of nitric oxide. Autopsy studies have shown elevated MGF-E8 expression in the aortic media of the elderly, along with markers of oxidative damage. Furthermore, inflammation and intimal/medial thickening were reported in wild-type (WT) aged mice, but were not observed in age-matched MFG-E8 KO mice, which did not express MFG-E8. Conversely, experimental studies on mice demonstrated that the levels of MFG-E8 protein and mRNA were increased in the dermis during wound healing [94]. Furthermore, in the same study, MFG-E8 was observed in granulation tissue surrounding blood vessels, whereas wound healing in MFG-E8 knockout mice was slower than in wild-type mice. The findings above indicate that the physiological effects of MFG-E8 are complex and incompletely understood.

In contrast to MFG-8, medin has only toxic effects. Medin causes human EC dysfunction by reducing EC viability, migration, and proliferation, possibly by inducing oxidative and nitrative stress via advanced glycation end-products (RAGE) [95,96]. Additionally, medin may stimulate EC inflammatory signaling, leading to the production of cytokines, including IL-6 [95]. In vitro studies have also demonstrated the direct toxicity of medin to ECs [97]. Endothelial dysfunction impairs VSMC function, further aggravating the effects of aging [98]. Finally, medin-induced alterations in the ECM disrupt EC-VSMC interactions, resulting in the production of a complex mixture of components derived from ECs and VSMCs [33,99], which may influence the function of neighboring cells [100].

Interestingly, medin in its amyloid state is decreased in patients with aortic aneurysm or dissection, whereas non-fibrillar medin levels are higher in the diseased aorta [101] and correlate with increased aortic stiffness [102], suggesting that non-fibrillar forms of medin may also be toxic.

2.2.3. Aortic Function

Inflammaging-induced alterations in the VSMC phenotype and in the elastin in the aortic wall cause vessel dilation, tortuosity, and stiffening that raise systolic and pulse blood pressure, enhance PWV, and lead to hyperpulsatile perfusion that damages the arterioles and promotes cognitive dysfunction (Figure 1) [103]. AMA contributes to this process by increasing inflammatory and oxidative stress, which accelerates age-related aortic dysfunction [95,97,104,105].

The aging-induced increase in AoSt occurs before the development of hypertension and may contribute to ethnic variations in cardiovascular and brain health [106–108]. Elevated AoSt impairs LV–arterial coupling and augments LV pulsatile load, resulting in LV hypertrophy, a significant risk factor for heart failure (HF) development [109,110]. In addition, increased AoSt impairs the optimization of the hemodynamic coupling between the LV, the aorta, and the brain [36,111]. This alters PWV dynamics, leading to increased transmission of pulsatile load, which is deleterious to the cerebral microvasculature and brain tissues. Eventually, the greater stiffness of the aorta compared to the carotid artery modifies wave reflections, facilitating the excessive transmission of pulsatile energy into the cerebrovascular network, contributing to cognitive impairments such as AD or another type of vascular dementia [112,113].

3. Medin, Aortic Stiffness, and Alzheimer's Disease

Cognitive impairment in the elderly is driven by both neurodegeneration and impairment of the cerebral vasculature [114]. Neurodegenerative and cerebrovascular interactions that adversely affect cognition are seen in AD and vascular dementia (VaD) [115], with medin playing an essential role within this context.

3.1. Cerebral Medin

3.1.1. Cerebral Medin Levels and Effects

There is evidence to suggest that cerebral medin facilitates AD development and progression. In a study that determined the medin levels in cerebral arterioles from brain donors without dementia (ND), AD, VaD, or combined AD and VaD, medin levels were higher in AD, VaD, or combined AD/VaD than in ND, with medin being the strongest predictor of AD diagnosis [116]. Further, in Aβ precursor protein (APP) transgenic mice and AD patients, medin was found in proximity to vascular Aβ deposits, and in mice, medin deficiency reduced vascular Aβ deposition by half [75]. Moreover, in both the mouse and human brain, MFG-E8 accumulated in the vessels, and expression of both MFG-E8 and medin increased in parallel with vascular Aβ. Additionally, analysis of data from the Religious Orders Study (ROS) and Rush Memory and Aging Project (MAP), referred to in combination as ROSMAP, patients with AD had higher MFGE8 vascular expression levels, which were associated with accelerated cognitive decline [75]. The same study reported that medin promoted Aβ aggregation by forming heterologous fibrils with Aβ. Thus, medin, together with vascular Aβ deposits, accelerates cognitive decline in Alzheimer's disease [117]. Finally, another study found that both Aβ and medin are aggregation-prone, and their mixture tends to form β-sheet-rich hetero-aggregates [118]. In the same study, the formation of Aβ-medin hetero-aggregates did not limit Aβ and medin from recruiting additional Aβ and medin peptides.

Medin exerts several adverse effects on the cerebral vasculature. Medin is capable of augmenting neuroinflammation by inducing endothelial dysfunction and immune activation, as well as cytotoxicity [95]. In human arterioles, both medin and Aβ induce endothelial dysfunction and oxidative stress [95,119], suggesting the potential for interactions between these two amyloids. In human brain VSMCs, medin increases the expression of their genes and the secretion of IL-6, IL-8, and MCP-1 [120], which are potent mediators of inflammatory atherogenesis, synergistically facilitating the interaction between VSMCs and monocytes [121]. Likewise, in studies in which human brain VSMCs (HBVSMCs) were exposed to medin, pro-inflammatory activation of HBVSMCs was observed [122].

Notably, medin in the vasculature of the human brain is not stained by amyloid dyes, suggesting that amyloids can exist as distinct "strains" with differing affinity for amyloid dyes [123,124]. It seems, therefore, that there may be local amplification of different medin strains that form in the mouse and human vasculature, explaining the differences in amyloid dye affinity. In this regard, it has also been demonstrated that soluble amyloid-β monomers and oligomers significantly impair vascular function in AD [125–127], and it is therefore essential to identify the specific medin aggregates that contribute to vascular dysfunction in the aging brain.

Medin is also involved in the activation of microglia and astrocytes in AD, and contributes to the release of pro-inflammatory cytokines and chemokines [128]. In culture, more astrocytes and microglia phagocytose synapses were derived from AD patients than from controls, with MFG-E8 having a protective effect [129]. Similar findings were observed in VaD, in which medin in the cerebral collateral arteries and parenchymal arterioles, white matter lesion scores, and astrocyte activity were higher in VaD patients versus cognitively normal donors [130]. EC immune activation and reduced EC viability, as well as IL-8 production in the astrocytes, were enhanced after exposure to medin-treated ECs. Further, medin induces aging-associated hypercoagulability by decreasing the synthesis of the anticoagulant protein thrombomodulin by human brain microvascular endothelial cells (HBMVECs) and increasing the expression of the procoagulant proteins PAI-1 and tissue factor [131].

Usually, an amyloidosis type is related to one or two amyloid peptides or proteins [77,132]. However, in clinical practice, different amyloid peptides are often present within the same tissue or organ [133]. The formation of "seeds", the nuclei generated by protein polymerization, promotes conversion of soluble proteins to fibrils [134]. Cross-seeding results from either seeds of the same protein (homologous) or seeds of one protein (heterologous) catalyzing the fibrillation of another protein [135]. Cross-seeding between different amyloid proteins may explain the simultaneous appearance of more than one misfolded protein in the same disease, as well as the simultaneous presence of more than one protein misfolding disorder (PMD) in the same individual [136,137]. Individuals suffering from a PMD are at high risk of developing another [138,139]. Finally, multiple amyloid diseases may be present simultaneously in the same patient (e.g., AD alongside Parkinson's disease [140] or AMA alongside AD [116]), increasing disease severity and accelerating progression.

3.1.2. Potential Mechanisms Underlying Elevated Cerebral Medin

The exact mechanism(s) underlying the elevated level of cerebral medin observed in AD are virtually unknown. However, there is some evidence implicating cerebral small vessel disease (cSVD) and the consequent hyperpulsatile flow occurring within the context of increased AoSt, which is associated both with an increase in arteriolar medin synthesis and a decrease in medin clearance via the glymphatic system.

cSVD and Increased Medin Synthesis

cSVD adversely affects small cells (40–200 μm) and contributes to 45% of dementias [141]. A white matter hyperintensity (WMH) burden, the predominant cSVD feature on brain magnetic resonance imaging (MRI), is tightly linked to increased AD risk [142].

The two primary forms of cSVD are arteriolosclerosis and cerebral amyloid angiopathy (CAA). Arteriolosclerosis refers to pathological processes affecting small vessels, which are susceptible to inadequately controlled hypertension or diabetes mellitus. CAA is characterized by the accumulation of Aβ and medin in the wall of small cortical and leptomeningeal arterioles and arteries [141]. Risk factors for CAA development include aging, hypertension, and carrying specific alleles of the apolipoprotein E (ApoE) gene [143]. CAA has a long preclinical phase starting decades before symptoms emerge, and it is incidentally found in approximately 16% of asymptomatic elderly individuals during MRI [144]. In ROSMAP, which included elderly men and women who underwent brain autopsy at the time of death, atherosclerosis was present in 39% and arteriolosclerosis in 35% of participants [145]. Further, the severity of both atherosclerosis and arteriolosclerosis was associated with a significantly higher risk of AD dementia [145].

In the same project, AD was the most prevalent pathology ($\approx$65% of brains), followed by ischaemic infarcts, CAA, atherosclerosis, and arteriolosclerosis [146]. Importantly, each vascular pathology was present in >30% of brains, and when present, each was related to 20–30% of an individual's cognitive decline associated with aging [18]. It seems that the overlap between neurodegenerative and vascular pathologies may contribute to the cognitive impairment associated with aging, as evidenced by the fact that over the last 3–4 decades, the dementia incidence in developed countries has declined, in parallel with the control of cardiovascular risk factors [147].

cSVD is characterized by increased vessel stiffness and pulsatility, white matter hypoperfusion, and reduced cerebrovascular reactivity (CVR) [148–152] occurring, as previously described, predominantly in elderly patients with increased AoSt.

In the Atherosclerosis Risk in Communities (ARIC)–Neurocognitive Study (n = 320, 76 (5) years, 45% black, 27% mild cognitive impairment (MCI)), the heart-carotid (hc)PWV was directly related to Aβ deposition, whereas the carotid-femoral (cf)PWV was inversely related to brain volumes in regions susceptible to AD and associated with a WMH burden [19]. A subanalysis of the ARIC Study revealed that intracranial atherosclerotic disease revealed by high-resolution vessel wall MRI was not associated with brain Aβ deposition [153].

In the Vanderbilt Memory & Aging Project, AoSt, assessed by aortic PWV measured with cardiac magnetic resonance (CMR), was related to lower regional cerebral blood flow

(CBF) and higher cerebrovascular resistance in cognitively normal elderly individuals, and this relationship was stronger among those with a genetic predisposition for AD (APOE ε4 carriers with mild cognitive impairment) [154].

Finally, a recent meta-analysis of 37 studies demonstrated that increased AoSt was associated with cSVD [155]. In the same study, subgroup analysis of AoSt studies demonstrated an association with WMH [155].

From a mechanistic point of view, the increased AoSt observed with advancing age is associated with pulse pressure (PP) widening throughout the arterial tree, including in microvessels, which, in the brain, have a thin wall and lack the structure required to confront the increased PP and hyperpulsatile flow [156,157]. The latter may eventually lead to VSMC damage, resulting in increased amyloid deposition (Aβ and presumably medin) and microvascular dysfunction. As agitation is a risk factor for nucleation, we may assume that excessive pressure and flow pulsatility in the cerebral vessels could also provide a highly mobile environment, facilitating the formation of Aβ and medin amyloid [158].

cSVD and Decreased Medin Clearance

The glymphatic system is a waste clearance system in the brain that utilizes a network of fluid-filled perivascular spaces (PVSs, also known as Virchow–Robin spaces) surrounding blood vessels, formed by astroglial cells (Figure 5) [159]. Moreover, the glymphatic system distributes essential compounds (e.g., glucose, lipids, amino acids, and neurotransmitters) throughout the brain. The parenchymal accumulation of Aβ plaques and tau protein tangles, which characterizes AD, has been attributed to several factors, including an overall impairment in Aβ clearance [160].

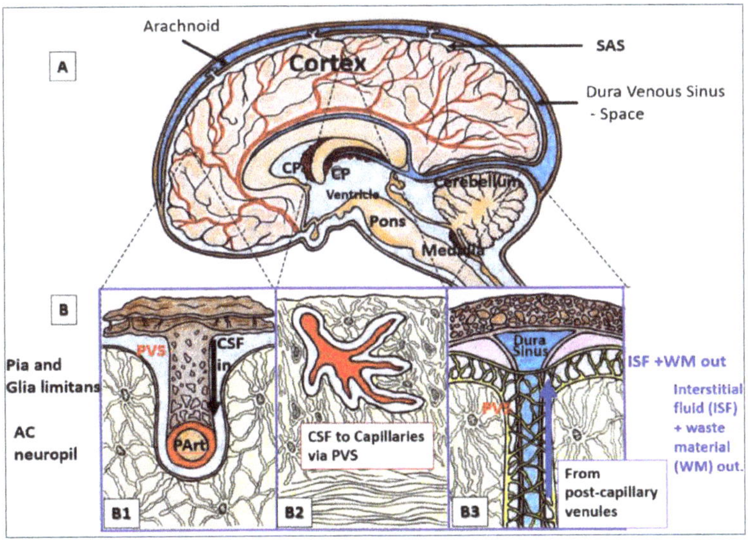

Figure 5. Brain perivascular spaces (PVSs). (**A**) Demarcations of specific brain regions (dashed lines). (**B**) The pia artery inside the subarachnoid space (SAS) penetrates deeper into the brain structure through adjacent PVSs, shown with blue coloration (**B1**), and horizontally–diagonally (**B2**). The PVSs allow the influx (black arrow) of cerebrospinal fluid (CSF) to the parenchymal interstitial fluid space (ISF) through the arteriolar PVS. Panel (**B3**) shows the efflux (blue arrow) of the interstitial fluid waste materials (WMs) of the pial venular PVS to the pial vein PVS that enters the subarachnoid space, which eventually drain into the dural venous sinus space. Adapted with permission from Ref. [161].

Glymphatic function depends on arterial pulsatility, which propels the cerebrospinal fluid (CSF) through the PVS and into the interstitial fluid (ISF) [162]. Cardiovascular disorders that impair cardiac output, such as heart failure or arrhythmias, as well as conditions that increase AoSt, impair CSF flow through the PVS [163]. In elderly hypertensive patients, glymphatic dysfunction is observed, as indicated by analysis of the perivascular space index, obtained with diffusion tensor imaging (DTI) [164]. Likewise, a decrease in glymphatic transport was observed in spontaneously hypertensive rats with cSVD [165]. Significantly, enlarged PVSs are associated with poor glymphatic clearance and cognitive performance in AD [166–168]. From a mechanistic point of view, it is reasonable to assume that increased AoSt, by generating excessive pressure and flow pulsatility, reduces glymphatic clearance of Aβ, tau protein, and, likely, medin [161,166,169]. Admittedly, however, medin clearance via the glymphatic system has not yet been convincingly demonstrated and remains under investigation.

4. Clinical Implications

Monoclonal antibodies (MABs) targeting toxic Aβ aggregates, such as oligomers, protofibrils, fibrils, and plaques (aducanumab, lecanemab, donanemab, and gantenerumab), have proved effective for the management of AD patients with Aβ pathology in the early stages of the disease [170–173]. In this regard, MABs have been shown to reduce Aβ plaques and slow cognitive decline by 30% [174]. However, MAB immunotherapy is far from curative [170,175], and is associated with an acceleration in brain volume loss and ventricular enlargement [176,177], which are objective measures of disease progression and severity of neurodegeneration in AD [178]. As the mechanisms and the long-term consequences of brain atrophy induced by anti-amyloid MABs in AD remain virtually unknown, concerns have been raised regarding the possible adverse effects of this treatment modality on the cognitive function and progression of the disease. Since anti-amyloid MABs clear away formed Aβ plaques and attenuate relevant toxicity, but do not delay plaque formation by inhibiting Aβ production, combination therapies have been proposed for AD [179].

AoSt increases moderately before 50 years of age, and markedly thereafter [180]. Although AD usually affects elderly individuals aged 60 and above [181], the accumulation of Aβ plaques usually precedes clinical manifestations by 10–20 years [182]. Thus, in many individuals, an increase in AoSt may precede or coincide with Aβ plaque and medin co-aggregation and clinical AD. Theoretically, therefore, both MFG-E8 and medin inhibition may attenuate age-induced AoSt and cognitive decline, as well as reducing the incidence of AD (Figure 6). However, MFG-E8 interferes with several biological functions, including immunomodulation, cell adhesion, angiogenesis, and tissue regeneration, which may be adversely affected by its inhibition [183]. On the other hand, medin has no known physiological functions, and its inhibition seems to be an attractive and feasible treatment modality to combat aortic stiffening and prevent AD. In addition, given its co-segregation with Aβ, targeting medin may have additional benefits in the management of established AD.

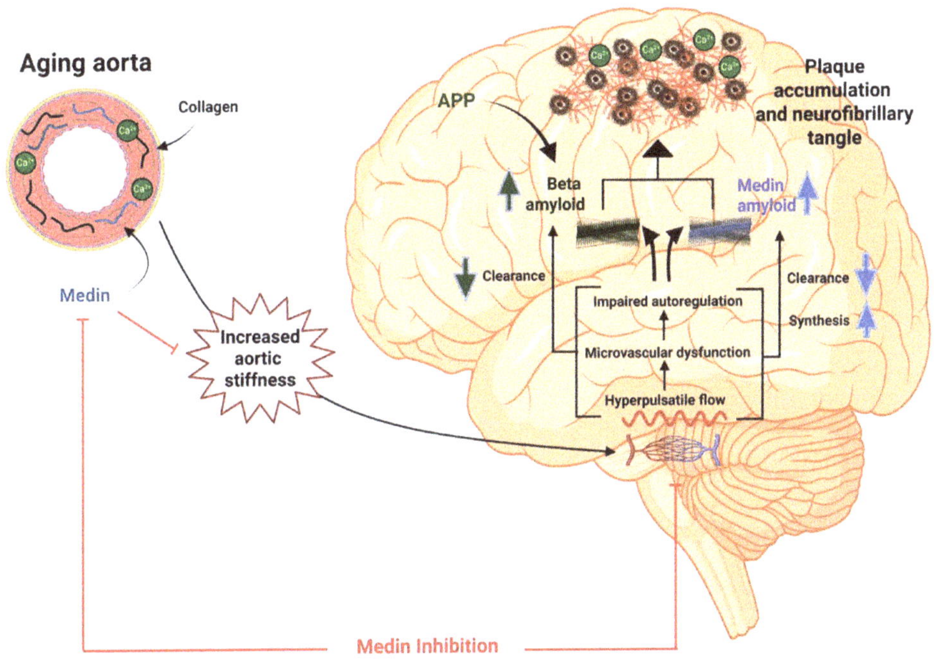

Figure 6. Hypothesis for the role of medin in linking aortic stiffness (AoSt) with Alzheimer's disease (AD) in the elderly. Medin amyloid is virtually a universal finding in the aging aorta, and contributes to an increase in AoSt. Increased AoSt is associated with an increased pulsatile load transmitted in the cerebral microvasculature, which, in turn, increases cerebral medin and medin amyloid by (a) increasing synthesis via vascular smooth muscle cell damage and agitation-enhanced nucleation, and (b) possibly decreasing glymphatic clearance. Medin synergizes with vascular Aβ deposits, contributing to an accelerated cognitive decline in AD. APP (Amyloid Precursor Protein) is a transmembrane protein that plays a crucial role in brain function and is particularly concentrated in synapses of neurons, and it is implicated in amyloid plaque formation in AD. Admittedly, however, this hypothesis, although rational, requires experimental and clinical validation. Created in BioRender.

5. Conclusions and Future Perspectives

There is converging evidence for a significant role of medin amyloid as a mediator of the age-related arterial pathology underlying both AoSt and AD. Nevertheless, crucial questions remain unanswered, such as (a) the source of medin modulating Aβ deposition leading to cognitive impairment, (b) the exact mechanism underlying clearance of medin in the brain, (c) the utility of medin as a biomarker for AD detection in the preclinical stage or the use of medin levels as a measure of AD severity, and d) whether medin aggregates increase AoSt due to physical interaction with elastin or indirectly through augmentation of inflammaging. Given the high prevalence of cardiovascular, cerebrovascular, and neurodegenerative diseases in the aging population associated with increased AoSt, the preservation of aortic function remains a major challenge in medical research. Targeting medin aggregation should therefore be investigated as a novel treatment modality to decelerate aortic stiffening and promote healthy aging.

Author Contributions: Conceptualization, F.T. and A.X.; software, F.T.; resources, F.T., A.X., H.B. and D.L.B.; data curation, F.T., A.X., H.B. and D.L.B.; writing—original draft preparation, F.T. and D.L.B.; writing—review and editing, A.X. and H.B.; visualization, F.T. and D.L.B.; supervision, F.T. All authors have read and agreed to the published version of the manuscript.

Funding: This work received no external funding.

Institutional Review Board Statement: Not applicable.

Informed Consent Statement: Not applicable.

Data Availability Statement: No new data were created or analyzed in this work. Data sharing is not applicable to this article.

Conflicts of Interest: The authors declare no conflict of interest.

References

1. O'Rourke, M.; Farnsworth, A.; O'Rourke, J. Aortic dimensions and stiffness in normal adults. *JACC Cardiovasc. Imaging* **2008**, *1*, 749–751. [CrossRef]
2. Kim, H.L. Arterial stiffness and hypertension. *Clin. Hypertens.* **2023**, *29*, 31. [CrossRef] [PubMed]
3. Mogi, M. Hypertension management to prevent dementia. *Hypertens. Res.* **2022**, *45*, 573–575. [CrossRef] [PubMed]
4. Triposkiadis, F.; Xanthopoulos, A.; Butler, J. Cardiovascular Aging and Heart Failure: JACC Review Topic of the Week. *J. Am. Coll. Cardiol.* **2019**, *74*, 804–813. [CrossRef] [PubMed]
5. Wolters, F.J.; Segufa, R.A.; Darweesh, S.K.L.; Bos, D.; Ikram, M.A.; Sabayan, B.; Hofman, A.; Sedaghat, S. Coronary heart disease, heart failure, and the risk of dementia: A systematic review and meta-analysis. *Alzheimer's Dement.* **2018**, *14*, 1493–1504. [CrossRef]
6. Afsar, B.; Sag, A.A.; Yalcin, C.E.; Kaya, E.; Siriopol, D.; Goldsmith, D.; Covic, A.; Kanbay, M. Brain-kidney cross-talk: Definition and emerging evidence. *Eur. J. Intern. Med.* **2016**, *36*, 7–12. [CrossRef]
7. Drew, D.A.; Weiner, D.E.; Sarnak, M.J. Cognitive Impairment in CKD: Pathophysiology, Management, and Prevention. *Am. J. Kidney Dis.* **2019**, *74*, 782–790. [CrossRef]
8. Triposkiadis, F.; Xanthopoulos, A.; Lampropoulos, K.; Briasoulis, A.; Sarafidis, P.; Skoularigis, J.; Boudoulas, H. Aortic Stiffness: A Major Risk Factor for Multimorbidity in the Elderly. *J. Clin. Med.* **2023**, *12*, 2321. [CrossRef]
9. Blumenthal, H.T. Amyloidosis: A universal disease of aging? *J. Gerontol. A Biol. Sci. Med. Sci.* **2004**, *59*, 361–369. [CrossRef]
10. Mucchiano, G.; Cornwell, G.G., 3rd; Westermark, P. Senile aortic amyloid. Evidence for two distinct forms of localized deposits. *Am. J. Pathol.* **1992**, *140*, 871–877.
11. Haggqvist, B.; Naslund, J.; Sletten, K.; Westermark, G.T.; Mucchiano, G.; Tjernberg, L.O.; Nordstedt, C.; Engstrom, U.; Westermark, P. Medin: An integral fragment of aortic smooth muscle cell-produced lactadherin forms the most common human amyloid. *Proc. Natl. Acad. Sci. USA* **1999**, *96*, 8669–8674. [CrossRef]
12. Davies, H.A.; Madine, J.; Middleton, D.A. Comparisons with amyloid-beta reveal an aspartate residue that stabilizes fibrils of the aortic amyloid peptide medin. *J. Biol. Chem.* **2015**, *290*, 7791–7803. [CrossRef]
13. Davies, H.A.; Rigden, D.J.; Phelan, M.M.; Madine, J. Probing Medin Monomer Structure and its Amyloid Nucleation Using (13)C-Direct Detection NMR in Combination with Structural Bioinformatics. *Sci. Rep.* **2017**, *7*, 45224. [CrossRef]
14. 2023 Alzheimer's disease facts and figures. *Alzheimer's Dement.* **2023**, *19*, 1598–1695. [CrossRef]
15. Gilbert, M.A.G.; Fatima, N.; Jenkins, J.; O'Sullivan, T.J.; Schertel, A.; Halfon, Y.; Wilkinson, M.; Morrema, T.H.J.; Geibel, M.; Read, R.J.; et al. CryoET of beta-amyloid and tau within postmortem Alzheimer's disease brain. *Nature* **2024**, *631*, 913–919. [CrossRef]
16. Walker, L.C. Abeta Plaques. *Free Neuropathol.* **2020**, *1*, 31. [CrossRef] [PubMed]
17. Orr, M.E.; Sullivan, A.C.; Frost, B. A Brief Overview of Tauopathy: Causes, Consequences, and Therapeutic Strategies. *Trends Pharmacol. Sci.* **2017**, *38*, 637–648. [CrossRef] [PubMed]
18. Greenberg, S.M.; Bacskai, B.J.; Hernandez-Guillamon, M.; Pruzin, J.; Sperling, R.; van Veluw, S.J. Cerebral amyloid angiopathy and Alzheimer disease—One peptide, two pathways. *Nat. Rev. Neurol.* **2020**, *16*, 30–42. [CrossRef] [PubMed]
19. Hughes, T.M.; Wagenknecht, L.E.; Craft, S.; Mintz, A.; Heiss, G.; Palta, P.; Wong, D.; Zhou, Y.; Knopman, D.; Mosley, T.H.; et al. Arterial stiffness and dementia pathology: Atherosclerosis Risk in Communities (ARIC)-PET Study. *Neurology* **2018**, *90*, e1248–e1256. [CrossRef]
20. Kim, H.; Kulshreshtha, A.; Alonso, A.; Goldstein, F.C.; Johnson, E.C.; Gold, M.E.; Quyyumi, A.A.; Lah, J.J. The association between pulse wave velocity and cerebrospinal fluid biomarkers for Alzheimer's disease. *J. Alzheimer's Dis.* **2025**, *105*, 729–735. [CrossRef]

21. Li, J.; Ren, Y.; Wang, L.; Zou, X.; Ding, X.; Hou, T.; Zhang, Q.; Tang, S.; Han, X.; Song, L.; et al. Estimated pulse wave velocity associated with cognitive phenotypes in a rural older population in China: A cohort study. *Alzheimer's Dement.* **2025**, *21*, e14491. [CrossRef]
22. Hughes, T.M.; Kuller, L.H.; Barinas-Mitchell, E.J.; Mackey, R.H.; McDade, E.M.; Klunk, W.E.; Aizenstein, H.J.; Cohen, A.D.; Snitz, B.E.; Mathis, C.A.; et al. Pulse wave velocity is associated with beta-amyloid deposition in the brains of very elderly adults. *Neurology* **2013**, *81*, 1711–1718. [CrossRef]
23. Heffernan, K.S.; Wilmoth, J.M.; London, A.S. Estimated Pulse Wave Velocity Is Associated With a Higher Risk of Dementia in the Health and Retirement Study. *Am. J. Hypertens.* **2024**, *37*, 909–915. [CrossRef]
24. Conlon, N.M. Stiff arteries, silent brain changes; a new diagnostic tool in Alzheimer's disease pathology. *J. Hypertens.* **2025**, *43*, 78–79. [CrossRef] [PubMed]
25. Collins, J.A.; Munoz, J.V.; Patel, T.R.; Loukas, M.; Tubbs, R.S. The anatomy of the aging aorta. *Clin. Anat.* **2014**, *27*, 463–466. [CrossRef] [PubMed]
26. Aird, W.C. Endothelial cell heterogeneity. *Cold Spring Harb. Perspect. Med.* **2012**, *2*, a006429. [CrossRef] [PubMed]
27. Segers, V.F.M.; Brutsaert, D.L.; De Keulenaer, G.W. Cardiac Remodeling: Endothelial Cells Have More to Say Than Just NO. *Front. Physiol.* **2018**, *9*, 382. [CrossRef]
28. Brutsaert, D.L. The indispensable role of cardiac endothelium in the structure and function of the heart. *Verh. K. Acad. Geneeskd. Belg.* **2003**, *65*, 75–116.
29. Brutsaert, D.L. Cardiac endothelial-myocardial signaling: Its role in cardiac growth, contractile performance, and rhythmicity. *Physiol. Rev.* **2003**, *83*, 59–115. [CrossRef]
30. Rafii, S.; Butler, J.M.; Ding, B.S. Angiocrine functions of organ-specific endothelial cells. *Nature* **2016**, *529*, 316–325. [CrossRef]
31. Mathieu, M.; Martin-Jaular, L.; Lavieu, G.; Thery, C. Specificities of secretion and uptake of exosomes and other extracellular vesicles for cell-to-cell communication. *Nat. Cell Biol.* **2019**, *21*, 9–17. [CrossRef]
32. Kruger-Genge, A.; Blocki, A.; Franke, R.P.; Jung, F. Vascular Endothelial Cell Biology: An Update. *Int. J. Mol. Sci.* **2019**, *20*, 4411. [CrossRef]
33. Wagenseil, J.E.; Mecham, R.P. Vascular extracellular matrix and arterial mechanics. *Physiol. Rev.* **2009**, *89*, 957–989. [CrossRef]
34. Lin, P.K.; Davis, G.E. Extracellular Matrix Remodeling in Vascular Disease: Defining Its Regulators and Pathological Influence. *Arter. Thromb. Vasc. Biol.* **2023**, *43*, 1599–1616. [CrossRef]
35. Emmott, A.; Garcia, J.; Chung, J.; Lachapelle, K.; El-Hamamsy, I.; Mongrain, R.; Cartier, R.; Leask, R.L. Biomechanics of the Ascending Thoracic Aorta: A Clinical Perspective on Engineering Data. *Can. J. Cardiol.* **2016**, *32*, 35–47. [CrossRef]
36. Mitchell, G.F.; van Buchem, M.A.; Sigurdsson, S.; Gotal, J.D.; Jonsdottir, M.K.; Kjartansson, O.; Garcia, M.; Aspelund, T.; Harris, T.B.; Gudnason, V.; et al. Arterial stiffness, pressure and flow pulsatility and brain structure and function: The Age, Gene/Environment Susceptibility--Reykjavik study. *Brain* **2011**, *134*, 3398–3407. [CrossRef]
37. Mitchell, G.F. Effects of central arterial aging on the structure and function of the peripheral vasculature: Implications for end-organ damage. *J. Appl. Physiol.* **2008**, *105*, 1652–1660. [CrossRef]
38. de Roos, A.; van der Grond, J.; Mitchell, G.; Westenberg, J. Magnetic Resonance Imaging of Cardiovascular Function and the Brain: Is Dementia a Cardiovascular-Driven Disease? *Circulation* **2017**, *135*, 2178–2195. [CrossRef] [PubMed]
39. Stone, J.; Johnstone, D.M.; Mitrofanis, J.; O'Rourke, M. The mechanical cause of age-related dementia (Alzheimer's disease): The brain is destroyed by the pulse. *J. Alzheimer's Dis.* **2015**, *44*, 355–373. [CrossRef] [PubMed]
40. Chirinos, J.A. Large Artery Stiffness, Microvascular Function, and Cardiovascular Risk. *Circ. Cardiovasc. Imaging* **2016**, *9*, e005903. [CrossRef] [PubMed]
41. Chirinos, J.A.; Segers, P.; Hughes, T.; Townsend, R. Large-Artery Stiffness in Health and Disease: JACC State-of-the-Art Review. *J. Am. Coll. Cardiol.* **2019**, *74*, 1237–1263. [CrossRef] [PubMed]
42. Aghilinejad, A.; Amlani, F.; Mazandarani, S.P.; King, K.S.; Pahlevan, N.M. Mechanistic insights on age-related changes in heart-aorta-brain hemodynamic coupling using a pulse wave model of the entire circulatory system. *Am. J. Physiol. Heart Circ. Physiol.* **2023**, *325*, H1193–H1209. [CrossRef]
43. Mammoto, A.; Matus, K.; Mammoto, T. Extracellular Matrix in Aging Aorta. *Front. Cell Dev. Biol.* **2022**, *10*, 822561. [CrossRef] [PubMed]
44. Peng, S.; Westermark, G.T.; Naslund, J.; Haggqvist, B.; Glennert, J.; Westermark, P. Medin and medin-amyloid in ageing inflamed and non-inflamed temporal arteries. *J. Pathol.* **2002**, *196*, 91–96. [CrossRef] [PubMed]
45. Wu, M.; Rementer, C.; Giachelli, C.M. Vascular calcification: An update on mechanisms and challenges in treatment. *Calcif. Tissue Int.* **2013**, *93*, 365–373. [CrossRef]
46. Wang, B.; Han, J.; Elisseeff, J.H.; Demaria, M. The senescence-associated secretory phenotype and its physiological and pathological implications. *Nat. Rev. Mol. Cell Biol.* **2024**, *25*, 958–978. [CrossRef]
47. Jia, G.; Aroor, A.R.; Jia, C.; Sowers, J.R. Endothelial cell senescence in aging-related vascular dysfunction. *Biochim. Biophys. Acta Mol. Basis Dis.* **2019**, *1865*, 1802–1809. [CrossRef]

48. Hwang, H.J.; Kim, N.; Herman, A.B.; Gorospe, M.; Lee, J.S. Factors and Pathways Modulating Endothelial Cell Senescence in Vascular Aging. *Int. J. Mol. Sci.* **2022**, *23*, 10135. [CrossRef]
49. Mehdizadeh, M.; Aguilar, M.; Thorin, E.; Ferbeyre, G.; Nattel, S. The role of cellular senescence in cardiac disease: Basic biology and clinical relevance. *Nat. Rev. Cardiol.* **2022**, *19*, 250–264. [CrossRef]
50. Franceschi, C.; Garagnani, P.; Parini, P.; Giuliani, C.; Santoro, A. Inflammaging: A new immune-metabolic viewpoint for age-related diseases. *Nat. Rev. Endocrinol.* **2018**, *14*, 576–590. [CrossRef]
51. Heinz, A. Elastic fibers during aging and disease. *Ageing Res. Rev.* **2021**, *66*, 101255. [CrossRef] [PubMed]
52. Tsamis, A.; Krawiec, J.T.; Vorp, D.A. Elastin and collagen fibre microstructure of the human aorta in ageing and disease: A review. *J. R. Soc. Interface* **2013**, *10*, 20121004. [CrossRef]
53. Angoff, R.; Mosarla, R.C.; Tsao, C.W. Aortic Stiffness: Epidemiology, Risk Factors, and Relevant Biomarkers. *Front. Cardiovasc. Med.* **2021**, *8*, 709396. [CrossRef] [PubMed]
54. Mas-Bargues, C.; Borras, C.; Alique, M. The Contribution of Extracellular Vesicles From Senescent Endothelial and Vascular Smooth Muscle Cells to Vascular Calcification. *Front. Cardiovasc. Med.* **2022**, *9*, 854726. [CrossRef] [PubMed]
55. Jellinger, K.A. Alzheimer disease and cerebrovascular pathology: An update. *J. Neural. Transm.* **2002**, *109*, 813–836. [CrossRef]
56. Itoh, Y.; Yamada, M.; Hayakawa, M.; Otomo, E.; Miyatake, T. Cerebral amyloid angiopathy: A significant cause of cerebellar as well as lobar cerebral hemorrhage in the elderly. *J. Neurol. Sci.* **1993**, *116*, 135–141. [CrossRef]
57. Chen, Y.; Zhao, X.; Wu, H. Arterial Stiffness: A Focus on Vascular Calcification and Its Link to Bone Mineralization. *Arter. Arterioscler. Thromb. Vasc. Biol.* **2020**, *40*, 1078–1093. [CrossRef]
58. Kapustin, A.N.; Shanahan, C.M. Emerging roles for vascular smooth muscle cell exosomes in calcification and coagulation. *J. Physiol.* **2016**, *594*, 2905–2914. [CrossRef]
59. Park, J.B.; Avolio, A. Arteriosclerosis and Atherosclerosis Assessment in Clinical Practice: Methods and Significance. *Pulse* **2023**, *11*, 1–8. [CrossRef]
60. Lin, X.; Li, S.; Wang, Y.J.; Wang, Y.; Zhong, J.Y.; He, J.Y.; Cui, X.J.; Zhan, J.K.; Liu, Y.S. Exosomal Notch3 from high glucose-stimulated endothelial cells regulates vascular smooth muscle cells calcification/aging. *Life Sci.* **2019**, *232*, 116582. [CrossRef]
61. Li, S.; Zhan, J.K.; Wang, Y.J.; Lin, X.; Zhong, J.Y.; Wang, Y.; Tan, P.; He, J.Y.; Cui, X.J.; Chen, Y.Y.; et al. Exosomes from hyperglycemia-stimulated vascular endothelial cells contain versican that regulate calcification/senescence in vascular smooth muscle cells. *Cell Biosci.* **2019**, *9*, 1. [CrossRef]
62. Alique, M.; Ruiz-Torres, M.P.; Bodega, G.; Noci, M.V.; Troyano, N.; Bohorquez, L.; Luna, C.; Luque, R.; Carmona, A.; Carracedo, J.; et al. Microvesicles from the plasma of elderly subjects and from senescent endothelial cells promote vascular calcification. *Aging* **2017**, *9*, 778–789. [CrossRef]
63. Buendia, P.; Montes de Oca, A.; Madueno, J.A.; Merino, A.; Martin-Malo, A.; Aljama, P.; Ramirez, R.; Rodriguez, M.; Carracedo, J. Endothelial microparticles mediate inflammation-induced vascular calcification. *FASEB J.* **2015**, *29*, 173–181. [CrossRef]
64. Henze, L.A.; Luong, T.T.D.; Boehme, B.; Masyout, J.; Schneider, M.P.; Brachs, S.; Lang, F.; Pieske, B.; Pasch, A.; Eckardt, K.U.; et al. Impact of C-reactive protein on osteo-/chondrogenic transdifferentiation and calcification of vascular smooth muscle cells. *Aging* **2019**, *11*, 5445–5462. [CrossRef]
65. Meerman, M.; Driessen, R.; van Engeland, N.C.A.; Bergsma, I.; Steenhuijsen, J.L.G.; Kozono, D.; Aikawa, E.; Hjortnaes, J.; Bouten, C.V.C. Radiation Induces Valvular Interstitial Cell Calcific Response in an in vitro Model of Calcific Aortic Valve Disease. *Front. Cardiovasc. Med.* **2021**, *8*, 687885. [CrossRef]
66. Hoeijmakers, J.H. DNA damage, aging, and cancer. *N. Engl. J. Med.* **2009**, *361*, 1475–1485. [CrossRef] [PubMed]
67. Chellan, B.; Sutton, N.R.; Hofmann Bowman, M.A. S100/RAGE-Mediated Inflammation and Modified Cholesterol Lipoproteins as Mediators of Osteoblastic Differentiation of Vascular Smooth Muscle Cells. *Front. Cardiovasc. Med.* **2018**, *5*, 163. [CrossRef] [PubMed]
68. Degenhardt, K.; Wagner, J.; Skodras, A.; Candlish, M.; Koppelmann, A.J.; Wild, K.; Maxwell, R.; Rotermund, C.; von Zweydorf, F.; Gloeckner, C.J.; et al. Medin aggregation causes cerebrovascular dysfunction in aging wild-type mice. *Proc. Natl. Acad. Sci. USA* **2020**, *117*, 23925–23931. [CrossRef] [PubMed]
69. Wang, M.; McGraw, K.R.; Monticone, R.E. Milk Fat Globule Epidermal Growth Factor VIII Fragment Medin in Age-Associated Arterial Adverse Remodeling and Arterial Disease. *Cells* **2023**, *12*, 253. [CrossRef]
70. Wang, M.; Jiang, L.; Monticone, R.E.; Lakatta, E.G. Proinflammation: The key to arterial aging. *Trends Endocrinol. Metab.* **2014**, *25*, 72–79. [CrossRef]
71. Uhlen, M.; Fagerberg, L.; Hallstrom, B.M.; Lindskog, C.; Oksvold, P.; Mardinoglu, A.; Sivertsson, A.; Kampf, C.; Sjostedt, E.; Asplund, A.; et al. Proteomics. Tissue-based map of the human proteome. *Science* **2015**, *347*, 1260419. [CrossRef] [PubMed]
72. Whitehead, M.; Yusoff, S.; Ahmad, S.; Schmidt, L.; Mayr, M.; Madine, J.; Middleton, D.; Shanahan, C.M. Vascular smooth muscle cell senescence accelerates medin aggregation via small extracellular vesicle secretion and extracellular matrix reorganization. *Aging Cell* **2023**, *22*, e13746. [CrossRef] [PubMed]

73. Larsson, A.; Malmstrom, S.; Westermark, P. Signs of cross-seeding: Aortic medin amyloid as a trigger for protein AA deposition. *Amyloid* **2011**, *18*, 229–234. [CrossRef] [PubMed]
74. Sack, G.H., Jr. Serum Amyloid A (SAA) Proteins. *Subcell. Biochem.* **2020**, *94*, 421–436. [CrossRef]
75. Wagner, J.; Degenhardt, K.; Veit, M.; Louros, N.; Konstantoulea, K.; Skodras, A.; Wild, K.; Liu, P.; Obermuller, U.; Bansal, V.; et al. Medin co-aggregates with vascular amyloid-beta in Alzheimer's disease. *Nature* **2022**, *612*, 123–131. [CrossRef]
76. Younger, S.; Jang, H.; Davies, H.A.; Niemiec, M.J.; Garcia, J.G.N.; Nussinov, R.; Migrino, R.Q.; Madine, J.; Arce, F.T. Medin Oligomer Membrane Pore Formation: A Potential Mechanism of Vascular Dysfunction. *Biophys. J.* **2020**, *118*, 2769–2782. [CrossRef]
77. Ke, P.C.; Sani, M.A.; Ding, F.; Kakinen, A.; Javed, I.; Separovic, F.; Davis, T.P.; Mezzenga, R. Implications of peptide assemblies in amyloid diseases. *Chem. Soc. Rev.* **2017**, *46*, 6492–6531. [CrossRef]
78. Sun, Y.; Kakinen, A.; Wan, X.; Moriarty, N.; Hunt, C.P.J.; Li, Y.; Andrikopoulos, N.; Nandakumar, A.; Davis, T.P.; Parish, C.L.; et al. Spontaneous Formation of beta-sheet Nano-barrels during the Early Aggregation of Alzheimer's Amyloid Beta. *Nano Today* **2021**, *38*, 101125. [CrossRef]
79. Olofsson, A.; Borowik, T.; Grobner, G.; Sauer-Eriksson, A.E. Negatively charged phospholipid membranes induce amyloid formation of medin via an alpha-helical intermediate. *J. Mol. Biol.* **2007**, *374*, 186–194. [CrossRef]
80. Davies, H.A.; Phelan, M.M.; Wilkinson, M.C.; Migrino, R.Q.; Truran, S.; Franco, D.A.; Liu, L.N.; Longmore, C.J.; Madine, J. Oxidative Stress Alters the Morphology and Toxicity of Aortic Medial Amyloid. *Biophys. J.* **2015**, *109*, 2363–2370. [CrossRef]
81. Larsson, A.; Peng, S.; Persson, H.; Rosenbloom, J.; Abrams, W.R.; Wassberg, E.; Thelin, S.; Sletten, K.; Gerwins, P.; Westermark, P. Lactadherin binds to elastin--a starting point for medin amyloid formation? *Amyloid* **2006**, *13*, 78–85. [CrossRef] [PubMed]
82. Huang, F.; Yan, J.; Zhang, X.; Xu, H.; Lian, J.; Yang, X.; Wang, C.; Ding, F.; Sun, Y. Computational insights into the aggregation mechanism and amyloidogenic core of aortic amyloid medin polypeptide. *Colloids Surf. B Biointerfaces* **2024**, *244*, 114192. [CrossRef] [PubMed]
83. Jackson, J.W.; Foster, J.S.; Martin, E.B.; Macy, S.; Wooliver, C.; Balachandran, M.; Richey, T.; Heidel, R.E.; Williams, A.D.; Kennel, S.J.; et al. Collagen inhibits phagocytosis of amyloid in vitro and in vivo and may act as a 'don't eat me' signal. *Amyloid* **2023**, *30*, 249–260. [CrossRef] [PubMed]
84. Schulte, T.; Chaves-Sanjuan, A.; Speranzini, V.; Sicking, K.; Milazzo, M.; Mazzini, G.; Rognoni, P.; Caminito, S.; Milani, P.; Marabelli, C.; et al. Helical superstructures between amyloid and collagen in cardiac fibrils from a patient with AL amyloidosis. *Nat. Commun.* **2024**, *15*, 6359. [CrossRef]
85. Sipe, J.D.; Cohen, A.S. Review: History of the amyloid fibril. *J. Struct. Biol.* **2000**, *130*, 88–98. [CrossRef]
86. Cline, E.N.; Bicca, M.A.; Viola, K.L.; Klein, W.L. The Amyloid-beta Oligomer Hypothesis: Beginning of the Third Decade. *J. Alzheimer's Dis.* **2018**, *64*, S567–S610. [CrossRef]
87. Williams, T.L.; Serpell, L.C. Membrane and surface interactions of Alzheimer's Abeta peptide--insights into the mechanism of cytotoxicity. *FEBS J.* **2011**, *278*, 3905–3917. [CrossRef]
88. Weibel, E.R.; Taylor, C.R.; Hoppeler, H. The concept of symmorphosis: A testable hypothesis of structure-function relationship. *Proc. Natl. Acad. Sci. USA* **1991**, *88*, 10357–10361. [CrossRef]
89. Lashuel, H.A.; Hartley, D.; Petre, B.M.; Walz, T.; Lansbury, P.T., Jr. Neurodegenerative disease: Amyloid pores from pathogenic mutations. *Nature* **2002**, *418*, 291. [CrossRef]
90. Chen, S.W.; Drakulic, S.; Deas, E.; Ouberai, M.; Aprile, F.A.; Arranz, R.; Ness, S.; Roodveldt, C.; Guilliams, T.; De-Genst, E.J.; et al. Structural characterization of toxic oligomers that are kinetically trapped during alpha-synuclein fibril formation. *Proc. Natl. Acad. Sci. USA* **2015**, *112*, E1994–E2003. [CrossRef]
91. Tang, H.; Andrikopoulos, N.; Li, Y.; Ke, S.; Sun, Y.; Ding, F.; Ke, P.C. Emerging biophysical origins and pathogenic implications of amyloid oligomers. *Nat. Commun.* **2025**, *16*, 2937. [CrossRef] [PubMed]
92. Stockl, M.T.; Zijlstra, N.; Subramaniam, V. alpha-Synuclein oligomers: An amyloid pore? Insights into mechanisms of alpha-synuclein oligomer-lipid interactions. *Mol. Neurobiol.* **2013**, *47*, 613–621. [CrossRef] [PubMed]
93. Sulatskaya, A.I.; Kosolapova, A.O.; Bobylev, A.G.; Belousov, M.V.; Antonets, K.S.; Sulatsky, M.I.; Kuznetsova, I.M.; Turoverov, K.K.; Stepanenko, O.V.; Nizhnikov, A.A. beta-Barrels and Amyloids: Structural Transitions, Biological Functions, and Pathogenesis. *Int. J. Mol. Sci.* **2021**, *22*, 11316. [CrossRef]
94. Uchiyama, A.; Yamada, K.; Ogino, S.; Yokoyama, Y.; Takeuchi, Y.; Udey, M.C.; Ishikawa, O.; Motegi, S. MFG-E8 regulates angiogenesis in cutaneous wound healing. *Am. J. Pathol.* **2014**, *184*, 1981–1990. [CrossRef] [PubMed]
95. Migrino, R.Q.; Davies, H.A.; Truran, S.; Karamanova, N.; Franco, D.A.; Beach, T.G.; Serrano, G.E.; Truong, D.; Nikkhah, M.; Madine, J. Amyloidogenic medin induces endothelial dysfunction and vascular inflammation through the receptor for advanced glycation endproducts. *Cardiovasc. Res.* **2017**, *113*, 1389–1402. [CrossRef]
96. Saki, G.; Eidi, A.; Mortazavi, P.; Panahi, N.; Vahdati, A. Effect of beta-asarone in normal and beta-amyloid-induced Alzheimeric rats. *Arch. Med. Sci.* **2020**, *16*, 699–706. [CrossRef]
97. Madine, J.; Middleton, D.A. Comparison of aggregation enhancement and inhibition as strategies for reducing the cytotoxicity of the aortic amyloid polypeptide medin. *Eur. Biophys. J.* **2010**, *39*, 1281–1288. [CrossRef]

98. Gong, S.; Li, Y.; Yan, K.; Shi, Z.; Leng, J.; Bao, Y.; Ning, K. The Crosstalk Between Endothelial Cells, Smooth Muscle Cells, and Macrophages in Atherosclerosis. *Int. J. Mol. Sci.* **2025**, *26*, 1457. [CrossRef]
99. Lutter, S.; Xie, S.; Tatin, F.; Makinen, T. Smooth muscle-endothelial cell communication activates Reelin signaling and regulates lymphatic vessel formation. *J. Cell Biol.* **2012**, *197*, 837–849. [CrossRef]
100. Davis, G.E.; Senger, D.R. Endothelial extracellular matrix: Biosynthesis, remodeling, and functions during vascular morphogenesis and neovessel stabilization. *Circ. Res.* **2005**, *97*, 1093–1107. [CrossRef]
101. Peng, S.; Larsson, A.; Wassberg, E.; Gerwins, P.; Thelin, S.; Fu, X.; Westermark, P. Role of aggregated medin in the pathogenesis of thoracic aortic aneurysm and dissection. *Lab. Invest.* **2007**, *87*, 1195–1205. [CrossRef] [PubMed]
102. Davies, H.A.; Caamano-Gutierrez, E.; Chim, Y.H.; Field, M.; Nawaytou, O.; Ressel, L.; Akhtar, R.; Madine, J. Idiopathic degenerative thoracic aneurysms are associated with increased aortic medial amyloid. *Amyloid* **2019**, *26*, 148–155. [CrossRef] [PubMed]
103. Safar, M.E. Arterial aging--hemodynamic changes and therapeutic options. *Nat. Rev. Cardiol.* **2010**, *7*, 442–449. [CrossRef] [PubMed]
104. Chiang, H.Y.; Chu, P.H.; Lee, T.H. MFG-E8 mediates arterial aging by promoting the proinflammatory phenotype of vascular smooth muscle cells. *J. Biomed. Sci.* **2019**, *26*, 61. [CrossRef]
105. Wang, M.; Wang, H.H.; Lakatta, E.G. Milk fat globule epidermal growth factor VIII signaling in arterial wall remodeling. *Curr. Vasc. Pharmacol.* **2013**, *11*, 768–776. [CrossRef]
106. Maroules, C.D.; Khera, A.; Ayers, C.; Goel, A.; Peshock, R.M.; Abbara, S.; King, K.S. Cardiovascular outcome associations among cardiovascular magnetic resonance measures of arterial stiffness: The Dallas heart study. *J. Cardiovasc. Magn. Reson.* **2014**, *16*, 33. [CrossRef]
107. King, K.S.; Sheng, M.; Liu, P.; Maroules, C.D.; Rubin, C.D.; Peshock, R.M.; McColl, R.W.; Lu, H. Detrimental effect of systemic vascular risk factors on brain hemodynamic function assessed with MRI. *Neuroradiol. J.* **2018**, *31*, 253–261. [CrossRef]
108. Goel, A.; Maroules, C.D.; Mitchell, G.F.; Peshock, R.; Ayers, C.; McColl, R.; Vongpatanasin, W.; King, K.S. Ethnic Difference in Proximal Aortic Stiffness: An Observation From the Dallas Heart Study. *JACC Cardiovasc. Imaging* **2017**, *10*, 54–61. [CrossRef]
109. Mitchell, G.F.; Tardif, J.C.; Arnold, J.M.; Marchiori, G.; O'Brien, T.X.; Dunlap, M.E.; Pfeffer, M.A. Pulsatile hemodynamics in congestive heart failure. *Hypertension* **2001**, *38*, 1433–1439. [CrossRef]
110. Pahlevan, N.M.; Gharib, M. Aortic wave dynamics and its influence on left ventricular workload. *PLoS ONE* **2011**, *6*, e23106. [CrossRef]
111. Maillard, P.; Mitchell, G.F.; Himali, J.J.; Beiser, A.; Tsao, C.W.; Pase, M.P.; Satizabal, C.L.; Vasan, R.S.; Seshadri, S.; DeCarli, C. Effects of Arterial Stiffness on Brain Integrity in Young Adults From the Framingham Heart Study. *Stroke* **2016**, *47*, 1030–1036. [CrossRef]
112. van Sloten, T.T.; Protogerou, A.D.; Henry, R.M.; Schram, M.T.; Launer, L.J.; Stehouwer, C.D. Association between arterial stiffness, cerebral small vessel disease and cognitive impairment: A systematic review and meta-analysis. *Neurosci. Biobehav. Rev.* **2015**, *53*, 121–130. [CrossRef] [PubMed]
113. King, K.S.; Chen, K.X.; Hulsey, K.M.; McColl, R.W.; Weiner, M.F.; Nakonezny, P.A.; Peshock, R.M. White matter hyperintensities: Use of aortic arch pulse wave velocity to predict volume independent of other cardiovascular risk factors. *Radiology* **2013**, *267*, 709–717. [CrossRef] [PubMed]
114. Corriveau, R.A.; Bosetti, F.; Emr, M.; Gladman, J.T.; Koenig, J.I.; Moy, C.S.; Pahigiannis, K.; Waddy, S.P.; Koroshetz, W. The Science of Vascular Contributions to Cognitive Impairment and Dementia (VCID): A Framework for Advancing Research Priorities in the Cerebrovascular Biology of Cognitive Decline. *Cell Mol. Neurobiol.* **2016**, *36*, 281–288. [CrossRef] [PubMed]
115. Madine, J.; Davies, H.A.; Migrino, R.Q.; Ruotsalainen, S.E.; Wagner, J.; Neher, J.J. Medin amyloid may drive arterial aging and disease in the periphery and brain. *Nat. Aging* **2023**, *3*, 1039–1041. [CrossRef]
116. Migrino, R.Q.; Karamanova, N.; Truran, S.; Serrano, G.E.; Davies, H.A.; Madine, J.; Beach, T.G. Cerebrovascular medin is associated with Alzheimer's disease and vascular dementia. *Alzheimer's Dement.* **2020**, *12*, e12072. [CrossRef]
117. Ge, X.; Li, L.; Xie, C. Medin synergized with vascular amyloid-beta deposits accelerates cognitive decline in Alzheimer's disease: A potential biomarker. *Neural Regen. Res.* **2024**, *19*, 1414. [CrossRef]
118. Huang, F.; Fan, X.; Wang, Y.; Zou, Y.; Lian, J.; Wang, C.; Ding, F.; Sun, Y. Computational insights into the cross-talk between medin and Abeta: Implications for age-related vascular risk factors in Alzheimer's disease. *Brief. Bioinform.* **2024**, *25*, bbad526. [CrossRef]
119. Migrino, R.Q.; Truran, S.; Karamanova, N.; Serrano, G.E.; Madrigal, C.; Davies, H.A.; Madine, J.; Reaven, P.; Beach, T.G. Human cerebral collateral arteriole function in subjects with normal cognition, mild cognitive impairment, and dementia. *Am. J. Physiol. Heart Circ. Physiol.* **2018**, *315*, H284–H290. [CrossRef]
120. Zhang, Y.; Karamanova, N.; Morrow, K.T.; Madine, J.; Truran, S.; Lozoya, M.; Weissig, V.; Li, M.; Nikkhah, M.; Park, J.G.; et al. Transcriptomic analyses reveal proinflammatory activation of human brain microvascular endothelial cells by aging-associated peptide medin and reversal by nanoliposomes. *Sci. Rep.* **2023**, *13*, 18802. [CrossRef]

121. Chen, L.; Frister, A.; Wang, S.; Ludwig, A.; Behr, H.; Pippig, S.; Li, B.; Simm, A.; Hofmann, B.; Pilowski, C.; et al. Interaction of vascular smooth muscle cells and monocytes by soluble factors synergistically enhances IL-6 and MCP-1 production. *Am. J. Physiol. Heart Circ. Physiol.* **2009**, *296*, H987–H996. [CrossRef] [PubMed]
122. Karamanova, N.; Morrow, K.T.; Maerivoet, A.; Madine, J.; Li, M.; Migrino, R.Q. Medin induces pro-inflammatory activation of human brain vascular smooth muscle cells. *Physiol. Rep.* **2025**, *13*, e70418. [CrossRef] [PubMed]
123. Condello, C.; Lemmin, T.; Stohr, J.; Nick, M.; Wu, Y.; Maxwell, A.M.; Watts, J.C.; Caro, C.D.; Oehler, A.; Keene, C.D.; et al. Structural heterogeneity and intersubject variability of Abeta in familial and sporadic Alzheimer's disease. *Proc. Natl. Acad. Sci. USA* **2018**, *115*, E782–E791. [CrossRef] [PubMed]
124. Jucker, M.; Walker, L.C. Propagation and spread of pathogenic protein assemblies in neurodegenerative diseases. *Nat. Neurosci.* **2018**, *21*, 1341–1349. [CrossRef]
125. Nortley, R.; Korte, N.; Izquierdo, P.; Hirunpattarasilp, C.; Mishra, A.; Jaunmuktane, Z.; Kyrargyri, V.; Pfeiffer, T.; Khennouf, L.; Madry, C.; et al. Amyloid beta oligomers constrict human capillaries in Alzheimer's disease via signaling to pericytes. *Science* **2019**, *365*, eaav9518. [CrossRef]
126. Suo, Z.; Humphrey, J.; Kundtz, A.; Sethi, F.; Placzek, A.; Crawford, F.; Mullan, M. Soluble Alzheimers beta-amyloid constricts the cerebral vasculature in vivo. *Neurosci. Lett.* **1998**, *257*, 77–80. [CrossRef]
127. Niwa, K.; Younkin, L.; Ebeling, C.; Turner, S.K.; Westaway, D.; Younkin, S.; Ashe, K.H.; Carlson, G.A.; Iadecola, C. Abeta 1-40-related reduction in functional hyperemia in mouse neocortex during somatosensory activation. *Proc. Natl. Acad. Sci. USA* **2000**, *97*, 9735–9740. [CrossRef]
128. Han, J.; Zhang, X.; Kang, L.; Guan, J. Extracellular vesicles as therapeutic modulators of neuroinflammation in Alzheimer's disease: A focus on signaling mechanisms. *J. Neuroinflamm.* **2025**, *22*, 120. [CrossRef]
129. Tzioras, M.; Daniels, M.J.D.; Davies, C.; Baxter, P.; King, D.; McKay, S.; Varga, B.; Popovic, K.; Hernandez, M.; Stevenson, A.J.; et al. Human astrocytes and microglia show augmented ingestion of synapses in Alzheimer's disease via MFG-E8. *Cell Rep. Med.* **2023**, *4*, 101175. [CrossRef]
130. Karamanova, N.; Truran, S.; Serrano, G.E.; Beach, T.G.; Madine, J.; Weissig, V.; Davies, H.A.; Veldhuizen, J.; Nikkhah, M.; Hansen, M.; et al. Endothelial Immune Activation by Medin: Potential Role in Cerebrovascular Disease and Reversal by Monosialoganglioside-Containing Nanoliposomes. *J. Am. Heart Assoc.* **2020**, *9*, e014810. [CrossRef]
131. Morrow, K.; Karamanova, N.; Nabaty, N.; Maerivoet, A.; Madine, J.; Li, M.; Migrino, R. Amyloidogenic medin induces pro-thrombotic activation in human brain microvascular endothelial cells. *Circulation* **2024**, *150* (Suppl. S1), 4146706. [CrossRef]
132. Iadanza, M.G.; Jackson, M.P.; Hewitt, E.W.; Ranson, N.A.; Radford, S.E. A new era for understanding amyloid structures and disease. *Nat. Rev. Mol. Cell Biol.* **2018**, *19*, 755–773. [CrossRef]
133. Ren, B.; Zhang, Y.; Zhang, M.; Liu, Y.; Zhang, D.; Gong, X.; Feng, Z.; Tang, J.; Chang, Y.; Zheng, J. Fundamentals of cross-seeding of amyloid proteins: An introduction. *J. Mater. Chem. B* **2019**, *7*, 7267–7282. [CrossRef] [PubMed]
134. Walsh, D.M.; Hartley, D.M.; Kusumoto, Y.; Fezoui, Y.; Condron, M.M.; Lomakin, A.; Benedek, G.B.; Selkoe, D.J.; Teplow, D.B. Amyloid beta-protein fibrillogenesis. Structure and biological activity of protofibrillar intermediates. *J. Biol. Chem.* **1999**, *274*, 25945–25952. [CrossRef] [PubMed]
135. Morales, R.; Green, K.M.; Soto, C. Cross currents in protein misfolding disorders: Interactions and therapy. *CNS Neurol. Disord. Drug Targets* **2009**, *8*, 363–371. [CrossRef] [PubMed]
136. Irwin, D.J.; Lee, V.M.; Trojanowski, J.Q. Parkinson's disease dementia: Convergence of alpha-synuclein, tau and amyloid-beta pathologies. *Nat. Rev. Neurosci.* **2013**, *14*, 626–636. [CrossRef]
137. Namkung, H.; Kim, S.H.; Sawa, A. The Insula: An Underestimated Brain Area in Clinical Neuroscience, Psychiatry, and Neurology: (Trends in Neuroscience 40, 200–207, 2017). *Trends Neurosci.* **2018**, *41*, 551–554. [CrossRef]
138. Biessels, G.J.; Staekenborg, S.; Brunner, E.; Brayne, C.; Scheltens, P. Risk of dementia in diabetes mellitus: A systematic review. *Lancet Neurol.* **2006**, *5*, 64–74. [CrossRef]
139. Luchsinger, J.A.; Reitz, C.; Patel, B.; Tang, M.X.; Manly, J.J.; Mayeux, R. Relation of diabetes to mild cognitive impairment. *Arch. Neurol.* **2007**, *64*, 570–575. [CrossRef]
140. Sengupta, U.; Kayed, R. Amyloid beta, Tau, and alpha-Synuclein aggregates in the pathogenesis, prognosis, and therapeutics for neurodegenerative diseases. *Prog. Neurobiol.* **2022**, *214*, 102270. [CrossRef]
141. Wardlaw, J.M.; Smith, C.; Dichgans, M. Small vessel disease: Mechanisms and clinical implications. *Lancet Neurol.* **2019**, *18*, 684–696. [CrossRef]
142. Merino, J.G. White Matter Hyperintensities on Magnetic Resonance Imaging: What Is a Clinician to Do? *Mayo Clin. Proc.* **2019**, *94*, 380–382. [CrossRef]
143. Cozza, M.; Amadori, L.; Boccardi, V. Exploring cerebral amyloid angiopathy: Insights into pathogenesis, diagnosis, and treatment. *J. Neurol. Sci.* **2023**, *454*, 120866. [CrossRef] [PubMed]
144. Walker, D.A.; Broderick, D.F.; Kotsenas, A.L.; Rubino, F.A. Routine use of gradient-echo MRI to screen for cerebral amyloid angiopathy in elderly patients. *AJR Am. J. Roentgenol.* **2004**, *182*, 1547–1550. [CrossRef] [PubMed]

145. Arvanitakis, Z.; Capuano, A.W.; Leurgans, S.E.; Bennett, D.A.; Schneider, J.A. Relation of cerebral vessel disease to Alzheimer's disease dementia and cognitive function in elderly people: A cross-sectional study. *Lancet Neurol.* **2016**, *15*, 934–943. [CrossRef] [PubMed]
146. Boyle, P.A.; Yu, L.; Wilson, R.S.; Leurgans, S.E.; Schneider, J.A.; Bennett, D.A. Person-specific contribution of neuropathologies to cognitive loss in old age. *Ann. Neurol.* **2018**, *83*, 74–83. [CrossRef]
147. Satizabal, C.L.; Beiser, A.S.; Chouraki, V.; Chene, G.; Dufouil, C.; Seshadri, S. Incidence of Dementia over Three Decades in the Framingham Heart Study. *N. Engl. J. Med.* **2016**, *374*, 523–532. [CrossRef]
148. Shi, Y.; Thrippleton, M.J.; Blair, G.W.; Dickie, D.A.; Marshall, I.; Hamilton, I.; Doubal, F.N.; Chappell, F.; Wardlaw, J.M. Small vessel disease is associated with altered cerebrovascular pulsatility but not resting cerebral blood flow. *J. Cereb. Blood Flow. Metab.* **2020**, *40*, 85–99. [CrossRef]
149. Shi, Y.; Thrippleton, M.J.; Marshall, I.; Wardlaw, J.M. Intracranial pulsatility in patients with cerebral small vessel disease: A systematic review. *Clin. Sci.* **2018**, *132*, 157–171. [CrossRef]
150. Webb, A.J.S.; Birks, J.S.; Feakins, K.A.; Lawson, A.; Dawson, J.; Rothman, A.M.K.; Werring, D.J.; Llwyd, O.; Stewart, C.R.; Thomas, J. Cerebrovascular Effects of Sildenafil in Small Vessel Disease: The OxHARP Trial. *Circ. Res.* **2024**, *135*, 320–331. [CrossRef]
151. Van Den Brink, H.; Pham, S.; Siero, J.C.; Arts, T.; Onkenhout, L.; Kuijf, H.; Hendrikse, J.; Wardlaw, J.M.; Dichgans, M.; Zwanenburg, J.J.; et al. Assessment of Small Vessel Function Using 7T MRI in Patients With Sporadic Cerebral Small Vessel Disease: The ZOOM@SVDs Study. *Neurology* **2024**, *102*, e209136. [CrossRef] [PubMed]
152. Ventura-Antunes, L.; Nackenoff, A.; Romero-Fernandez, W.; Wang, Y.; Bosworth, A.M.; Prusky, A.; Wang, E.; Carvajal-Tapia, C.; Shostak, A.; Harmsen, H.; et al. Arteriolar degeneration and stiffness in cerebral amyloid angiopathy are linked to Abeta deposition and lysyl oxidase. *Alzheimer's Dement.* **2025**, *21*, e70254. [CrossRef] [PubMed]
153. Gottesman, R.F.; Mosley, T.H.; Knopman, D.S.; Hao, Q.; Wong, D.; Wagenknecht, L.E.; Hughes, T.M.; Qiao, Y.; Dearborn, J.; Wasserman, B.A. Association of Intracranial Atherosclerotic Disease With Brain beta-Amyloid Deposition: Secondary Analysis of the ARIC Study. *JAMA Neurol.* **2020**, *77*, 350–357. [CrossRef] [PubMed]
154. Jefferson, A.L.; Cambronero, F.E.; Liu, D.; Moore, E.E.; Neal, J.E.; Terry, J.G.; Nair, S.; Pechman, K.R.; Rane, S.; Davis, L.T.; et al. Higher Aortic Stiffness Is Related to Lower Cerebral Blood Flow and Preserved Cerebrovascular Reactivity in Older Adults. *Circulation* **2018**, *138*, 1951–1962. [CrossRef]
155. Scheuermann, B.C.; Parr, S.K.; Schulze, K.M.; Kunkel, O.N.; Turpin, V.G.; Liang, J.; Ade, C.J. Associations of Cerebrovascular Regulation and Arterial Stiffness With Cerebral Small Vessel Disease: A Systematic Review and Meta-Analysis. *J. Am. Heart Assoc.* **2023**, *12*, e032616. [CrossRef]
156. Winder, N.R.; Reeve, E.H.; Walker, A.E. Large artery stiffness and brain health: Insights from animal models. *Am. J. Physiol. Heart Circ. Physiol.* **2021**, *320*, H424–H431. [CrossRef]
157. Mitchell, G.F. Aortic stiffness, pressure and flow pulsatility, and target organ damage. *J. Appl. Physiol. (1985)* **2018**, *125*, 1871–1880. [CrossRef]
158. Davies, H.A.; Lee, C.F.; Miller, L.; Liu, L.N.; Madine, J. Insights into the Origin of Distinct Medin Fibril Morphologies Induced by Incubation Conditions and Seeding. *Int. J. Mol. Sci.* **2018**, *19*, 1357. [CrossRef]
159. Jessen, N.A.; Munk, A.S.; Lundgaard, I.; Nedergaard, M. The Glymphatic System: A Beginner's Guide. *Neurochem. Res.* **2015**, *40*, 2583–2599. [CrossRef]
160. Mawuenyega, K.G.; Sigurdson, W.; Ovod, V.; Munsell, L.; Kasten, T.; Morris, J.C.; Yarasheski, K.E.; Bateman, R.J. Decreased clearance of CNS beta-amyloid in Alzheimer's disease. *Science* **2010**, *330*, 1774. [CrossRef]
161. Shulyatnikova, T.; Hayden, M.R. Why Are Perivascular Spaces Important? *Medicina* **2023**, *59*, 917. [CrossRef] [PubMed]
162. Corbali, O.; Levey, A.I. Glymphatic system in neurological disorders and implications for brain health. *Front. Neurol.* **2025**, *16*, 1543725. [CrossRef] [PubMed]
163. Nedergaard, M.; Goldman, S.A. Glymphatic failure as a final common pathway to dementia. *Science* **2020**, *370*, 50–56. [CrossRef] [PubMed]
164. Kikuta, J.; Kamagata, K.; Takabayashi, K.; Taoka, T.; Yokota, H.; Andica, C.; Wada, A.; Someya, Y.; Tamura, Y.; Kawamori, R.; et al. An Investigation of Water Diffusivity Changes along the Perivascular Space in Elderly Subjects with Hypertension. *AJNR Am. J. Neuroradiol.* **2022**, *43*, 48–55. [CrossRef]
165. Xue, Y.; Liu, N.; Zhang, M.; Ren, X.; Tang, J.; Fu, J. Concomitant enlargement of perivascular spaces and decrease in glymphatic transport in an animal model of cerebral small vessel disease. *Brain Res. Bull.* **2020**, *161*, 78–83. [CrossRef]
166. Bown, C.W.; Khan, O.A.; Liu, D.; Remedios, S.W.; Pechman, K.R.; Terry, J.G.; Nair, S.; Davis, L.T.; Landman, B.A.; Gifford, K.A.; et al. Enlarged perivascular space burden associations with arterial stiffness and cognition. *Neurobiol. Aging* **2023**, *124*, 85–97. [CrossRef]
167. Maclullich, A.M.; Wardlaw, J.M.; Ferguson, K.J.; Starr, J.M.; Seckl, J.R.; Deary, I.J. Enlarged perivascular spaces are associated with cognitive function in healthy elderly men. *J. Neurol. Neurosurg. Psychiatry* **2004**, *75*, 1519–1523. [CrossRef]

168. Passiak, B.S.; Liu, D.; Kresge, H.A.; Cambronero, F.E.; Pechman, K.R.; Osborn, K.E.; Gifford, K.A.; Hohman, T.J.; Schrag, M.S.; Davis, L.T.; et al. Perivascular spaces contribute to cognition beyond other small vessel disease markers. *Neurology* **2019**, *92*, e1309–e1321. [CrossRef]
169. Zhu, Y.C.; Tzourio, C.; Soumare, A.; Mazoyer, B.; Dufouil, C.; Chabriat, H. Severity of dilated Virchow-Robin spaces is associated with age, blood pressure, and MRI markers of small vessel disease: A population-based study. *Stroke* **2010**, *41*, 2483–2490. [CrossRef]
170. Sims, J.R.; Zimmer, J.A.; Evans, C.D.; Lu, M.; Ardayfio, P.; Sparks, J.; Wessels, A.M.; Shcherbinin, S.; Wang, H.; Monkul Nery, E.S.; et al. Donanemab in Early Symptomatic Alzheimer Disease: The TRAILBLAZER-ALZ 2 Randomized Clinical Trial. *JAMA* **2023**, *330*, 512–527. [CrossRef]
171. Bateman, R.J.; Smith, J.; Donohue, M.C.; Delmar, P.; Abbas, R.; Salloway, S.; Wojtowicz, J.; Blennow, K.; Bittner, T.; Black, S.E.; et al. Two Phase 3 Trials of Gantenerumab in Early Alzheimer's Disease. *N. Engl. J. Med.* **2023**, *389*, 1862–1876. [CrossRef] [PubMed]
172. Sevigny, J.; Chiao, P.; Bussiere, T.; Weinreb, P.H.; Williams, L.; Maier, M.; Dunstan, R.; Salloway, S.; Chen, T.; Ling, Y.; et al. The antibody aducanumab reduces Abeta plaques in Alzheimer's disease. *Nature* **2016**, *537*, 50–56. [CrossRef] [PubMed]
173. van Dyck, C.H.; Swanson, C.J.; Aisen, P.; Bateman, R.J.; Chen, C.; Gee, M.; Kanekiyo, M.; Li, D.; Reyderman, L.; Cohen, S.; et al. Lecanemab in Early Alzheimer's Disease. *N. Engl. J. Med.* **2023**, *388*, 9–21. [CrossRef] [PubMed]
174. Cummings, J.; Osse, A.M.L.; Cammann, D.; Powell, J.; Chen, J. Anti-Amyloid Monoclonal Antibodies for the Treatment of Alzheimer's Disease. *BioDrugs* **2024**, *38*, 5–22. [CrossRef]
175. Sperling, R.A.; Donohue, M.C.; Raman, R.; Rafii, M.S.; Johnson, K.; Masters, C.L.; van Dyck, C.H.; Iwatsubo, T.; Marshall, G.A.; Yaari, R.; et al. Trial of Solanezumab in Preclinical Alzheimer's Disease. *N. Engl. J. Med.* **2023**, *389*, 1096–1107. [CrossRef]
176. Alves, F.; Kalinowski, P.; Ayton, S. Accelerated Brain Volume Loss Caused by Anti-beta-Amyloid Drugs: A Systematic Review and Meta-analysis. *Neurology* **2023**, *100*, e2114–e2124. [CrossRef]
177. Ayton, S. Brain volume loss due to donanemab. *Eur. J. Neurol.* **2021**, *28*, e67–e68. [CrossRef]
178. Royle, N.A.; Booth, T.; Valdes Hernandez, M.C.; Penke, L.; Murray, C.; Gow, A.J.; Maniega, S.M.; Starr, J.; Bastin, M.E.; Deary, I.J.; et al. Estimated maximal and current brain volume predict cognitive ability in old age. *Neurobiol. Aging* **2013**, *34*, 2726–2733. [CrossRef]
179. Kim, B.H.; Kim, S.; Nam, Y.; Park, Y.H.; Shin, S.M.; Moon, M. Second-generation anti-amyloid monoclonal antibodies for Alzheimer's disease: Current landscape and future perspectives. *Transl. Neurodegener.* **2025**, *14*, 6. [CrossRef]
180. Mitchell, G.F. Arterial Stiffness in Aging: Does It Have a Place in Clinical Practice?: Recent Advances in Hypertension. *Hypertension* **2021**, *77*, 768–780. [CrossRef]
181. Qiu, C.; Kivipelto, M.; von Strauss, E. Epidemiology of Alzheimer's disease: Occurrence, determinants, and strategies toward intervention. *Dialogues Clin. Neurosci.* **2009**, *11*, 111–128. [CrossRef] [PubMed]
182. Chapleau, M.; Iaccarino, L.; Soleimani-Meigooni, D.; Rabinovici, G.D. The Role of Amyloid PET in Imaging Neurodegenerative Disorders: A Review. *J. Nucl. Med.* **2022**, *63*, 13S–19S. [CrossRef] [PubMed]
183. Duran-Jara, E.; Vera-Tobar, T.; Lobos-Gonzalez, L.L. Lactadherin: From a Well-Known Breast Tumor Marker to a Possible Player in Extracellular Vesicle-Mediated Cancer Progression. *Int. J. Mol. Sci.* **2022**, *23*, 3855. [CrossRef] [PubMed]

biomolecules

MDPI

Review

Cardiac Myosin Inhibitors in the Treatment of Hypertrophic Cardiomyopathy: Clinical Trials and Future Challenges

Arnold Kukowka [1] and Marek Droździk [2,*]

[1] Department of Pharmacokinetics and Therapeutic Drug Monitoring, Pomeranian Medical University, 72 Powstańców Wielkopolskich Avenue, 70-111 Szczecin, Poland; arnold.kukowka@pum.edu.pl
[2] Department of Pharmacology, Pomeranian Medical University, 72 Powstańców Wielkopolskich Avenue, 70-111 Szczecin, Poland
* Correspondence: marek.drozdzik@pum.edu.pl

Abstract

Hypertrophic cardiomyopathy (HCM) is a prevalent and often underdiagnosed genetic cardiac disorder characterized by left ventricular hypertrophy and, in many cases, dynamic left ventricular outflow tract obstruction (LVOTO). The development of cardiac myosin inhibitors (CMIs) represents an emerging therapeutic approach in the pharmacological management of obstructive HCM (oHCM). This review offers an integrated and up-to-date synthesis of the cardiac myosin inhibitor class, with a focus on mavacamten, aficamten, and the broader landscape of emerging agents. It also highlights recent clinical trial outcomes, pharmacokinetic and pharmacogenetic considerations, and potential future directions in therapy. Furthermore, we incorporate the most recent data up to May 2025, including late-breaking trial results and real-world safety findings, aiming to provide clinicians with a practical and comprehensive perspective on this evolving drug class. A narrative review was conducted by systematically searching PubMed, Scopus, Google Scholar, and ClinicalTrials.gov for English-language articles and trials published between January 2016 and May 2025. Keywords included "cardiac myosin inhibitor", mavacamten", "aficamten", "MYK-224", and "hypertrophic cardiomyopathy." Inclusion criteria encompassed clinical trials and comprehensive reviews specifically addressing CMIs in cardiac applications. CMIs such as mavacamten and aficamten have demonstrated significant clinical benefits in reducing LVOT gradients, improving exercise capacity, and alleviating symptoms in patients with oHCM. Mavacamten is currently approved for clinical use, while aficamten is in advanced regulatory review. Comparative data suggest potential advantages of aficamten in the onset of action, pharmacokinetic profile, and tolerability. Emerging evidence supports the exploration of CMIs in pediatric populations, heart failure with preserved ejection fraction (HFpEF), and non-obstructive HCM (nHCM), although results are still preliminary. Cardiac myosin inhibitors offer a novel, pathophysiology-targeted approach to managing oHCM. While mavacamten has established efficacy, next-generation agents like aficamten may offer improved safety and versatility. Further long-term studies are needed to clarify their role across broader patient populations.

Keywords: cardiac myosin inhibitors; hypertrophic cardiomyopathy; mavacamten; aficamten; LVOT obstruction; heart failure; pharmacotherapy

1. Introduction

According to the guidelines of the European Society of Cardiology (ESC), hypertrophic cardiomyopathy (HCM) is diagnosed when an increased thickness of the left ventricular

(LV) wall is present, in the absence of secondary causes (such as hypertension or tachyarrhythmias). In order to diagnose HCM in adult patients, the LV wall thickness should be equal to or higher than 15mm in any myocardial segment. For children, LV wall thickness is required to be greater than two standard deviations (SD) from the predicted mean (z-score more than 2) [1].

From a statistical standpoint, it is estimated that 1 in 500 individuals may be affected by HCM. However, this figure may be underestimated, as only 10–20% of cases are currently identified [2]. Recent data suggest that the prevalence of HCM may be approximately 1 in 200 individuals, highlighting a growing clinical problem. More widespread use of echocardiographic diagnostics in the general population is the primary reason for the increased HCM diagnosis [3,4].

For example, in Poland, it is estimated that approximately 180,000 people are affected by HCM. Data on the prevalence of hypertrophic cardiomyopathy in Poland between 2016 and 2020 revealed a yearly increase in newly diagnosed HCM cases, except in 2020, which is attributed to the COVID-19 pandemic [5].

In addition to secondary causes of hypertrophic cardiomyopathy, two primary etiologies of HCM can be distinguished: genetic mutations and diseases that mimic HCM in imaging studies (e.g., echocardiography and cardiac magnetic resonance), referred to as phenocopies. These phenocopies exhibit imaging features similar to those seen in genetically determined HCM but have different underlying causes. They are most commonly either storage diseases, such as Fabry disease and Danon disease, or infiltrative conditions, such as amyloidosis.

Regarding genetic mutations, HCM is most frequently associated with mutations in genes encoding cardiac sarcomere proteins. The most commonly affected genes are `MYH7`, `MYBPC3` and `TNNT2`. Mutations in `MYH7` and `MYBPC3` account for approximately 60–70% of HCM cases. In contrast, mutations in `TNNT2` represent around 3–5% [1,6,7]

The treatment of hypertrophic cardiomyopathy is a complex clinical challenge. Current therapeutic algorithms distinguish two main treatment pathways, and the choice between them is closely linked to two key aspects: The severity of heart failure symptoms, classified using the New York Heart Association (NYHA) functional classification [8], and the presence or absence of left ventricular outflow tract obstruction (LVOTO).

The parameter used to assess LVOTO is the pressure gradient across the left ventricular outflow tract (LVOT), as measured by echocardiography. A normal LVOT pressure gradient should not exceed 30 mmHg, whereas a value equal to or greater than 50 mmHg is considered hemodynamically significant, leading to marked impairment of blood flow. This threshold (50 mmHg) serves as a critical decision point in both pharmacological and invasive treatment strategies according to current clinical guidelines. Yet, it is the value of 30 mmHg of LVOT gradient that is the value used to differentiate patients into two clinical subtypes: those with obstructive hypertrophic cardiomyopathy (oHCM), characterized by LVOT obstruction, and those with non-obstructive hypertrophic cardiomyopathy (nHCM), in whom obstruction is absent. European Society of Cardiology guidelines underline that most patients in the range of 30 mmHg to 49 mmHg LVOT gradient can be treated according to the non-obstructive hypertrophic cardiomyopathy pathway. Cardiac myosin inhibitors (CMIs) are currently used exclusively in the treatment of oHCM. The LVOT gradient is highly relevant in the context of heart failure symptoms. Its elevation can lead to worsening of heart failure manifestations, such as fatigue, increased dyspnea, and a significant reduction in exercise capacity, due to impeded delivery of oxygenated blood to peripheral tissues [1,9–11].

Treatment of LVOTO is guided by the presence of heart failure symptoms. In symptomatic patients with HCM, the first-line therapy is based on beta-blockers. If symptoms persist, non-dihydropyridine calcium channel blockers (such as verapamil or diltiazem) may be added. Subsequently, cardiac myosin inhibitors or disopyramide can be introduced. If symptoms remain refractory to pharmacotherapy, invasive procedures are recommended, either alcohol septal ablation or surgical myectomy (see Figure 1) [1].

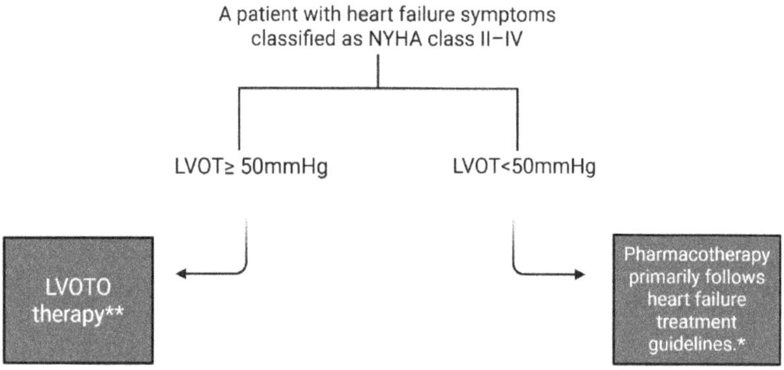

Figure 1. Simplified diagram of the therapeutic pathway for hypertrophic cardiomyopathy based on the guidelines of the European Society of Cardiology [1]. * depending on ejection fraction; ** described in the main text. Created in BioRender: https://BioRender.com/wzb0fkf (accessed on 30 May 2025).

The introduction of a new class of medications, i.e., cardiac myosin inhibitors, is a major breakthrough in the treatment of HCM [12]. This article provides a detailed discussion of CMIs. The emergence of this novel pharmacological option not only enables improved symptom control but also contributes to delaying the need for invasive intervention. The development and clinical benefits of CMIs have been acknowledged by the leading cardiology societies, i.e., the European Society of Cardiology (ESC) and the American College of Cardiology (ACC), both of which have issued positive recommendations for their use, in 2023 and 2024, respectively [1,13–15].

Currently, the most widely used CMIs is mavacamten, which was approved by the U.S. Food and Drug Administration (FDA) in 2022 [16] and by the European Medicines Agency (EMA) in 2023 [17]. Other agents in this class currently undergoing clinical investigation include the following: aficamten, which has completed Phase III clinical trials [18], and MYK-224, which is currently in Phase II clinical trials [19]. It is noteworthy that, in parallel with the growing population of individuals affected by HCM, there is a steady increase in the number of clinical trials focused on the pharmacological treatment of this condition [20]. Although several recent reviews have focused primarily on mavacamten or aficamten, the present article provides a broader, comparative overview of the entire class of cardiac myosin inhibitors, including mavacamten, aficamten, and investigational agents such as MYK-224 and MYK-581. This review also incorporates the most up-to-date clinical trial data available as of May 2025. In addition, we highlight important aspects such as CYP2C19 pharmacogenetic considerations, regulatory milestones, and the potential future use of CMIs in non-obstructive HCM, heart failure with preserved ejection fraction, and pediatric populations. By addressing these elements, this review aims to deliver a comprehensive and clinically relevant summary of current evidence and future directions in the pharmacological management of HCM with CMIs.

Physiology of Cardiac Muscle Contraction and Pathophysiology of HCM

Under physiological conditions, cardiac muscle contraction occurs as a result of interactions between myosin filaments (thick filaments) and actin filaments (thin filaments) within a structural unit known as the sarcomere. A key component of myosin is the myosin head, which contains two crucial binding sites: one for actin and one for ATP (adenosine triphosphate).

When ATP binds to the myosin head, it undergoes hydrolysis catalyzed by cardiac myosin ATPase, releasing energy. This energy drives a conformational change in the myosin head, enabling it to bind to actin, forming a cross-bridge. Following the release of inorganic phosphate and then the ADP (adenosine diphosphate), which are products of ATP hydrolysis, the thin filament is pulled toward the center of the sarcomere, resulting in muscle contraction. A new ATP molecule then binds to the myosin head, causing it to detach from actin, allowing the cycle to repeat [21,22]. The basic physiology of cardiac muscle contraction is shown in Figure 2.

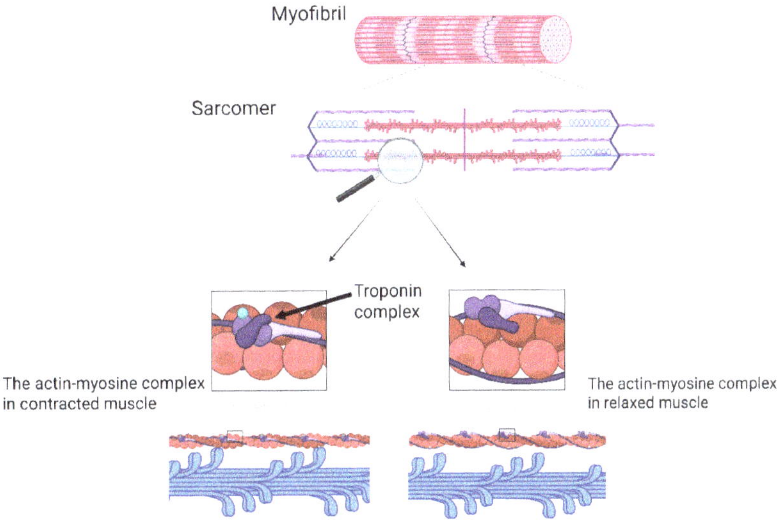

Figure 2. Basic physiology of cardiac muscle contraction: from myofibril structure to sarcomere and actin–myosin interaction. Troponin complex describes the combination of troponin C, I, and T. The colors used in this figure: blue describes myosin (thick filament), red describes actin (thin filament), purple describes tropomyosin, and green describes ion calcium (Ca^{2+}). Created in BioRender: https://BioRender.com/y3xg904 (accessed on 30 May 2025).

It is also important to note that, physiologically, two resting states of myosin can be distinguished: the SRX (super-relaxed) state and the DRX (disordered-relaxed) state. These states differ significantly in their ATP consumption rates: SRX consumes far less ATP compared to DRX (0.003 s^{-1} vs. 0.03 s^{-1}). Furthermore, the DRX state represents a condition in which the myosin head is primed for activity, i.e., capable of forming cross-bridges and initiating contraction. In contrast, in the SRX state, the myosin head cannot bind to actin, thereby inhibiting contraction [23]. Under pathophysiological conditions, the underlying cause of HCM is often genetic mutations, such as those affecting the β-myosin heavy chain gene (MYH7). This mutation may lead to an increased number of myosin heads in the DRX state, which in turn causes a higher number of cross-bridges between myosin and actin in cardiac sarcomeres [24]. Figure 3 shows the pathophysiological mechanism of HCM. It also includes cardiac myosin inhibitors, which are described more specifically later in this article.

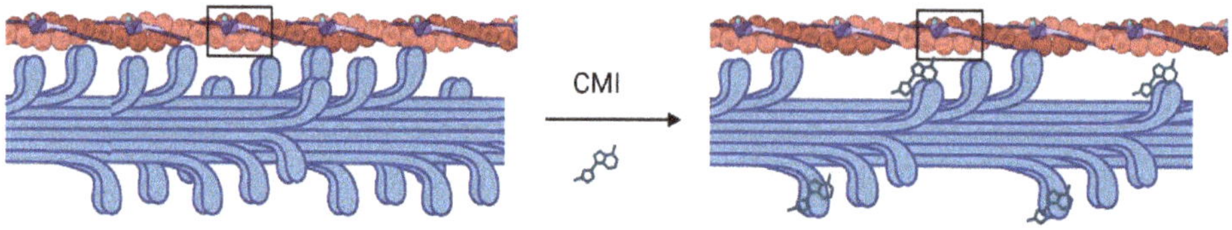

Figure 3. Pathological actin–myosin interaction in hypertrophic cardiomyopathy and its modulation by cardiac myosin inhibitors. The left side of Figure 3 presents the pathological state of cardiac myosin-actin complex in HCM with a larger number of cross-bridges, which results in hypercontraction (as described in the text). The right side shows a state of myosin–actin complex in the presence of CMI: the cardiac myosin inhibitors block excessive myosin heads, and as a result, there is a lower number of cross-bridges. The colors used in this figure: blue describes myosin (thick filament), red describes actin (thin filament), and purple describes tropomyosin. CMI: cardiac myosin inhibitor. The chemical structure shown is symbolic and does not represent any actual compound. Created in BioRender (2025): https://BioRender.com/pie1hc8 (accessed on 30 May 2025).

The clinical consequences of the dysfunction include the following: increased myocardial contractility, impaired diastolic relaxation, narrowing of the left ventricular outflow tract (due to increased myocardial fiber mass), and myocardial fibrosis [12,25–27].

2. Materials and Methods

2.1. Data Sources

We searched PubMed, Scopus, and Google Scholar to gather relevant literature. We selected a data frame from January 2016 to May 2025. These platforms were chosen simply because they offer a wide range of peer-reviewed biomedical research and are well-established databases in biomedical sciences. We also took a thorough look at ClinicalTrials.gov to find any relevant clinical trials, mainly 3-phase clinical trials for CMIs.

2.2. Search Strategy

The search was performed using a narrative desk review approach. The following keywords and combinations were used: "cardiac myosin inhibitor", "mavacamten," "aficamten", "myosin ATPase", "hypertrophic cardiomyopathy", "HCM", and "hyperertrophic cardiomyopathy treatment". Boolean operators (AND/OR) were applied to enhance search precision. Only articles published in English were considered. For additional depth, manual screening of reference lists from key articles was performed. The final search was completed on 15 May 2025.

2.3. Inclusion and Exclusion Criteria

Included materials encompassed original research articles, clinical trials, pharmacological studies, and comprehensive review articles specifically addressing cardiac myosin inhibitors. Studies were excluded if they provided only general discussions on sarcomeric proteins without specific reference to pharmacological inhibitors, were editorials, commentaries, or opinion pieces without original data or comprehensive synthesis, or were non-English publications or non-peer-reviewed sources (e.g., blogs or preprints without peer review).

2.4. Article Selection and Quality Assessment

Chosen articles were investigated manually for relevance and quality. Emphasis was placed on peer-reviewed studies from reputable journals with transparent methodology and well-described objectives. Clinical trials were evaluated based on the design phase, registration status, and reported outcomes. No formal scoring system was applied, but a qualitative assessment was conducted to prioritize clinically meaningful publications.

2.5. Bias Assessment

Bias was minimized through the use of multiple databases and a defined inclusion protocol. Critical source analysis was applied to assess the scientific validity, originality, and contextual relevance of each article. These measures aimed to accurately represent the current state of knowledge on cardiac myosin inhibition.

2.6. Transparency and Reproducibility

The methodology employed in this review follows a qualitative narrative approach, ensuring reproducibility through defined search parameters and source documentation. No new experiments, datasets, or protocols were created. All data cited are publicly available or accessible via institutional subscriptions. No ethical approval was required. Generative AI was not used for content generation, data analysis, or synthesis in this manuscript.

3. Cardiac Myosin Inhibitors

Cardiac myosin ATPase inhibitors are more commonly referred to in the scientific literature as cardiac myosin inhibitors. They are small-molecule drugs that act by inhibiting a key enzyme responsible for cardiac muscle contraction, namely cardiac myosin ATPase. This inhibition results in a reduced formation of actin–myosin cross-bridges. Additionally, drugs in this class help maintain myosin in its super-relaxed state (SRX). Collectively, these mechanisms counteract the pathophysiological basis of HCM by decreasing myocardial contractility, improving diastolic function, and reducing left ventricular outflow tract obstruction. These effects translate into clinical benefits for patients, including symptom relief, reduction in the LVOT pressure gradient, and improvement in exercise capacity. Importantly, unlike other drug classes used in HCM, CMIs act not only on symptoms but also target the underlying pathophysiology of the disease [28–30].

The primary agents in this class include the following: mavacamten (Camzyos ®), which currently approved by both the FDA [16] and the European Medicines Agency [17]; aficamten (CK-274), which is currently undergoing clinical investigation [18]; MYK-224, which is in Phase II clinical trials [19]; MYK-581, which is currently in preclinical studies [31]; Blebbastatin, which is used for research purposes [32]. All of these compounds are small-molecule drugs [33]. A summary of the compounds classified as cardiac myosin inhibitors is presented in Table 1.

Table 1. Regulatory status and clinical development stage of selected cardiac myosin inhibitors (references in the text).

Drug Name	FDA Approval	EMA Approval	Clinical Development Stage
Mavacamten	Approved (2022)	Approved (2023)	Available on the market
Aficamten	Pending approval	Pending approval	Phase III completed
MYK-224	Not approved	Not approved	Phase II
MYK-581	Not approved	Not approved	Preclinical phase
Blebbistatin	Not approved	Not approved	Research use only

FDA—Food and Drug Administration; EMA—European Medicines Agency.

3.1. Mavacamten

Mavacamten, also known as MYK-461, is an allosteric inhibitor of cardiac myosin ATPase and the first approved drug in this class [30], making it also the most extensively studied. The structure of mavacamten is shown in Figure 4, and its molecular weight is 273.33 g/mol [34].

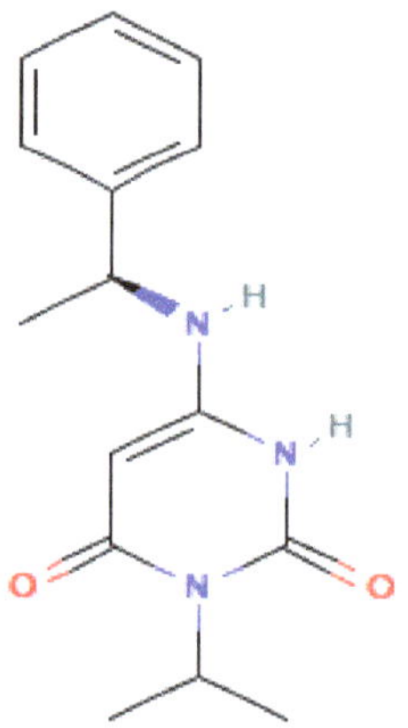

Figure 4. Structure of Mavacamten [34].

Its mechanism of action, which involves reducing the formation of actin–myosin cross-bridges, has been described previously [30]. From a pharmacokinetic perspective, mavacamten demonstrates high oral bioavailability (>85%), allowing for once-daily oral administration. It has a long half-life, estimated at 6–9 days. Mavacamten is metabolized primarily by two cytochrome P450 isoenzymes: CYP2C19 (≈74%), for which genotyping is recommended by the EMA to identify poor metabolizers [17], and to a lesser extent, CYP3A4 (≈18%) and CYP2C9 (≈8%). Excretion is primarily via the biliary route [35,36]. In preclinical studies on mice, the early administration of mavacamten significantly reduced the progression of hypertrophic cardiomyopathy. Moreover, the same study showed that mavacamten could reverse cardiac hypertrophy in subjects where hypertrophy was already established [37]. In Phase II clinical trials, the PIONEER-HCM study and its long-term extension PIONEER-OLE ($n = 13$) demonstrated that mavacamten significantly reduced LVOT gradients, improved exercise capacity, and enhanced quality of life [38,39]. In Phase III, the EXPLORER-HCM trial ($n = 251$) confirmed its clinical efficacy by showing a mean reduction of 37 mmHg in the post-exercise LVOT gradient (compared to 14 mmHg in the placebo group), improvement in peak oxygen consumption (pVO_2), an improvement of ≥1 NYHA class in 80% of patients, and a 9.1-point increase in quality-of-life scores versus 5.1 points in the placebo group. Mavacamten demonstrated a safety profile comparable to the placebo [40]. Another clinical trial, The HORIZON-HCM, was conducted on the Japanese population. The study had similar outcomes regarding efficacy and safety of mavacamten when compared to EXPLORER-HCM [41].

The MAVA-LTE trial (ongoing through 2029), a long-term extension of EXPLORER-HCM, aims to assess long-term efficacy and safety. Of the 231 patients completing EXPLORER-HCM, 211 were enrolled in MAVA-LTE. The initial dose was 5 mg, with titration based on ejection fraction and LVOT gradient. Preliminary data show sustained LVOT gradient reduction, left atrial volume reduction (~5.5 mL/m^2), which is a marker of improved diastolic function, at week 180, 66.3% of patients were in NYHA class I, a low discontinuation rate (5.6%), with main adverse effects being atrial fibrillation (14.3%), and worsening heart failure (6.1%) [42]. The VALOR-HCM, an important clinical trial regarding mavacamten and its impact on the treatment of HCM, evaluated its potential to delay or prevent invasive procedures (alcohol septal ablation or surgical myectomy) in patients with NYHA class III/IV of HCM already scheduled for such interventions. At week 16, only 17.9% of patients in the mavacamten group still qualified for invasive treatment, offering a significant clinical benefit [14,43]. However, an important limitation of mavacamten is its potential to reduce left ventricular ejection fraction (LVEF). It is not recommended for patients with LVEF < 55%, and echocardiographic monitoring is essential, especially during the initial treatment phase [17,35]. Following the approval of mavacamten by the FDA and EMA, the number of real-world experience studies evaluating this drug has increased. For example, a study conducted by Ramonfaur et al. [44] in a racially diverse population over an 18-month period demonstrated efficacy and safety outcomes consistent with those observed in clinical trials. The study also found that patients with HCM and obesity were less likely to experience symptom improvement with mavacamten. Another real-world study by Lim et al. [45] demonstrated that mavacamten can be a safe and effective treatment in the Korean population diagnosed with oHCM. Acknowledging mavacamten metabolism, it should be avoided or used with special caution at lower doses in patients co-administered with moderate or strong CYP2C19 or CYP3A4 inhibitors/inducers due to its metabolic profile. Because of the polymorphic nature of the gene coding for CYP2C19, patients should be genotyped for the *CYP2C19* gene to determine the appropriate dose of mavacamten. Patients with a poor metabolizer phenotype may experience up to a threefold increase in mavacamten exposure, which may lead to an increased risk of systolic dysfunction compared to normal metabolizers. If treatment is initiated before the CYP2C19 phenotype is determined, patients should follow the dosing recommendations for poor metabolizers until their CYP2C19 status is known [17,35,46]. This issue has been especially relevant in Asian populations, where the prevalence of poor CYP2C19 metabolism is higher, as observed in the HORIZON-HCM trial (Japanese population) [39], as well as in Chinese [47] and Korean cohorts [45]. In summary, regarding CYP2C19 polymorphism, patients who are poor metabolizers should initiate treatment at a lower dose than normal metabolizers (2.5 mg vs. 5 mg), with subsequent dose adjustment based on echocardiographic imaging [17,46]. The most common adverse events during mavacamten treatment were dizziness, syncope, and atrial fibrillation (AF) [35,36,40]. Due to the occurrence of AF during mavacamten therapy, there is a need for clinical guidelines regarding the safety of anticoagulants and antiarrhythmic (AAD) drugs. A review by Ricci et al. [48] addressed this issue, noting that commonly used anticoagulants, such as vitamin K antagonists and non-vitamin K oral anticoagulants (NOACs), are generally considered safe. However, given the lack of large-scale studies, patients should be closely monitored for potential complications when combining anticoagulation therapy with mavacamten. Regarding AADs, it is important to consider replacing flecainide or propafenone with alternative agents due to their proarrhythmic properties in the HCM population [48,49]. Additionally, mavacamten use has been associated with other arrhythmias, such as premature atrial contractions (PACs), premature ventricular contractions (PVCs), and supraventricular tachycardia (SVT). These arrhythmias were observed primarily at the start of treatment, but the correlation was transient and not present over the long term [50]. To conclude, mavacamten has shown beneficial effects not only on clinical symptoms but also on echocardiographic parameters, with a favorable safety profile. It also offers a non-invasive alternative in treatment pathways.

3.2. *Aficamten*

Aficamten, currently under FDA review [51], is a structural analog of mavacamten with a molecular weight equal to 337.4 g/mol (see Figure 5) [52].

Overall, aficamten's mechanism of action is similar to mavacamten, but it exhibits greater selectivity for the cardiac myosin active site [53]. Thanks to its pharmacokinetic properties, the steady-state drug concentration is set within 2 weeks of treatment, which translates to a faster clinical effect, including NYHA class improvement and LVOT gradient reduction, and is typically achieved in this period of two weeks of treatment [54], compared to 6–9 weeks for mavacamten [40]. Pharmacokinetically, aficamten is metabolized by multiple cytochrome P450 enzymes, reducing the risk of drug–drug interactions and the need for pharmacogenetics testing. It has a half-life of ~4 days, reaches a steady state in ~2 weeks, shows less plasma protein binding (compared to mavacamten), and is renally excreted (~32%), unlike mavacamten, which undergoes minimal renal elimination [55–57]. The SEQUOIA-HCM Phase III trial (n = 282) demonstrated that aficamten improved pVO_2 by +1.8 mL/kg/min, reduced resting LVOT gradient by 34.7mmHg, improved NYHA class in 59% of patients, and enhanced overall quality of life [54,58,59]. Earlier trials also showed improvements in LVOT gradient and NYHA class, though some did not reach statistical significance [60–62]. In terms of safety, aficamten appears better tolerated than mavacamten. The most common adverse effects of aficamten are dizziness and LVEF reduction below 50%, which occurred in only 3.5% of patients (n = 144) after 28 weeks vs. 8.7% in the EXPLORE-HCM study with mavacamten (n = 180) [39,52,54,63]. A safety meta-analysis by Davis et al. [64] found that atrial fibrillation occurred in 4.1/100 patients with aficamten vs. 11.5/100 for mavacamten. LVEF-related discontinuation occurred only once for aficamten (n = 188), and was transient; in contrast, mavacamten was discontinued completely in eight patients and temporarily in 28 out of 414 [64]. In the context of the cardiac conduction system, aficamten did not prolong the QTc interval and did not influence heart rate [65]. Moreover, regarding aficamten pharmacokinetics and safety, it is worth mentioning that patients with mild to moderate kidney and/or liver impairment may not require dose adjustment, as the pharmacokinetics of aficamten were not significantly different from those observed in individuals with normal liver and kidney function [66]. These findings suggest pharmacokinetic and pharmacodynamic differences between aficamten and mavacamten. Overall, it seems that aficamten acts faster, is cleared more rapidly, and may offer a safer profile, though larger studies are needed. FDA approval is expected by late 2025, and EMA approval by mid-2026 [51,64,67].

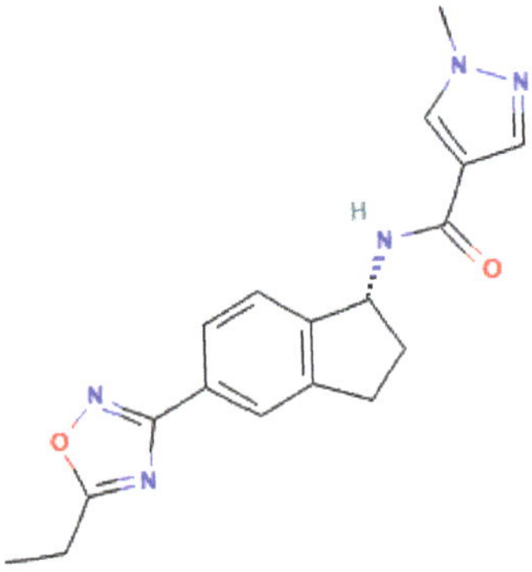

Figure 5. Structure of aficamten [52].

3.3. Experimental and Emerging CMIs

Blebbistatin is a prototype myosin II inhibitor widely used in experimental research, including studies of muscle physiology and cancer models, due to its fluorescent properties [32,68]. Despite its value as a research tool, its utility in vivo is limited by phototoxicity, poor solubility, and cytotoxicity [69]. To overcome these drawbacks, several derivatives, such as para-nitroblebbistatin and para-aminoblebbistatin, have been developed, offering improved solubility and reduced phototoxicity while retaining inhibitory activity. However, these analogs still exhibit residual cytotoxicity and genotoxicity, limiting their suitability for preclinical development [68,70]. While blebbistatin itself is not clinically viable, its molecular scaffold remains a valuable template for the development of future isoform-selective myosin inhibitors.

Building on these early insights, newer cardiac myosin inhibitors have progressed into clinical development. MYK-224, a structural analog of mavacamten, is currently in Phase II trials. It is designed with a shorter half-life to offer greater dosing flexibility, although detailed preclinical data remain limited [71]. The MERCUTIO trial evaluating MYK-224 in obstructive HCM was terminated for business-related reasons [72]; however, the AURORA-HFpEF trial remains ongoing, assessing its safety and tolerability in patients with heart failure with preserved ejection fraction (HFpEF) [31]. Notably, mavacamten has also shown promise in this population: in the EMBARK-HFpEF trial ($n = 30$), treatment led to a 13% reduction in troponin I and a 26% reduction in NT-proBNP levels, suggesting potential efficacy in HFpEF [73]. In parallel, MYK-581, another CMI analog, is currently in preclinical development and has demonstrated inhibition of HCM progression in animal models [74].

4. Discussion

The future of pharmacotherapy for hypertrophic cardiomyopathy using cardiac myosin inhibitors remains filled with open questions. Furthermore, ongoing studies continue to investigate new agents in this class, such as the aforementioned MYK-224 or MYK-581.

To determine the definitive role of CMIs in oHCM, comparative trials with beta-blockers are necessary. The MAPLE-HCM trial, comparing aficamten vs. metoprolol, is completed, but results are pending [75]. However, valuable insights have already emerged from post hoc analyses of existing trial data. In a detailed subgroup analysis of the EXPLORER-HCM and MAVA-LTE trials, Wheeler et al. [76] investigated the impact of background beta-blocker use on the efficacy of mavacamten. Their findings revealed that while the improvement in peak oxygen consumption (VO_2) was attenuated in patients on beta-blockers due to chronotropic incompetence, the overall clinical benefit of mavacamten was preserved. Key endpoints such as reductions in LVOT gradients, improvements in NYHA functional class, and decreases in NT-proBNP levels were largely unaffected by beta-blocker use. Notably, the data raise the hypothesis that in certain patients, beta-blocker withdrawal or dose reduction could potentially unmask additional benefits of CMI therapy, particularly in terms of exercise capacity. Nevertheless, given the multifactorial indications for beta-blockers in HCM (e.g., arrhythmia management, coronary disease, and comorbid hypertension), this approach must be individualized. The upcoming results from head-to-head trials such as MAPLE-HCM are therefore essential to establish whether CMIs could supplant beta-blockers as first-line pharmacotherapy in symptomatic obstructive HCM.

The role of CMIs in nHCM is under investigation in the ACACIA-HCM trial (aficamten). This is the first aficamten study in the nHCM population [77]. Regarding aficamten, an interesting study was conducted by Masri et al. [78], in which they evaluated the combined efficacy of aficamten administered concurrently with disopyramide, a drug with a well-established role in the treatment of hypertrophic cardiomyopathy. In this study, researchers divided patients from clinical trials of aficamten (REDWOOD-HCM, SEQUOIA-HCM, and FOREST-HCM) into four groups: patients receiving continuous treatment with both aficamten and disopyramide; patients receiving both aficamten and disopyramide, with subsequent discontinuation of disopyramide; patients receiving disopyramide and the placebo; patients receiving disopyramide and aficamten, with subsequent discontinuation of aficamten. The results were very positive for aficamten. The study showed that the combination of disopyramide and aficamten did not demonstrate superiority compared to aficamten alone. Moreover, the discontinuation of aficamten (which left patients treated only with disopyramide) resulted in a higher left ventricular outflow tract gradient, recurrence of symptoms, and NT-proBNP levels returning to baseline levels (before starting aficamten treatment). Thus, the use of aficamten alone may be sufficient to alleviate symptoms of obstructive HCM, potentially allowing for the discontinuation of disopyramide when patients begin treatment with aficamten. In addition, these results may lead to a reduced number of drugs in the patient's treatment, which offers benefits such as a lower risk of drug interactions and fewer treatment complications. [79,80]. Mavacamten has already been evaluated in nHCM through the MAVERICK-HCM trial [81], which showed improved biomarkers (NT-proBNP and troponin I), but no significant improvements in pVO_2 or NYHA class, which was limited by the small sample size (n = 59), short duration (16 weeks), and reliance on surrogate endpoints. To address these limitations, the ODYSSEY-HCM trial (n = 580) was launched in 2022, with extended follow-up through 2029. However, interim results were negative, showing no improvement in pVO_2 or KCCQ-23 CSS scores, leading to early termination. Investigators suggested that nHCM may represent a distinct disease requiring different pharmacologic strategies [82,83]. A study by Amr et al. [84] also addressed a similar issue, but from a different perspective. Among patients with obstructive HCM, only a portion of this population was included in clinical trials of cardiac myosin inhibitors (CMIs) during phases 2 and 3. According to the authors' observations, the most common reason for exclusion was an insufficient LVOT gradient. This highlights an important issue, particularly for patients with a resting LVOT gradient in the 30–49 mmHg range [1,84]. For example, in the SEQUOIA-HCM trial (a clinical trial for aficamten), the inclusion criteria were a resting LVOT gradient $\geq$30 mmHg and <50 mmHg and a post-Valsalva (provoked) LVOT gradient $\geq$50 mmHg [13]. In contrast, the EXPLORER-HCM trial (a clinical trial for mavacamten) used stricter criteria, requiring a peak LVOT gradient of at least 50 mmHg (either at rest or after Valsalva maneuver/exercise) [40]. These differences raise an important question: from what LVOT gradient should CMIs be considered? Currently, the general consensus is that 50 mmHg serves as the primary threshold. Another key area of CMIs is their pediatric use (<18 years). The CEDAR-HCM trial, currently recruiting, is testing aficamten in this population [85]. Given promising animal model data with mavacamten [35] and aficamten's potentially superior safety [64], this study may be highly impactful in reshaping HCM therapy. The clinical development of cardiac myosin inhibitors continues to evolve, with next-generation agents such as MYK-224 and MYK-581 under investigation and expanding interest in broader indications (for well-established CMIs such as mavacamten and aficamten), including HFpEF, non-obstructive HCM, and pediatric populations. Despite some setbacks and unanswered questions, ongoing trials reflect a hopeful trajectory for the future of CMI-based pharmacotherapy in hypertrophic cardiomyopathy.

5. Conclusions

Cardiac myosin inhibitors (CMIs) reduce myocardial contractility and are currently indicated for a narrow subset of patients with obstructive hypertrophic cardiomyopathy (oHCM). As of now, mavacamten is the only CMI approved for clinical use. Its newer analog, aficamten, may reach the market by late 2025, pending regulatory approval. Both agents have demonstrated favorable clinical and hemodynamic effects, along with acceptable safety profiles in clinical trials. CMIs may expand their therapeutic scope in the future to include pediatric populations and patients with HFpEF or nHCM. However, as a relatively new drug class, CMIs will require further large-scale clinical trials to establish their long-term efficacy, safety, and optimal place in therapy.

Author Contributions: Conceptualization, A.K.; methodology, A.K.; validation, A.K. and M.D.; formal analysis, A.K.; investigation, A.K.; resources, A.K.; data curation, A.K.; writing—original draft preparation, A.K.; writing—review and editing, A.K. and M.D.; visualization, A.K.; supervision,

M.D.; project administration, A.K.; funding acquisition, M.D. All authors have read and agreed to the published version of the manuscript.

Funding: This research received no external funding.

Institutional Review Board Statement: Not applicable.

Informed Consent Statement: Not applicable.

Data Availability Statement: No new data were created or analyzed in this study. Data sharing is not applicable to this article.

Acknowledgments: All of the figures (except Figures 4 and 5) in this article were created on the BioRender.com website. The authors declare that for each figure, a publication license was obtained, generated via the official site (BioRender.com). ChatGPT (GPT-4 omni) was only used to assess grammatical and spelling errors, it was not used for any other reason.

Conflicts of Interest: The authors declare no conflicts of interest.

Abbreviations

The following abbreviations are used in this manuscript:

HFpEF	Heart Failure with Preserved Ejection Fraction
HCM	Hypertrophic Cardiomyopathy
oHCM	Obstructive Hypertrophic Cardiomyopathy
nHCM	Non-Obstructive Hypertrophic Cardiomyopathy
CMI	Cardiac Myosin Inhibitor

References

1. Arbelo, E.; Protonotarios, A.; Gimeno, J.R.; Arbustini, E.; Barriales-Villa, R.; Basso, C.; Bezzina, C.R.; Biagini, E.; Blom, N.A.; de Boer, R.A.; et al. 2023 ESC Guidelines for the management of cardiomyopathies. *Eur. Heart J.* **2023**, *44*, 3503–3626. [CrossRef]
2. Maron, B.J.; Desai, M.Y.; Nishimura, R.A.; Spirito, P.; Rakowski, H.; Towbin, J.A.; Rowin, E.J.; Maron, M.S.; Sherrid, M.V. Diagnosis and Evaluation of Hypertrophic Cardiomyopathy: JACC State-of-the-Art Review. *J. Am. Coll. Cardiol.* **2022**, *79*, 372–389. [CrossRef] [PubMed]
3. Semsarian, C.; Ingles, J.; Maron, M.S.; Maron, B.J. New perspectives on the prevalence of hypertrophic cardiomyopathy. *J. Am. Coll. Cardiol.* **2015**, *65*, 1249–1254. [CrossRef] [PubMed]
4. Mandeş, L.; Roşca, M.; Ciupercă, D.; Popescu, B.A. The role of echocardiography for diagnosis and prognostic stratification in hypertrophic cardiomyopathy. *J. Echocardiogr.* **2020**, *18*, 137–148. [CrossRef] [PubMed]
5. Mizia-Stec, K.; Leszek, P.; Cegłowska, U.; Wiśniewska, A.; Hałgas, K.; Wybraniec, M.; Pachciński, O.; Stec, M.; Cieśla, D.; Gąsior, M.; et al. Incidence and prevalence of cardiomyopathies in Poland and outcomes for patients in the years 2016–2020. *Kardiol. Pol.* **2024**, *82*, 217–219. [CrossRef]
6. Chumakova, O.S.; Baklanova, T.N.; Zateyshchikov, D.A. Clinical Features and Prospective Outcomes of Thin-Filament Hypertrophic Cardiomyopathy: Intrinsic Data and Comparative Insights from Other Cohorts. *J. Clin. Med.* **2025**, *14*, 866. [CrossRef]
7. Hong, Y.; Xi, H.T.; Yang, X.Y.; Su, W.W.; Li, X.P. Pathogenic genes and clinical prognosis in hypertrophic cardiomyopathy. *World J. Cardiol.* **2025**, *17*, 99595. [CrossRef]
8. The Criteria Committee of the New York Heart Association. *Nomenclature and Criteria for Diagnosis of Diseases of the Heart and Great Vessels*, 9th ed.; Little, Brown & Co.: Boston, MA, USA, 1994; pp. 253–256.
9. Vilcant, V.; Hai, O. Left Ventricular Outflow Tract Obstruction. In *StatPearls*; StatPearls Publishing: Treasure Island, FL, USA, 2022.
10. Huffman, M.D.; Roth, G.A.; Sliwa, K.; Yancy, C.W.; Prabhakaran, D. Heart Failure. In *Cardiovascular, Respiratory, and Related Disorders*, 3rd ed.; Prabhakaran, D., Anand, S., Gaziano, T.A., Mbanya, J.C., Wu, Y., Nugent, R., Eds.; The International Bank for Reconstruction and Development/The World Bank: Washington, DC, USA, 2017.
11. Elliott, P.M.; Gimeno, J.R.; Tomé, M.T.; Shah, J.; Ward, D.; Thaman, R.; Mogensen, J.; McKenna, W.J. Left ventricular outflow tract obstruction and sudden death risk in patients with hypertrophic cardiomyopathy. *Eur. Heart J.* **2006**, *27*, 1933–1941. [CrossRef]
12. Kalinski, J.K.; Xu, B.; Boyd, R.; Tasseff, N.; Rutkowski, K.; Ospina, S.; Smedira, N.; Thamilarasan, M.; Popovic, Z.B.; Desai, M.Y. Novel Cardiac Myosin Inhibitor Therapy for Hypertrophic Cardiomyopathy in Adults: A Contemporary Review. *Am. J. Cardiovasc. Drugs* **2024**, *24*, 591–602. [CrossRef]

13. Sherrod Iv, C.F.; Saberi, S.; Nassif, M.E.; Claggett, B.L.; Coats, C.J.; Garcia-Pavia, P.; Januzzi, J.L.; Lewis, G.D.; Ma, C.; Maron, M.S.; et al. Effect of Aficamten on Health Status Outcomes in Obstructive Hypertrophic Cardiomyopathy: Results From SEQUOIA-HCM. *J. Am. Coll. Cardiol.* **2024**, *84*, 1773–1785. [CrossRef]
14. Desai, M.Y.; Owens, A.; Wolski, K.; Geske, J.B.; Saberi, S.; Wang, A.; Sherrid, M.; Cremer, P.C.; Lakdawala, N.K.; Tower-Rader, A.; et al. Mavacamten in Patients With Hypertrophic Cardiomyopathy Referred for Septal Reduction: Week 56 Results From the VALOR-HCM Randomized Clinical Trial. *JAMA Cardiol.* **2023**, *8*, 968–977. [CrossRef]
15. Ommen, S.R.; Ho, C.Y.; Asif, I.M.; Balaji, S.; Burke, M.A.; Day, S.M.; Dearani, J.A.; Epps, K.C.; Evanovich, L.; Ferrari, V.A.; et al. 2024 AHA/ACC/AMSSM/HRS/PACES/SCMR Guideline for the Management of Hypertrophic Cardiomyopathy: A Report of the American Heart Association/American College of Cardiology Joint Committee on Clinical Practice Guidelines. *Circulation* **2024**, *149*, e1239–e1311. [CrossRef] [PubMed]
16. U.S. Foods and Drugs Administration (FDA). Available online: https://www.fda.gov/drugs/news-events-human-drugs/fda-approves-new-drug-improve-heart-function-adults-rare-heart-condition (accessed on 15 May 2025).
17. European Medicine Agency. Available online: https://www.ema.europa.eu/en/medicines/human/EPAR/camzyos (accessed on 15 May 2025).
18. Clinical Trails. Available online: https://clinicaltrials.gov/study/NCT05186818 (accessed on 15 May 2025).
19. Clinical Trails. Available online: https://clinicaltrials.eu/inn/bms-986435/ (accessed on 15 May 2025).
20. Zhang, H.; Yu, C.; Cheng, Y.; Chen, Z.; Chen, M.; He, W.; Jin, Z.; Cai, S.; Yu, L. Clinical Trials in Hypertrophic Cardiomyopathy Therapy: A Comprehensive Analysis of Trials Registered in Global Clinical Databases. *Drug Des. Devel. Ther.* **2023**, *17*, 1863–1877. [CrossRef] [PubMed]
21. Gordon, A.M.; Homsher, E.; Regnier, M. Regulation of contraction in striated muscle. *Physiol. Rev.* **2000**, *80*, 853–924. [CrossRef] [PubMed]
22. Squire, J. Special Issue: The Actin-Myosin Interaction in Muscle: Background and Overview. *Int. J. Mol. Sci.* **2019**, *20*, 5715. [CrossRef] [PubMed]
23. Jani, V.P.; Song, T.; Gao, C.; Gong, H.; Sadayappan, S.; Kass, D.A.; Irving, T.C.; Ma, W. The structural OFF and ON states of myosin can be decoupled from the biochemical super- and disordered-relaxed states. *PNAS Nexus* **2024**, *3*, 39. [CrossRef] [PubMed]
24. Cheng, Z.; Fang, T.; Huang, J.; Guo, Y.; Alam, M.; Qian, H. Hypertrophic cardiomyopathy: From phenotype and pathogenesis to treatment. *Front. Cardiovasc. Med.* **2021**, *8*, 722340. [CrossRef]
25. Trivedi, D.V.; Adhikari, A.S.; Sarkar, S.S.; Ruppel, K.M.; Spudich, J.A. Hypertrophic cardiomyopathy and the myosin mesa: Viewing an old disease in a new light. *Biophys. Rev.* **2018**, *10*, 27–48. [CrossRef]
26. Kim, K.H.; Kim, H.K.; Hwang, I.C.; Lee, S.P.; Park, E.A.; Lee, W.; Kim, Y.J.; Park, J.H.; Sohn, D.W. Myocardial scarring on cardiovascular magnetic resonance in asymptomatic or minimally symptomatic patients with "pure" apical hypertrophic cardiomyopathy. *J. Cardiovasc. Magn. Reson.* **2012**, *14*, 52. [CrossRef]
27. Ingles, J.; Burns, C.; Bagnall, R.D.; Lam, L.; Yeates, L.; Sarina, T.; Puranik, R.; Briffa, T.; Atherton, J.J.; Driscoll, T.; et al. Nonfamilial Hypertrophic Cardiomyopathy: Prevalence, Natural History, and Clinical Implications. *Circ. Cardiovasc. Genet.* **2017**, *10*, e001620. [CrossRef]
28. Turer, A.T.; Wang, A. Cardiac Myosin Inhibitors: Unlocking Potential to Improve Treatment in Hypertrophic Cardiomyopathy. *Circulation* **2023**, *147*, 700–702. [CrossRef] [PubMed]
29. Aman, A.; Akram, A.; Akram, B.; Maham, M.; Bokhari, M.Z.; Akram, A.; Akram, S.; Yaqub, F. Efficacy of cardiac myosin inhibitors mavacamten and aficamten in hypertrophic cardiomyopathy: A systematic review and meta-analysis of randomised controlled trials. *Open Heart* **2025**, *12*, e003215. [CrossRef] [PubMed]
30. Braunwald, E.; Saberi, S.; Abraham, T.P.; Elliott, P.M.; Olivotto, I. Mavacamten: A first-in-class myosin inhibitor for obstructive hypertrophic cardiomyopathy. *Eur. Heart J.* **2023**, *44*, 4622–4633. [CrossRef] [PubMed]
31. Del Rio, C.L.; Aprajita, Y.; Ferguson, B.S.; Christopher, Z.; Trisha, S.; Frank, R.; John, S.; Lee-Jae, G.; Allison, H.; Julie, G.; et al. Abstract 14585: Chronic Treatment with a Mavacamten-Like Myosin-Modulator (MYK-581) Blunts Disease Progression in a Mini-Pig Genetic Model of Non-Obstructed Hypertrophic Cardiomyopathy: In Vivo Evidence for Improved Relaxation and Functional Reserve. *Circulation* **2019**, *140*, A14585.
32. Swift, L.M.; Asfour, H.; Posnack, N.G.; Arutunyan, A.; Kay, M.W.; Sarvazyan, N. Properties of blebbistatin for cardiac optical mapping and other imaging applications. *Pflug. Arch.* **2012**, *464*, 503–512. [CrossRef]
33. Lehman, S.J.; Crocini, C.; Leinwand, L.A. Targeting the sarcomere in inherited cardiomyopathies. *Nat. Rev. Cardiol.* **2022**, *19*, 353–363. [CrossRef]
34. National Center for Biotechnology Information. PubChem Compound Summary for CID 117761397, Mavacamten. PubChem Identifier. Available online: https://pubchem.ncbi.nlm.nih.gov/compound/Mavacamten#section=2D-Structure&fullscreen=true (accessed on 21 May 2025).
35. Keam, S.J. Mavacamten: First Approval. *Drugs* **2022**, *82*, 1127–1135, Erratum in *Drugs* **2022**, *82*, 1235. [CrossRef]
36. Available online: https://www.accessdata.fda.gov/drugsatfda_docs/label/2022/214998s000lbl.pdf (accessed on 15 May 2025).

37. Green, E.M.; Wakimoto, H.; Anderson, R.L.; Evanchik, M.J.; Gorham, J.M.; Harrison, B.C.; Henze, M.; Kawas, R.; Oslob, J.D.; Rodriguez, H.M.; et al. A small-molecule inhibitor of sarcomere contractility suppresses hypertrophic cardiomyopathy in mice. *Science* **2016**, *351*, 617–621. [CrossRef]
38. Heitner, S.B.; Jacoby, D.; Lester, S.J.; Owens, A.; Wang, A.; Zhang, D.; Lambing, J.; Lee, J.; Semigran, M.; Sehnert, A.J. Mavacamten Treatment for Obstructive Hypertrophic Cardiomyopathy: A Clinical Trial. *Ann. Intern. Med.* **2019**, *170*, 741–748. [CrossRef]
39. Masri, A.; Lester, S.J.; Stendahl, J.C.; Hegde, S.M.; Sehnert, A.J.; Balaratnam, G.; Shah, A.; Fox, S.; Wang, A. Long-Term Safety and Efficacy of Mavacamten in Symptomatic Obstructive Hypertrophic Cardiomyopathy: Interim Results of the PIONEER-OLE Study. *J. Am. Heart Assoc.* **2024**, *13*, e030607. [CrossRef]
40. Olivotto, I.; Oreziak, A.; Barriales-Villa, R.; Abraham, T.P.; Masri, A.; Garcia-Pavia, P.; Saberi, S.; Lakdawala, N.K.; Wheeler, M.T.; Owens, A.; et al. Mavacamten for treatment of symptomatic obstructive hypertrophic cardiomyopathy (EXPLORER-HCM): A randomised, double-blind, placebo-controlled, phase 3 trial. *Lancet* **2020**, *396*, 759–769, Erratum in *Lancet* **2020**, *396*, 758. [CrossRef]
41. Kitaoka, H.; Ieda, M.; Ebato, M.; Kozuma, K.; Takayama, M.; Tanno, K.; Komiyama, N.; Sakata, Y.; Maekawa, Y.; Minami, Y.; et al. Phase 3 Open-Label Study Evaluating the Efficacy and Safety of Mavacamten in Japanese Adults with Obstructive Hypertrophic Cardiomyopathy—The HORIZON-HCM Study. *Circ. J.* **2024**, *89*, 130–138. [CrossRef]
42. Rader, F.; Oręziak, A.; Choudhury, L.; Saberi, S.; Fermin, D.; Wheeler, M.T.; Abraham, T.P.; Garcia-Pavia, P.; Zwas, D.R.; Masri, A.; et al. Mavacamten Treatment for Symptomatic Obstructive Hypertrophic Cardiomyopathy: Interim Results From the MAVA-LTE Study, EXPLORER-LTE Cohort. *JACC Heart Fail.* **2024**, *12*, 164–177. [CrossRef]
43. Inestroza, K.; Mijares-Rojas, I.; Matute-Martínez, C.; Ergui, L.; Albosta, M.; Vergara-Sanchez, C.; Dangl, M.; Hernandez, R.J.; Ebner, B.; Vincent, L.T.; et al. In-hospital outcomes of septal myectomy vs. alcohol septal ablation for hypertrophic cardiomyopathy with outflow tract obstruction: An update and insights from the national inpatient sample from 2011 to 2019. *J. Investig. Med.* **2024**, *72*, 262–269. [CrossRef]
44. Ramonfaur, D.; Gasperetti, A.; Blake, V.E.; Rivers, B.; Kassamali, A.A.; Kasper, E.K.; Barouch, L.A.; Wu, K.C.; Madrazo, J.A.; Carrick, R.T. Eighteen-Month Real-World Experience Using Mavacamten for Treatment of Obstructive Hypertrophic Cardiomyopathy in a Racially Diverse Population. *J. Am. Heart Assoc.* **2024**, *13*, e034069. [CrossRef] [PubMed]
45. Lim, J.; Cho, J.Y.; Kwak, S.; Park, C.S.; Park, J.; Choi, H.M.; Cho, G.Y.; Choi, G.H.; Kim, J.; Na, J.O. Real-World Experience of Mavacamten for Patients With Obstructive Hypertrophic Cardiomyopathy in South Korea: A Prospective Multi-Center Observational Study. *Korean Circ. J.* **2025**, *55*, 339–354. [CrossRef] [PubMed]
46. Scholtz, S.; Coppée, C.; Mohemed, K.; Potratz, M.; Sequeira, V.; Rudolph, V.; Scholtz, W.; Reil, J.C. Mavacamten maintenance dose determination: Insights into individualised therapy for hypertrophic cardiomyopathy. *Open Heart* **2025**, *12*, e003192. [CrossRef] [PubMed]
47. Wu, X.; Chen, N.; Hsu, P.; Sun, J.; Li, W.T.; Wang, Q.; Samira, M.; Wei, Q.; Yu, J.C.; Cao, G.Y.; et al. Pharmacokinetics and safety of mavacamten in healthy Chinese participants with different CYP2C19 phenotypes. *Clin. Transl. Sci.* **2024**, *17*, e13877. [CrossRef]
48. Ricci, F.; Molinari, L.V.; Mansour, D.; Galanti, K.; Vagnarelli, F.; Renda, G.; Gallina, S.; Owens, A.; Luzum, J.A.; Olivotto, I.; et al. Managing drug-drug interactions with mavacamten: A focus on combined use of antiarrhythmic drugs and anticoagulants. *Heart Rhythm* **2025**, *22*, 510–525. [CrossRef]
49. Ammirati, E.; Contri, R.; Coppini, R.; Cecchi, F.; Frigerio, M.; Olivotto, I. Pharmacological treatment of hypertrophic cardiomyopathy: Current practice and novel perspec-tives. *Eur. J. Heart Fail.* **2016**, *18*, 1106–1118. [CrossRef]
50. Badr, A.; Roehl, K.; Suppah, M.; Abdullah, H.A.; Arsanjani, R.; Siontis, K.C.; Geske, J.B.; Ommen, S.R.; Giudicessi, J.R.; Alsidawi, R. Temporal Patterns of Holter-Detected Arrhythmias in Hypertrophic Cardiomyopathy Patients Treated with Mavacamten. *Biomedicines* **2025**, *13*, 1005. [CrossRef]
51. Cytokinetics. Available online: https://pnr-files.pro1.gus.wdc.dianum.io/globenewswire/articles/2989688/en/cytokinetics-announces-fda-acceptance-of-new-drug.pdf (accessed on 15 May 2025).
52. National Center for Biotechnology Information. PubChem Compound Summary for CID 139331495, Aficamten. Available online: https://pubchem.ncbi.nlm.nih.gov/compound/Aficamten (accessed on 21 May 2025).
53. Hartman, J.J.; Hwee, D.T.; Robert-Paganin, J.; Chuang, C.; Chin, E.R.; Edell, S.; Lee, K.H.; Madhvani, R.; Paliwal, P.; Pernier, J.; et al. Aficamten is a small-molecule cardiac myosin inhibitor designed to treat hypertrophic cardiomyopathy. *Nat. Cardiovasc. Res.* **2024**, *3*, 1003–1016. [CrossRef]
54. Maron, M.S.; Masri, A.; Nassif, M.E.; Barriales-Villa, R.; Arad, M.; Cardim, N.; Choudhury, L.; Claggett, B.; Coats, C.J.; Düngen, H.D.; et al. Aficamten for Symptomatic Obstructive Hypertrophic Cardiomyopathy. *N. Engl. J. Med.* **2024**, *390*, 1849–1861. [CrossRef] [PubMed]
55. Chuang, C.; Collibee, S.; Ashcraft, L.; Wang, W.; Wal, M.V.; Wang, X.L.; Hwee, D.T.; Wu, Y.S.; Wang, J.Y.; Chin, E.R.; et al. Discovery of Aficamten (CK-274), a Next-Generation Cardiac Myosin Inhibitor for the Treatment of Hypertrophic Cardiomyopathy. *J. Med. Chem.* **2021**, *64*, 14142–14152. [CrossRef] [PubMed]

56. Malik, F.I.; Robertson, L.A.; Armas, D.R.; Robbie, E.P.; Osmukhina, A.; Xu, D.; Li, H.; Solomon, S.D. A Phase 1 Dose-Escalation Study of the Cardiac Myosin Inhibitor Aficamten in Healthy Participants. *JACC Basic Transl. Sci.* **2022**, *7*, 763–775. [CrossRef] [PubMed]
57. Xu, D.; Divanji, P.; Griffith, A.; Sukhun, R.; Cheplo, K.; Li, J.; German, P. Pharmacokinetics, disposition, and biotransformation of the cardiac myosin inhibitor aficamten in humans. *Pharmacol. Res. Perspect.* **2024**, *12*, e70006. [CrossRef]
58. Maron, M.S.; Masri, A.; Nassif, M.E.; Barriales-Villa, R.; Abraham, T.P.; Arad, M.; Cardim, N.; Choudhury, L.; Claggett, B.; Coats, C.J.; et al. Impact of Aficamten on Disease and Symptom Burden in Obstructive Hypertrophic Cardiomyopathy: Results From SEQUOIA-HCM. *J. Am. Coll. Cardiol.* **2024**, *84*, 1821–1831. [CrossRef]
59. Schulze, C.; Abraham, T.P.; Barriales-Villa, R.; Claggett, B.; Coats, C.; García-Pavía, P.; Hagège, A.; Januzzi, J.L.; Kulac, I.J.; Kwong, R.Y.; et al. Abstract 4140102: Changes in EQ-5D-5L with Aficamten in Obstructive Hypertrophic Cardiomyopathy (oHCM): The SEQUOIA-HCM Trial. *Circulation* **2024**, *150*, Suppl_1. [CrossRef]
60. Saberi, S.; Abraham, T.P.; Choudhury, L.; Owens, A.T.; Tower-Rader, A.; Rader, F.; Pavia, P.G.; Olivotto, I.; Coats, C.; Fifer, M.A.; et al. Efficacy and safety of Aficamten in the first cohort of patients with symptomatic obstructive hypertrophic cardiomyopathy completing 1-year follow-up: Findings from the FOREST-HCM study. *J. Am. Coll. Cardiol.* **2024**, *83*, 356. [CrossRef]
61. Owens, A.T.; Masri, A.; Abraham, T.P.; Choudhury, L.; Rader, F.; Symanski, J.D.; Turer, A.T.; Wong, T.C.; Tower-Rader, A.; Coats, C.J.; et al. Aficamten for Drug-Refractory Severe Obstructive Hypertrophic Cardiomyopathy in Patients Receiving Disopyramide: REDWOOD-HCM Cohort 3. *J. Card. Fail.* **2023**, *29*, 1576–1582. [CrossRef]
62. Maron, M.S.; Masri, A.; Choudhury, L.; Olivotto, I.; Saberi, S.; Wang, A.; Garcia-Pavia, P.; Lakdawala, N.K.; Nagueh, S.F.; Rader, F.; et al. Phase 2 Study of Aficamten in Patients With Obstructive Hypertrophic Cardiomyopathy. *J. Am. Coll. Cardiol.* **2023**, *81*, 34–45. [CrossRef]
63. Coats, C.J.; Masri, A.; Nassif, M.E.; Barriales-Villa, R.; Arad, M.; Cardim, N.; Choudhury, L.; Claggett, B.; Düngen, H.; Garcia-Pavia, P.; et al. Dosing and Safety Profile of Aficamten in Symptomatic Obstructive Hypertrophic Cardiomyopathy: Results From SEQUOIA-HCM. *J. Am. Heart Assoc.* **2024**, *13*, e035993. [CrossRef] [PubMed]
64. Davis, B.J.; Volk, H.; Nguyen, O.; Kamna, D.; Chen, H.; Barriales-Villa, R.; Garcia-Pavia, P.; Olivotto, I.; Owens, A.T.; Coats, C.J.; et al. Safety and Efficacy of Mavacamten and Aficamten in Patients With Hypertrophic Cardiomyopathy. *J. Am. Heart Assoc.* **2025**, *14*, e038758. [CrossRef] [PubMed]
65. Maharao, N.; Xu, D.; Simkins, T.J.; Bowles, O.; Liu, G.; Benattia, Y.; Griffith, A.; Heitner, S.B.; Kupfer, S.; German, P. Clinical Evaluation of the Effect of Aficamten on QT/QTc Interval in Healthy Participants. *Clin. Transl. Sci.* **2025**, *18*, e70218. [CrossRef] [PubMed]
66. Xu, D.; Lutz, J.D.; Divanji, P.; Li, J.; Benattia, Y.; Griffith, A.; Heitner, S.B.; Kupfer, S.; German, P. Effect of Hepatic Impairment or Renal Impairment on the Pharmacokinetics of Aficamten. *Clin. Pharmacokinet.* **2025**, *64*, 397–406. [CrossRef]
67. Cytokinetics. Available online: https://pnr-files.pro1.gus.wdc.dianum.io/globenewswire/articles/3001161/en/cytokinetics-announces-european-medicines-agency.pdf (accessed on 15 May 2025).
68. Rauscher, A.Á.; Gyimesi, M.; Kovács, M.; Málnási-Csizmadia, A. Targeting Myosin by Blebbistatin Derivatives: Optimization and Pharmacological Potential. *Trends. Biochem. Sci.* **2018**, *43*, 700–713. [CrossRef]
69. Roman, B.I.; Verhasselt, S.; Stevens, C.V. Medicinal Chemistry and Use of Myosin II Inhibitor (S)-Blebbistatin and Its Derivatives. *J. Med. Chem.* **2018**, *61*, 9410–9428. [CrossRef]
70. Gyimesi, M.; Rauscher, A.Á.; Suthar, S.K.; Hamow, K.Á.; Oravecz, K.; Lőrincz, I.; Borhegyi, Z.; Déri, M.T.; Kiss, Á.F.; Monostory, K.; et al. Improved Inhibitory and Absorption, Distribution, Metabolism, Excretion, and Toxicology (ADMET) Properties of Blebbistatin Derivatives Indicate That Blebbistatin Scaffold Is Ideal for drug Development Targeting Myosin-2. *J. Pharmacol. Exp. Ther.* **2021**, *376*, 358–373. [CrossRef]
71. Alsulami, K.; Marston, S. Small Molecules acting on Myofilaments as Treatments for Heart and Skeletal Muscle Diseases. *Int. J. Mol. Sci.* **2020**, *21*, 9599. [CrossRef]
72. AdisInsight. Available online: https://adisinsight.springer.com/drugs/800056091 (accessed on 15 May 2025).
73. Clinical Trails. Available online: https://clinicaltrials.gov/study/NCT06122779 (accessed on 15 May 2025).
74. Shah, S.J.; Rigolli, M.; Javidialsaadi, A.; Patel, R.B.; Khadra, S.; Goyal, P.; Little, S.; Wever-Pinzon, O.; Owens, A.T.; Skali, H.; et al. Cardiac Myosin Inhibition in Heart Failure with Normal and Supranormal Ejection Fraction: Primary Results of the EMBARK-HFpEF Trial. *JAMA Cardiol.* **2025**, *10*, 170–175. [CrossRef]
75. Garcia-Pavia, P.; Bilen, O.; Burroughs, M.; Costabel, J.P.; Correia, E.d.B.; Dybro, A.M.; Elliott, P.; Lakdawala, N.K.; Mann, A.; Nair, A.; et al. Aficamten vs Metoprolol for Obstructive Hypertrophic Cardiomyopathy: MAPLE-HCM Rationale, Study Design, and Baseline Characteristics. *JACC Heart Fail.* **2025**, *13*, 346–357. [CrossRef]
76. Wheeler, M.T.; Jacoby, D.; Elliott, P.M.; Saberi, S.; Hegde, S.M.; Lakdawala, N.K.; Myers, J.; Sehnert, A.J.; Edelberg, J.M.; Li, W.; et al. Effect of beta-blocker therapy on the response to mavacamten in patients with symptomatic obstructive hypertrophic cardiomyopathy. *Eur. J. Heart Fail.* **2023**, *25*, 260–270. [CrossRef] [PubMed]
77. Clinical Trails. Available online: https://clinicaltrials.gov/study/NCT06081894 (accessed on 15 May 2025).

78. Masri, A.; Maron, M.S.; Abraham, T.P.; Nassif, M.E.; Barriales-Villa, R.; Bilen, O.; Coats, C.J.; Elliott, P.; Garcia-Pavia, P.; Massera, D.; et al. Concomitant Aficamten and Disopyramide in Symptomatic Obstructive Hypertrophic Cardiomyopathy. *JACC Heart Fail.* **2025**. [CrossRef] [PubMed]
79. Hughes, C. Appropriate and inappropriate polypharmacy-Choosing the right strategy. *Br. J. Clin. Pharmacol.* **2021**, *87*, 84–86. [CrossRef] [PubMed]
80. Delara, M.; Murray, L.; Jafari, B.; Bahji, A.; Goodarzi, Z.; Kirkham, J.; Chowdhury, M.; Seitz, D.P. Prevalence and factors associated with polypharmacy: A systematic review and Meta-analysis. *BMC Geriatr.* **2022**, *22*, 601, Erratum in *BMC Geriatr.* **2022**, *22*, 742. [CrossRef]
81. Ho, C.Y.; Mealiffe, M.E.; Bach, R.G.; Bhattacharya, M.; Choudhury, L.; Edelberg, J.M.; Hegde, S.M.; Jacoby, D.; Lakdawala, N.K.; Lester, S.J.; et al. Evaluation of Mavacamten in Symptomatic Patients With Nonobstructive Hypertrophic Cardiomyopathy. *J. Am. Coll. Cardiol.* **2020**, *75*, 2649–2660. [CrossRef]
82. Desai, M.Y.; Nissen, S.E.; Abraham, T.; Olivotto, I.; Garcia-Pavia, P.; Lopes, R.D.; Verheyen, N.; Wever-Pinzon, O.; Wolski, K.; Jaber, W.; et al. Mavacamten in Symptomatic Nonobstructive Hypertrophic Cardiomyopathy: Design, Rationale, and Baseline Characteristics of ODYSSEY-HCM. *JACC Heart Fail.* **2025**, *13*, 358–370. [CrossRef]
83. Bristol Myers Squibb. Available online: https://news.bms.com/news/details/2025/Bristol-Myers-Squibb-Provides-Update-on-Phase-3-ODYSSEY-HCM-Trial/ (accessed on 15 May 2025).
84. Amr, A.; Kayvanpour, E.; Reich, C.; Koelemen, J.; Asokan, S.; Frey, N.; Meder, B.; Sedaghat-Hamedani, F. Assessing the Applicability of Cardiac Myosin Inhibitors for Hypertrophic Cardiomyopathy Management in a Large Single Center Cohort. *Rev. Cardiovasc. Med.* **2024**, *25*, 225. [CrossRef]
85. Clinical Trails. Available online: https://clinicaltrials.gov/study/NCT06412666 (accessed on 15 May 2025).

MDPI AG
Grosspeteranlage 5
4052 Basel
Switzerland
Tel.: +41 61 683 77 34

Biomolecules Editorial Office
E-mail: biomolecules@mdpi.com
www.mdpi.com/journal/biomolecules

www.ingramcontent.com/pod-product-compliance
Lightning Source LLC
LaVergne TN
LVHW070450061126
851858LV00011B/199